Troubleshooting and Repairing Major Appliances, Second Edition

About the Author

Eric Kleinert is a professional with 35 years of experience in commercial and domestic major appliance, refrigeration, and HVAC sales, service, and installation. He owned and operated a major appliance and air conditioning sales and service corporation.

As an instructor on behalf of the Palm Beach County School District, he taught adults aspects of preventative and diagnostic services and techniques, which enhanced their ability to better evaluate products and services necessary to maintain residential climate-control systems and major appliances.

Troubleshooting and Repairing Major Appliances, Second Edition

Eric Kleinert

New York Chicago San Francisco
Lisbon London Madrid Mexico City
Milan New Delhi San Juan
Seoul Singapore Sydney Toronto

The McGraw·Hill Companies

Library of Congress Cataloging-in-Publication Data

Kleinert, Eric.
  Troubleshooting and repairing major appliances / Eric Kleinert.—2nd ed.
    p. cm.
  ISBN 0-07-148148-6 (alk. paper)
  1. Household appliances, Electric—Maintenance and repair. I. Title.
TK7018.K56 2007
683'.80288—dc22

2007018878

McGraw-Hill books are available at special quantity discounts to use as premiums and sales promotions, or for use in corporate training programs. For more information, please write to the Director of Special Sales, Professional Publishing, McGraw-Hill, Two Penn Plaza, New York, NY 10121-2298. Or contact your local bookstore.

Troubleshooting and Repairing Major Appliances, Second Edition

1234567890 DOC DOC 01987

ISBN-13: 978-0-07-148148-9
ISBN-10:   0-07-148148-6

Sponsoring Editor
  Judy Bass

Editorial Supervisor
  Patty Mon

Project Manager
  Madhu Bhardwaj

Acquisitions Coordinator
  Laura Hahn

Copy Editor
  Lisa McCoy

Proofreader
  Karen Jane Simon

Indexer
  Kevin Broccoli

Production Supervisor
  Jean Bodeaux

Composition
  International Typesetting
  and Composition

Illustration
  International Typesetting
  and Composition

Art Director, Cover
  Jeff Weeks

Cover Designer
  Ty Nowicki

Contents at a Glance

Contents

Part V   Appliance Service, Installation, and Preventive Maintenance Procedures

Acknowledgments

I would like to convey my utmost appreciation to the following companies for their valuable contributions to this improved second edition:

- BrandsmartUSA
- Dwyer Instruments, Inc.
- Klein Tools, Inc.
- Major Appliance Consumer Action Panel (MACAP)
- National Appliance Service Technicians Certification (NASTeC)
- Tecumseh Compressor Company

I would also like to thank Judy Bass and the staff of McGraw-Hill for giving me the opportunity to write this second edition.

I would like to thank my wife, Eileen, for putting up with me while I disappeared in my office all day and night over the course of writing this second edition. Your sacrifices will not go unrewarded.

Lastly, again, I would like to thank my daughter, Brandi, for her assistance and her ability to repair her own appliances.

Introduction

A few months ago, my grown-up daughter called and asked me to help repair her dishwasher; it was leaking water on the floor. I was glad to assist, but I was 1800 miles away. I knew she had a copy of the first edition of this book. I told her to turn to the section on dishwashers as I explained the probable cause for the water leak. She began reading the section about the common washability problems associated with dishwashers. "Oh, I see why it's doing that," she said. "If I just follow the illustrations, I should be able to repair it." She e-mailed me pictures as she worked. The final photo was of her giving me the "thumbs-up" sign and the working dishwasher in the background. I was happy to save her money on a service call, but I began thinking about all the changes in appliances over the last decade.

New products, energy conservation, and electronics have made the industry what it is today. The second edition of *Troubleshooting and Repairing Major Appliances* will assist you in servicing these products. All that's needed is the ability to think clearly, to use hand tools, and to use test meters.

I can only expose you to the knowledge that's needed for preventative, diagnostic, and service techniques in the major appliance and air conditioner industry. This will help you evaluate these products and services. If you are a homeowner, this book can help you decide if it's a task you can fix yourself or if it's one that's best left to the experts. For the beginning technician, what you learn from this book will give you the keys for a rewarding service career. And for the experienced professional, this book will serve as a good resource and refresher on the latest techniques and technologies.

As a novice servicer, you may have to look up terms in the glossary or consult the component descriptions and diagnostic charts. However, you'll be amply rewarded for this effort when a major appliance or air conditioner is fixed for the first time. This book will also be informative and useful for the experienced servicer and repair technician.

This edition is comprised of five parts: general, types of fuels, theory, parts, and specific information. Part One, "The Fundamentals of Service," identifies the problems of selecting, purchasing, and installing new major appliances and air conditioners. Also included is information on manufacturer warranties, use and care manuals, and where consumers can get help if needed.

Chapters 2 through 4 map out a complete guide for repairing major appliances and air conditioners. We'll discuss the safety precautions for installation, operation, and repair of major appliances and air conditioners, as well as the tools needed for the right job and the basic approach before beginning to repair these products.

Part Two, "Electricity, Electronics, and Gas," covers everything you ever needed to know about electricity, gas, and electronics in major appliances and air conditioners. I will teach you how to use test meters, read circuit diagrams, test electronic parts, gas appliance venting, and much more.

In Part Three, "Refrigeration and Air Conditioning," you'll learn about the refrigeration cycle and its components.

In Chapters 8, 9, and 10, we'll focus on the description, operation, and locations of the parts that are in today's major appliances and air conditioners. In Part Five, "Appliance Service, Installation, and Preventive Maintenance Procedures," we'll explore in-depth the different types of appliances and air conditioners you'll be expected to fix. Chapters 11 through 23 cover domestic dishwashers, garbage disposals, electric and gas water heaters, automatic washers, automatic electric and gas dryers, electric and gas ranges/ovens, microwave ovens, refrigerators, freezers, automatic ice makers, and air conditioners. We'll go over step-by-step procedures for testing and replacing parts, chart troubleshooting, preventative maintenance procedures, and appliance and air conditioner installation procedures.

To present a broad overview of service techniques, I've provided general troubleshooting information about the various types of major home appliances and air conditioners, rather than specific models that may become outdated. The pictures and illustrations are for demonstration purposes only to clarify the description of how to service them; they do not reflect a particular product's reliability.

Troubleshooting and Repairing Major Appliances aims to provide you with a basic understanding of the operation and common problems of major appliances and air conditioners as you become your own technician. Soon, you'll have the know-how to fix practically any appliance in the home that breaks down.

PART

Fundamentals of Service

Selecting, Purchasing, and Installing Major Home Appliances and Air Conditioners

Both the technician and the consumer will appreciate this chapter because it addresses some of the problems of selecting, acquiring, and installing major home appliances and air conditioners. It is also pertinent that technicians familiarize themselves with the features and functions of the appliances and air conditioners, which are also stated in every new appliance and air conditioner use and care manual. This will be essential when diagnosing problems with the diverse types of appliances and air conditioners.

This information will provide proper planning techniques and a better understanding of major home appliances and air conditioners, including appliance and air conditioner (A/C) warranties and where to get help when it is needed.

In today's market, major home appliances and air conditioners are manufactured to meet the needs of the average person. Remember that price should not be the most influential factor when choosing an appliance or air conditioner. Physical and mental limitations should also be considered when selecting the product that will be juxtaposed with the consumer's needs.

Electric and Gas Ranges, Cooktops, and Ovens

The domestic range was designed as a multipurpose cooking appliance. It consists of a surface area with heating elements on the top to cook the food. The oven cavity is used for baking food at a set temperature. Within the same oven cavity, the broiling of food is also incorporated.

Domestic ranges are available in either electric or gas, in sizes ranging from 20 inches to 48 inches in width, and with a wide selection of configurations and colors. The following sections describe a few of the common configurations available to the consumer.

Freestanding Ranges

The freestanding range stands between two base cabinets, or sometimes at the end of a cabinet line. The consumer has a choice of gas or electric. These ranges are available in 20-, 24-, 30-, 36-, 40-, and 48-inch widths (Figure 1-1). Some 40- and 48-inch models have two ovens. Designs include gas burners and gas ovens, or standard electric heating elements with electric ovens, and a glass cooktop with concealed electric elements underneath. Also available is a combination of electric and gas cooking. The controls might be located on the rear console or across the front.

Oven cleaning systems include self-cleaning,[1] continuous cleaning,[2] and standard manual cleaning. Freestanding ranges have the following advantages and disadvantages:

Advantages

- Freestanding ranges are generally less expensive than other types. Prices vary with features.
- Freestanding ranges can be moved when the family moves.
- Most models have some center space for placing utensils.
- Front controls can be reached easily from a seated position.
- Bottom drawer adds to kitchen storage space.

Disadvantages

- Rear console controls are virtually impossible for a seated person to reach.
- Low broiler in gas ranges is less accessible from a seated position than the oven broiler in electric ranges.

Figure 1-1   The freestanding electric or gas range is available in 20-, 24-, 30-, 36-, 40-, and 48-inch widths and a variety of colors.

FIGURE 1-2
The slide-in range with a 30-inch width is available in a variety of colors.

Slide-in Ranges and Drop-in Ranges

Slide-in ranges and drop-in ranges might not always be available in all sizes: 30-inch width is the most common (Figure 1-2). However, drop-in ranges are also available in a 27-inch width. Designs include gas burners, standard electric heating elements, and a glass cooktop with concealed electric elements underneath. The oven and cooktop controls are usually located across the front of the range; however, some models have the cooktop controls along the side. They are an excellent choice for an island or peninsula-shaped counter because they are flush with the surrounding counter. These ranges can also be installed overlapping the adjacent countertop edges, thereby eliminating dirt-catching gaps. A drop-in range either hangs from a countertop or sits on a low cabinet base; a slide-in range sits on the floor.

Slide-in ranges and drop-in ranges (Figure 1-3) have the following advantages and disadvantages:

FIGURE 1-3
The drop-in range is most commonly available with a 30-inch width. It is available in a variety of colors.

Advantages

- Can be installed a few inches higher or lower than a freestanding range.
- Controls can be reached by most cooks.

Disadvantages

- Built into kitchen and generally cannot be moved when the family relocates.
- Requires installation by a carpenter and an electrician.

Built-in Cooktops

Built-in cooktops are set into a countertop and are made in various sizes, from 15 to 48 inches wide. Built-in cooktop designs include gas burners, standard electric heating elements, and a glass cooktop with concealed electric elements underneath (Figures 1-4 and 1-5). They might have side or front controls and might be of modular design. Special plug-in cooking accessories are also available. Built-in cooktops have the following advantages and disadvantages:

Advantages

- Can be installed at the most convenient height for the cook.
- Side or front controls are easily reached by most cooks.
- Counter installation provides open space below the cooktop.

The slide-in range with a 30-inch width is available in a variety of colors.

FIGURE 1-4 Electric or gas built-in cooktops are available with two, four, or five heating elements. They are available in a variety of colors.

FIGURE 1-5
The electric built-in cooktop (downdraft model). The unit has a fan in between the surface elements that draws the cooking fumes into it and out of the kitchen.

Disadvantages

- Built into kitchen and generally cannot be moved when the family relocates.
- Requires installation by a carpenter and an electrician.

Built-in Ovens

Built-in ovens usually have one oven cavity, but models with two oven cavities are also available. In the double oven cavity model, the second oven might be a conventional oven, a microwave oven, or a combination of both (Figure 1-6). Built-in ovens are available in 24- , 27- , and 30-inch widths. Their height varies, depending upon whether they are single- or double-oven units. Oven cleaning systems available include self-cleaning, continuous cleaning, and standard manual cleaning. Two-oven models also offer conventional ovens with two different cleaning systems. Built-in ovens with the microwave feature are available with either solid-state microcomputer or electromechanical controls.

Built-in ovens have the following advantages and disadvantages:

Advantages

- Can be installed at the most convenient height for the user, putting controls within reach for a standing or a seated cook.
- Automatic cleaning systems virtually eliminate the task of cleaning the oven manually.

Disadvantages

- Built into the kitchen and generally cannot be moved when the family relocates.
- Installation might involve structural and wiring changes, requiring a carpenter and an electrician.
- Two-oven combinations with microwave ovens often require learning some new cooking techniques.

FIGURE 1-6   The built-in oven is available in different configurations and in a variety of colors.

Eye-Level Two-Oven Ranges

Eye-level ranges[3] are 30-inch-wide ranges that have a second oven mounted over the cooktop. The upper oven will be a microwave oven. Designs include features like:

- A selection of either gas burners or electric elements
- Choice between gas or electric oven
- Glass cooktop with concealed electric elements underneath

The lower ovens are available with self-cleaning or continuous cleaning systems.

Microwave oven combinations are available with solid-state microcomputer or electromechanical controls. This type of range gives the maximum cooking capacity in a minimum amount of floor space.

Eye-level one- and two-oven ranges (Figure 1-7) have the following advantages and disadvantages:

Advantages

- Microwave controls at eye level are convenient for cooks who use walkers, crutches, or a cane.
- Automatic cleaning systems virtually eliminate the task of manually cleaning the oven.
- Can be moved when the family relocates.

Disadvantages

- Controls located at eye level are not usable by seated cooks.

FIGURE 1-7
Eye-level two-oven range with or without microwave oven combination. Available in a variety of colors.

Microwave Ovens

Do not go out and purchase a microwave oven until you do your homework. Brand, type, size, power, features, warranty, and location are just some of the features you need to consider. You can also go on the Internet and conduct research.

When comparing microwaves, fill out the following checklist. Check all that apply.

Measure the area available for the microwave oven:

_____width
_____depth
_____height

Before purchasing, read the installation instructions for the built-in or over-the-range model you are considering. These instructions will provide you with information such as cutout dimensions, venting, voltage, location, and more.

What is the voltage?

❑ 120 volts   ❑ 240 volts

What type of microwave?

❑ Countertop   ❑ Over the range

❑ Combination   ❑ Convection
(range and microwave)

❑ Under the counter   ❑ Built-in

❑ Other (list the features)

How much room do I need to fit the microwave oven on the countertop?

❑ Large: 22 to 27 inches wide by 13 to16 inches high by 15 to 21 inches deep

❑ Medium: 20 to 24 inches wide by 16 to 16 inches high by 13 to 18 inches deep

❑ Small: 18 to 20 inches wide by 10 to 12 inches high by 11 to 12 inches deep

❑ Touchpad controls:

❑ Are the numbers large and easy to read?

❑ Is the oven easy to program?

❑ Does the model display prompters that will guide you as you set the controls?

❑ Mechanical controls:

❑ Does the mechanical timer have a wide time range setting?

❑ Seconds?   ❑ Minutes?   ❑ Hours?

❑ Is there an on/off switch to operate the oven after setting the time function?

❑ Does the timer knob turn on the oven after setting the time function?

How many watts of cooking?

❑ 400 to 700 watts   ❑ 600 to 800 watts

❑ 800 to 1000 watts   ❑ 1000 to 1600 watts

What additional features do you want?

❑ Auto defrost

❑ Pre-programmed cooking

Does the oven have a light inside the cavity?

❑ yes   ❑ no

Is there a window in the door?

❑ yes   ❑ no

Does the oven have a rack?

❑ yes   ❑ no

Do you want a turntable for the food to rotate on?

❑ yes   ❑ no

Does the model have multi-stage cooking?

❑ yes   ❑ no

❑ Temperature probe

❑ Cooking sensors

Will my plates fit the oven cavity?

❑ yes   ❑ no

Other (list here)

Additional information on warranties, safety, recalls, and maintenance must be taken into account when purchasing a microwave oven. Ask the salesperson for a demonstration of the microwave you are considering.

Checklist for Cooking Products

Prior to selecting and purchasing an appliance, read this section. Fill out the following checklist. Check all that apply.

Measure the area available for the range:

_____width

_____depth

_____height

These measurements are the cut-out measurements, not the range measurements. Also, be sure that the range can fit through the doorways of the house.

The type of range desired:

❑ Freestanding with one oven   ❑ Eye-level (high or low) with two ovens

❑ Built-in   ❑ Drop-in

❑ Slide-in   ❑ Electric

❑ Gas

❑ LP   ❑ Natural

Type of oven needed:

☐ Single   ☐ Double
☐ Conventional   ☐ Microwave
☐ Combination oven   ☐ Convection oven

Location desired:

☐ Below cooktop   ☐ One over, one under
☐ Separate built-in oven(s)

Oven controls:

☐ On back console   ☐ On range front console
☐ On hood   ☐ Dial type
☐ Touch pads   ☐ Automatic oven clock/timer

Cleaning system:

☐ Self-cleaning (pyrolytic)   ☐ Continuous
☐ Manual

Broiler type:

☐ Top of oven   ☐ Low broiler
☐ Variable heat

Cooktop style:

☐ Conventional   ☐ Glass-ceramic
☐ Grill/griddle convertible   ☐ Induction

Cooktop controls:

☐ Thermostatic control   ☐ Eye-level controls
☐ Eye-level controls on hood ☐ Controls on backsplash
☐ Controls at front of cooktop ☐ Controls at side of cooktop

Accessories:

☐ Rotisserie   ☐ Roast temperature probe
☐ Griddle

Venting system:

☐ Separate hood   ☐ Built-in down draft
☐ Vent-microwave oven combination ☐ Hood attached to upper oven
☐ Vent over regular cooktop ☐ Vent over grill/griddle cooktop

The preferred color of the range _____

Who will use the range? _____

How many people are being cooked for? _____

Is the range design convenient for all family members? ☐ yes ☐ no

Price range: $_____

Warranty and service information:_____

Refrigerators and Freezers

As with other kitchen appliances, refrigerators and freezers come in a wide variety of styles, sizes, and colors (Figure 1-8). Some designs might meet the needs of a family member who has a physical or mental limitation better than others.

Some questions to consider when choosing a refrigerator are:

- Does it have a true no-frost system to do away with manual defrosting?

- Does it have an automatic ice maker, which produces ice without trays to fill or empty?

FIGURE 1-8 Refrigerators and freezers are available in a wide variety of sizes and colors. (*continued*)

FIGURE 1-8   Refrigerators and freezers are available in a wide variety of sizes and colors.

- Does it have in-the-door dispensers to deliver ice and cold water without opening the door?
- Does it have shelves, bins, and drawers that pull out to make reachable those foods stored in the back?

Five basic types of refrigerators are on the market today:

- Compacts or portables: 11.2 to 6 cubic feet (12 inches to 24 inches wide)
- Single-door models: 9 to 14 cubic feet (23 inches to 30 inches wide)
- Top-mount refrigerator-freezer combination: 10 to 23 cubic feet (24 inches to 33 inches wide)
- Bottom-mount refrigerator-freezer: 18 to 20 cubic feet (32 to 36 inches wide)
- Side-by-side refrigerator-freezer: 17 to 30 cubic feet (30 to 48 inches wide)

Excluding compacts and portables, refrigerators range in height from 56 to 84 inches and in depth from 24 to 31 inches.

When selecting the capacity of a refrigerator, the following guidelines should be considered:

- Allow a minimum of 12 cubic feet for the first two persons in the household.
- Add 2 cubic feet for each additional member. The most popular size for an average family is 18 cubic feet.
- Subtract from this rule if many meals are eaten away from home, using the number of meals eaten outside the household as a basis.
- Add to the rule if the customer entertains often, if the family is expanding, if there is a vegetable garden growing, or if cooking is enjoyable.

Compact and Portable Refrigerators

Compact and portable refrigerators are often used as a supplementary model in family rooms, offices, dorms, vacation homes, campers, and other convenient places. Many fit on or under a countertop. They might be freestanding or built-in models, and they come in a variety of colors and finishes.

Compact and portable refrigerators have the following advantages and disadvantages:

Advantages

- Refrigerators of this type have a smaller capacity—less than 6 cubic feet. They can be installed at any height.
- Has a small freezer compartment for ice trays.
- Some models have an optional ice maker.

Disadvantages

- Frozen food storage is limited to a few days—a week at most—depending on the temperature.
- A few compact models have automatic defrosting. However, most are manually defrosted.
- Small size provides limited storage capacity.

Single-Door Refrigerators

Single-door refrigerators provide both fresh and frozen food storage. Frozen food compartments are located on top of the inside of the refrigerator and usually contain ice cube trays.

Single-door refrigerators have the following advantages and disadvantages:

Advantages

- Refrigerators of this type generally have a small capacity—less than 14 cubic feet. Most shelf areas are within reach of a seated person.
- Freezer compartments have side-opening doors, rather than drop-down doors, for easier accessibility.

Disadvantages

- Refrigerators of this type usually require manual defrosting, a difficult chore for disabled persons.
- Freezer compartments with drop-down doors are inaccessible from a seated position.
- Limited storage capacity.
- Freezer compartment can only be used for short-term storage of commercially frozen food and for making ice. High-sugar foods, such as ice cream, might not stay frozen.

Two-Door Refrigerators

Two-door refrigerators with top freezers provide storage for fresh and frozen food. The freezer maintains a temperature of 0 to 10 degrees Fahrenheit. These models come with and without automatic ice makers. Some models have an ice or water dispenser on the outside of the doors.

Two-door refrigerators with top freezers have the following advantages and disadvantages:

Advantages

- Provide proper storage conditions for both fresh and frozen foods. Keep ice cream frozen.

Disadvantages

- Top freezer is not accessible from a seated position.
- Foods stored near the rear of deep shelves might be difficult to reach without using special aids.

Two-Door Refrigerators with Bottom Freezers

Two-door refrigerators with bottom freezers provide storage for both fresh and frozen foods. The freezer maintains a temperature of 0 to 10 degrees Fahrenheit. Models with and without automatic ice makers are available.

Two-door refrigerators with bottom freezers have the following advantages and disadvantages:

Advantages

- Freezer shelf and basket slide out for easy accessibility.
- Lower shelves of the fresh food storage area are within easy reach from a seated position.

Disadvantages

- Although this design might meet the needs of some users with disabilities, the bottom-freezer refrigerator is generally not convenient for those in wheelchairs. The low freezer is also inconvenient for users who have trouble stooping or bending.

Side-by-Side Refrigerators

Side-by-side refrigerators have deep, vertical fresh and frozen food compartments. They require less room for the doors to open than other types.

Side-by-side refrigerators have the following advantages and disadvantages:

Advantages

- Provide universal access to the majority of shelves in both refrigerator and freezer compartments.
- Models with in-door dispensers give easy access to ice and ice water without opening the door.
- Models with a shallow, third-door compartment have the option of putting frequently used foods within easy reach without opening the main refrigerator door.
- Pull-out shelves, drawers, bins, and baskets provide easier reach for foods stored at the back.

Disadvantages

- Special features add to the total cost.
- Might require more space than available in existing kitchens due to a wider design.

Freezers

A compact, upright freezer will best meet the needs of the disabled person. Installing the freezer on a box or raised platform can help make its contents more easily accessible. Freezers are conveniences for people who do not frequent the supermarket. They are especially useful in homes with smaller refrigerators or refrigerators having only an ice cube tray compartment. Home freezers come in chest and upright models. Two designs of upright models are available on the market today: manual defrost and automatic defrost. Home freezers are available with wire shelves and baskets, and with storage shelves on the doors in upright models.

Freezers have the following advantages and disadvantages:

Advantages

- Make it possible to keep a supply of all kinds of frozen food on hand.
- By stocking up on food at sale prices and storing them for later use, consumers can easily take advantage of price specials.
- Require fewer shopping trips.

Disadvantages

- Kitchen might not allow space for the freezer.

Checklist for Refrigerator and Freezer Products

Prior to selecting and purchasing a refrigerator or freezer, fill out the following checklist. Check all that apply.

Size

- How large is the space for the freezer?
  _____ width _____ depth _____ height
- How much room does the door need to swing open? _____
- Will there be enough room to open the doors completely so as to remove the storage bins?_____
- Direction of door swing: handle on the left side ❏ right side ❏
- How many people will be using the refrigerator?
  ❏ One to two people (need at least 14 cubic feet)
  ❏ Three to four people (need 16 to 18 cubic feet)
  ❏ More than four people (add 2 cubic feet per additional person). To accommodate for later expansion needs, plan for additional refrigerator space, especially if the family is growing larger or to accommodate peak loads.
- Is there a regular stock of cold beverages in the refrigerator?
  ❏ yes   ❏ no
- How often does the customer go food shopping?
  ❏ two to three times a week
  ❏ weekly   ❏ daily
- Does the consumer host large holiday dinners?
  ❏ yes   ❏ no

Each "yes" answer will add to the refrigerator size requirements, as specified earlier.

Model

- Are there any handicapped or disabled members in the household?
  ❏ yes   ❏ no

A side-by-side model allows easy access to both freezer and fresh food compartments for those who have limited activity requirements.

- Choose the model desired:
  - ❑ Side-by-side
  - ❑ Bottom-mount freezer
  - ❑ Compact
  - ❑ Top-mount freezer
  - ❑ Three door
  - ❑ Single door (no long-term freezer compartment)
- Which features are important?
  - ❑ Automatic defrost
  - ❑ Cycle defrost (requires manual defrosting of freezer)
  - ❑ Manual defrost (requires manual defrosting of both refrigerator and freezer)
  - ❑ Reversible doors
- Automatic ice maker:
  - ❑ Factory installed
  - ❑ Equipped for later installation
- Through-the-door dispenser:
  - ❑ Ice cubes only
  - ❑ Cubes/crushed ice/chilled water
  - ❑ Cubes/crushed ice
- Storage drawers:
  - ❑ See-through
  - ❑ Adjustable humidity
  - ❑ Adjustable temperature
  - ❑ Sealed snack pack for lunch meats/cheese
- Refrigerator shelves:
  - ❑ Glass
  - ❑ Wire
  - ❑ Adjustable
  - ❑ Nonadjustable
  - ❑ Full width only
  - ❑ Half width only
  - ❑ Combination full and half width
- Door storage:
  - ❑ Egg compartment
  - ❑ Covered
  - ❑ Removable
- Dairy compartment:
  - ❑ Butter only
  - ❑ Deep enough for liter-size bottles or six packs
  - ❑ Butter and cheese
  - ❑ Removable storage/servers
- Freezer compartment:
  - ❑ Needed only for one to two days' storage of frozen food
  - ❑ Needed for storage of ice cream, meats, and other frozen food over longer periods
- Freezer shelves:
  - ❑ No shelves needed
  - ❑ Need adjustable shelves
  - ❑ Nonadjustable shelves acceptable

- Convenience features:
  - ❑ Juice/can dispenser
  - ❑ Ice cube bin
  - ❑ Ice tray shelves
  - ❑ Wine bottle holder
- Price range: $_____
- Warranty and service information:_____

Dishwashers

During the past decades, dishwashers have proven their value and usefulness in reducing the cleanup tasks in the kitchen. They not only save time, energy, and labor, but they also deliver dishes cleaner than those washed by hand.

The most common type is the built-in dishwasher. However, other styles are available for special situations. When selecting a new dishwasher (Figure 1-9), consider the following:

- Is it a front-loading undercounter model? Portables are less convenient because they must be moved into position and hooked up to a faucet every time they are used.
- Does it have a self-cleaning filter, rather than one that has to be removed and cleaned?
- Does it have dispensers and silverware baskets in an easy-to-reach location?
- Are the silverware baskets on the door or in the bottom rack?
- Are the racks designed with flexibility for easy loading of tall or bulky dishes?

FIGURE 1-9   Undercounter dishwasher. Available in 18- or 24-inch width. Portable models are also available.

Built-in Dishwashers

Built-in dishwashers are designed to fit into a 24-inch-wide space between two kitchen cabinets and under the countertop. Also available are smaller-capacity models that fit into an 18-inch space. The consumer now has a choice between a drop-down door (standard) and a single or double slide-out drawer.

Convertible/Portable Dishwashers

Convertible/portable models are essentially the same as built-ins, but they have finished sides and tops, drain and fill hoses with a faucet connector, and casters for easy rolling to the sink. These can be installed later as built-ins, if desired.

Undersink Dishwashers

An undersink model is designed to fit under a special six-inch-deep sink in just 24 inches of space; or under a special double-bowl sink, with a disposer under the second bowl, in 36 inches of space. These dishwashers were designed for small kitchens with limited cabinet space for storage.

Dishwasher-Sink Combination

A dishwasher-sink combination unit is also available. It includes a stainless steel sink with drainboard, an enameled metal undersink cabinet, and a dishwasher in 48 inches of space. Some "Pullman-type" combination units include a dishwasher as well as a sink, range, and/or, in some, a refrigerator—all in one unit. Dishwashers have the following advantages and disadvantages:

Advantages

- Can save the physical labor of washing dishes.
- Provide out-of-the-way storage for dirty or clean dishes.

Disadvantages

- Require an 18-inch or 24-inch space in the kitchen, so a cabinet might have to be removed.
- Are nearly always designed for use with standard 36-inch-high countertops.

This might require a two-level counter if the sink is installed at a 30-inch height for a wheelchair-bound person.

Checklist for Dishwashers

Before selecting and purchasing a dishwasher, complete the following checklist. Check all that apply.

- Model:
    - ❑ Built-in
    - ❑ Slide-out drawer
    - ❑ Dishwasher-sink combination
    - ❑ Standard drop-down door
    - ❑ Convertible/portable

- Loading convenience:
  - ❏ Racks designed to handle large dishes
  - ❏ Adjustable racks
  - ❏ Special baskets:_____
- Wash system:
  - ❏ Water temperature booster
  - ❏ Two- or three-level spray arms
  - ❏ Rinse aid dispenser
  - ❏ Self-cleaning filter
  - ❏ Soft food dispenser
- Control panel:
  - ❏ Dials
  - ❏ Push buttons
  - ❏ Touch pads
  - ❏ Cycle time indicator
  - ❏ Energy level indicator
  - ❏ System status
- Cycles:
  - ❏ Normal
  - ❏ Rinse only
  - ❏ Economy/water saver
  - ❏ Super-heated rinse
  - ❏ No-heat drying
  - ❏ Light duty
  - ❏ China and glass
  - ❏ Pots and pans
  - ❏ Delay start
- Finishes:
  - Tub _____
  - Racks_____
  - Color panels_____
  - Trim kits_____
- Preferred color_____
- Price range: $_____
- Warranty and service information:_____

Laundry Equipment

Today's laundry equipment, along with changes in fibers and fashions, have eliminated the need for hand laundering, clothesline drying, and routine ironing. There features can often be preprogrammed into the appliance for any type of laundry load. These all give excellent results. From diapers and jeans, to delicate silks and knits, today's laundry system is equipped for all fabric needs.

The typical laundry pair, a standard washer and dryer, will stand side by side in 4-1/2 to 5 feet of wall space, depending on the brand and model (Figure 1-10). Some questions to consider when choosing laundry appliances are as follows:

- Is there enough space available for laundry appliances?
- What control location will be best for the principal user? Some models offer front or rear controls.
- What capacity will best satisfy family needs?
- Will built-in dispensers for bleach and fabric softener increase the washer's utility?

FIGURE 1-10
FIGURE 1-10
The automatic washer and automatic dryer. The dryer is available in electric or gas.

- Will the dryer need to be vented outside?
- Which is preferred (gas or electric) for drying clothes?
- How many different washing cycles are needed?
- How many different drying cycles are needed?

Automatic Washers

Basically, all automatic washers will wash the clothes in the same manner, but there are some key differences in design and special features from model to model and manufacturer to manufacturer.

Top-Loading Automatic Washers

Top-loading models vary in width from about 24 to 30 inches (Figure 1-11). They are available in a variety of load capacities. Standard-capacity washers are built for the average two- to four-person household. However, a large-capacity model reduces the number of loads washed, saving time. Some models offer front panel controls, and many models have dispensers. Top-loading washers have the following advantages and disadvantages:

Advantages

- Provide a convenient, at-home way to do laundry.
- Models installed in a small space, only 24 inches wide, are available.
- Models with front controls can be reached and operated easily from a seated position.
- Provide a variety of designs and control positions to meet varying user needs.

Disadvantages

- Models with rear console controls are virtually impossible to operate from a seated position.

FIGURE 1-11
The top-loading washer with dryer built in above is available in a variety of colors.

- Some designs might require special aids to remove loads, set controls, and clean filters from a seated position.
- Compact models—although they provide the greatest accessibility from a wheelchair—have a smaller load capacity than other designs.

Front-Loading Automatic Washers

Front-loading models might have drop-down or side-opening doors (Figure 1-12 and 1-13). In these models, the entire wash basket revolves. As clothes tumble, they are lifted by vanes on the sides of the basket. Front loaders use less water than top loaders, but they will not handle very large loads because they must have empty space in the drum to tumble clothes. Front-loading washers have the following advantages and disadvantages:

Advantages

- Front controls can be reached and operated easily from a seated position.
- Front opening makes loading and unloading easier for users with limitations.

Disadvantages

- Drop-down door might create wheelchair barrier.
- Door opening might be too low for some wheelchair users or those who cannot stoop or bend.

Figure 1-12   Front-load automatic washer.

Figure 1-13   The front-loading washer can be stacked, placed under the counter, or installed in the laundry area. Available in a variety of colors.

Figure 1-14
Compact or portable washer.

Compact Automatic Washers

These compact, "apartment-sized" washers range from 24 to 27 inches wide to fit spatial needs. They are available in two forms: built-in or on casters so that they can be rolled to the kitchen sink for use. Matching dryers can be installed next to the washer, stacked on a special rack, wall-hung, or purchased as a one-piece unit with the washer (Figure 1-14).

Automatic Dryers

Automatic dryers perform in the same tumbling manner as front-loading automatic washing machines. However,

FIGURE 1-15   The automatic dryer is available with a drop-down or side-opening door.

there are some key differences in design and special features from model to model and manufacturer to manufacturer.

Dryers are available in electric or gas. They vary in load capacity. Some models offer front or rear controls and side-opening or drop-down doors (Figure 1-15). For optimum efficiency, an electric dryer should have a minimum rating of 4400 watts. Gas dryers require a 120-volt outlet for such features as the motor, lights, and ignition. The gas heater should have a rating of at least 20,000 BTU/hour for top performance. Dryers should be installed in an area that permits proper venting.

Compact dryers are electric (either 120- or 240-volt). The 120-volt dryer takes at least twice as long to dry clothes as the 240-volt model does. While venting is recommended for all dryers, some 120-volt models can be used without venting if they are not in an enclosed space. The 240-volt dryers must be vented to prevent damage from moisture buildup in the home. Compact dryers can be installed next to the washer, stacked above the washer on a special rack, wall hung, or purchased as a one-piece unit with the washer.

Automatic dryers have the following advantages and disadvantages:

Advantages

- Eliminate the difficulties inherent in line drying.
- Give modern fabrics proper care, practically eliminating the ironing chore.
- Designs available to meet the needs of most disabled persons.

Disadvantages

- Models with rear controls are virtually impossible to operate from a seated position.
- Dryer door might be too low without a raised installation.
- As with washers, compact dryers have a smaller load capacity than other designs.

Checklist for Washers and Dryers

Before selecting and purchasing laundry equipment, fill out this checklist on washers and dryers. Check all that apply.

Washer
- What size is wanted?
  - ❑ Compact/portable ❑ Standard capacity
  - ❑ Large capacity ❑ One-piece washer/dryer combination
- Style:
  - ❑ Front-loading ❑ Top-loading
- Cycle selections:
  - ❑ Permanent press ❑ Delicate
  - ❑ Knits ❑ Pre-wash
  - ❑ Soak ❑ Extra-clean
- Options:
  - ❑ Variable water level ❑ Water temperature control
  - ❑ Extra rinse cycle ❑ Electronic controls
  - ❑ Water saver ❑ Small load basket
    (to reuse wash water)
  - ❑ Bleach dispenser ❑ Fabric softener dispenser
  - ❑ Detergent dispenser
- Color_____

Dryer
- Which type is preferred?
  - ❑ Electric ❑ Gas
  - ❑ LP or natural
- What size is needed?
  - ❑ One-piece washer/ ❑ Standard capacity
    dryer combination
  - ❑ Large capacity
- Cycle selections:
  - ❑ Permanent press/medium heat ❑ Delicate/low heat
  - ❑ No heat ❑ Timed cycles
  - ❑ Automatic drying ❑ No-heat tumbling at end of the drying cycle

- Other options:
    - ❏ Electronic controls
    - ❏ Drying shelf
    - ❏ Drop-down door
    - ❏ End-of-cycle signal
    - ❏ Side-opening door
- Color_____
- Drum (stainless or porcelain)_____
- Price range: $_____
- Warranty and service information:_____

Air Conditioners

Room air conditioners are self-contained units that can be installed in a window, through a wall, or moved around on wheels (Figure 1-16). The smaller air conditioners are designed to cool the immediate area of a room. However, manufacturers are designing room air conditioners from 4000 BTU/hour to over 30,000 BTU/hour. These larger-sized units will cool a larger area or cool multiple rooms in a home or office at one time. The standard features are manual or electronic controls, straight cool only, electric heat and cooling, or

FIGURE 1-16   Room air conditioners are available in a variety of sizes and configurations.

reverse-cycle air conditioning (heat pump). Air flow circulating from the air conditioners discharge grill might have fixed, directional, or motorized louvers. Some models are available with wireless remote control handheld units. The remote control makes it easier to control the on/off, fan speed, and temperature of the air conditioner from across the room. Room air conditioners are available in 120 volts or 230 volts. To properly install a room air conditioner, it is strongly recommended that you follow the manufacturer's recommendations. Factors that should be taken into account when purchasing an air conditioner include the following:

- Size
- Energy rating
- Location of air conditioner within the home or office
- Location of the electrical outlet and voltage needed to run the air conditioner

Room air conditioners have the following advantages and disadvantages:

Advantages

- Save money on electric bill by only cooling rooms needed.
- Easy installation.
- No ductwork needed.
- You can have different temperature settings for each room air conditioner in a home or office.

Disadvantages

- Room air conditioners are a little noisier than central air conditioning.
- Window air conditioners are drafty in winter time and most likely have to be removed.

Sizing Room Air Conditioners

To properly size an air conditioner that is needed for a particular room, you must determine the square footage of that room. If the room is rectangular or square (Figure 1-17), multiply the length by the width. This will provide you with the square footage of the area to be cooled. Some rooms may be oddly shaped, consisting of both a rectangular or square area and a triangular area (Figure 1-18). First determine the square footage of a triangle—measure the base of the triangle and the height of the triangle in feet. Now multiply both amounts, and divide that figure by two. The result will be the square footage of that triangle. Then determine the square footage of the rectangle or square as described previously. When you have determined the square footage of the area to be cooled, see Table 1-1 to find out what air conditioner capacity is needed.

   Once you have determined what size air conditioner is needed for the room, there are other circumstances that have to be taken into account before you purchase an air conditioner:

- Most air conditioners are designed to cool a room based on two occupants. For every additional person in the room, you will have to add 600 BTUs to the basic cooling capacity.

PART I

FIGURE 1-17
In a square or rectangular room, multiply the length by the width in feet to determine the square footage.

FIGURE 1-18
In oddly shaped room, determine the square footage of the rectangle. Then determine the square footage of the triangle. Subtract the square footage of the triangle from the total square footage of the rectangle to determine the square footage of the oddly shaped room.

Area To Be Cooled (square feet)	Capacity Needed (BTUs per hour)
100 to 150	5000
150 to 250	6000
250 to 300	7000
300 to 350	8000
350 to 400	9000
400 to 450	10,000
450 to 550	12,000
550 to 700	14,000
700 to 1000	18,000
1000 to 1200	21,000
1200 to 1400	23,000
1400 to 1500	24,000
1500 to 2000	30,000
2000 to 2500	34,000

TABLE 1-1   Room Air Conditioner Sizing Chart

- If the room to be cooled is not insulated, you will have to add 15 percent to the basic cooling capacity.
- You will have to add to the basic cooling capacity by 10 percent if the room has a west and/or southwest exposure.
- If the room to be cooled is facing north or northeast or is heavily shaded, the cooling capacity will be reduced by 10 percent.
- If you are only going to use this air conditioner at night, deduct 30 percent from the basic cooling capacity.
- If you have high ceilings, or if the ceilings are not insulated, increase the basic cooling capacity by 10 percent.

Figures are based on sizing an air conditioner in a moderate climate around 75 to 80 degrees Fahrenheit with an average humidity of 50 percent. If the climate is warmer and the humidity is higher, you might have to increase the basic cooling capacity by 10 to 20 percent.

How to Purchase the Best Air Contitioner or Appliance Value Using the EnergyGuide Label

Proper planning and evaluation before buying can save time, trouble, and money each step of the way. Take the time to determine the air conditioner or appliance's annual cost of operation.

Remember, while some energy-efficient products have higher purchase prices than less efficient ones, they will cost less in the long run because they require less electricity to operate. Calculate and evaluate the product's annual cost of operation and carefully read the EnergyGuide label, which appears on most appliances and air conditioners sold today, to get the best buy.

The EnergyGuide label (Figure 1-19) is required by the U.S. government on many home appliances and air conditioners, and gives information to help customers select and save. The information gained from the EnergyGuide label is important because is helps determine the following:

- It will help you compare the estimated annual operating costs of one model versus another.
- It will give information about the size range of the models being compared.
- It will tell how each model compares in terms of its energy costs to other models in the same size range.

FIGURE 1-19
The EnergyGuide label that appears on most appliances.

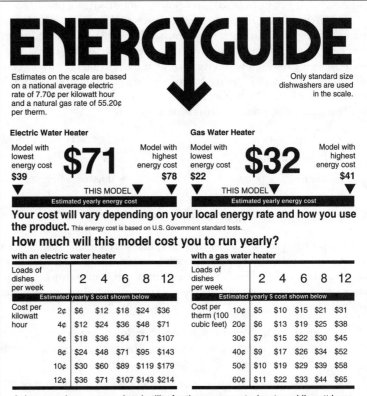

ENERGYGUIDE

Estimates on the scale are based on a national average electric rate of 7.70¢ per kilowatt hour and a natural gas rate of 55.20¢ per therm.

Only standard size dishwashers are used in the scale.

Electric Water Heater

Model with lowest energy cost **$71** Model with highest energy cost
$39 $78

THIS MODEL ▼

Estimated yearly energy cost

Gas Water Heater

Model with lowest energy cost **$32** Model with highest energy cost
$22 $41

THIS MODEL ▼

Estimated yearly energy cost

Your cost will vary depending on your local energy rate and how you use the product. This energy cost is based on U.S. Government standard tests.

How much will this model cost you to run yearly?

with an electric water heater

Loads of dishes per week		2	4	6	8	12
Estimated yearly $ cost shown below						
Cost per kilowatt hour	2¢	$6	$12	$18	$24	$36
	4¢	$12	$24	$36	$48	$71
	6¢	$18	$36	$54	$71	$107
	8¢	$24	$48	$71	$95	$143
	10¢	$30	$60	$89	$119	$179
	12¢	$36	$71	$107	$143	$214

with a gas water heater

Loads of dishes per week		2	4	6	8	12
Estimated yearly $ cost shown below						
Cost per therm (100 cubic feet)	10¢	$5	$10	$15	$21	$31
	20¢	$6	$13	$19	$25	$38
	30¢	$7	$15	$22	$30	$45
	40¢	$9	$17	$26	$34	$52
	50¢	$10	$19	$29	$39	$58
	60¢	$11	$22	$33	$44	$65

Ask your salesperson or local utility for the energy rate (cost per kilowatt hour or therm) in your area, and for estimated costs if you have a propane or oil water heater.

Important Removal of this label before consumer purchaser is a violation of federal law (42 U.S.C. 6302)

To read the EnergyGuide label, first look for the estimated annual energy cost in the center. To either side are energy costs of lower-rated and higher-rated models. These costs are derived from national average electricity rates, so knowledge of the local rate for electricity is helpful.

Major Appliance Consumer Action Panel (MACAP)

The Major Appliance Consumer Action Panel, or MACAP, is an independent, complaint-mediation group made up of professionals with expertise in textiles, equipment, consumer law, and engineering who volunteer their time. Panelists cannot be connected with the appliance industry. They receive no financial remuneration other than the reimbursement of their travel and living expenses while attending meetings. MACAP receives comments and complaints from appliance owners; excessive charges, delays in parts and service, and alleged unnecessary repairs are frequently mentioned. Other complaints are related to product performance, such as operating noise, temperature, maintenance, and running time. Non-responsiveness of dealers and manufacturers, warranty coverage, food loss claims, imperfect finishes, improper installation, and purchasing dissatisfaction are also alleged.

MACAP also studies industry practices and advises industries of ways to improve their services to consumers. It recommends how to educate consumers on proper appliance purchase, use, and care. The panel develops and distributes educational publications and periodic news releases when its review of individual consumer complaints pinpoints information that would be useful to consumers. The panel is sponsored by the Association of Home Appliance Manufacturers.

The types of appliances represented include:

- Compactors
- Dehumidifiers
- Microwave ovens
- Refrigerators and freezers
- Room air conditioners

- Laundry equipment
- Ranges
- Dishwashers
- Disposers

After a complaint reaches MACAP, the staff screens it so as to ascertain whether the consumer has already requested assistance from the local dealer and brand-name owner's headquarters office. If this has been done, the complaint is sent to the sponsoring association and thus enters MACAP's communications phase. This phase comprises the following actions:

- Copies of the consumer's correspondence or a summary of telephone comments are sent to the brand-name owner involved, requesting a report on proposed action within two weeks.

- A letter is sent to the consumer acknowledging receipt of the complaint, reporting action taken, and asking for any additional information, if necessary.

- When an answer is received from the brand-name owner, the staff writes a letter to the consumer confirming any company action or information. The consumer is asked to return a card verifying this action.

- The file is then resolved, unless the consumer otherwise advises MACAP.
- If the complaint reaches an impasse, the file moves to the study phase.

In the study phase, the panel discusses the file at a meeting. Preparation for this review includes the gathering of an exact and detailed background of the complaint from the consumer and from the brand-name owner, if such information is not already on file. If conflicting reports are received, the panel might ask for an independent, on-site evaluation by a utility, extension, university home economist, or an engineer.

All information is included in a confidential summary prepared by the staff and discussed by MACAP. The panel might make a recommendation to the company and/or consumer, ask for additional information, or close the file on the basis of the information presented.

Purchasing Decisions

The purchase of a major appliance is one of the most important investments made for the home. MACAP's experience in handling consumer appliance complaints has shown that poor purchasing decisions lead to unhappy appliance owners.

Pointers from MACAP to help consumers make wise decisions include the following:

- Ask the dealer for specification sheets from several manufacturers of the appliance types you plan to purchase. Study them carefully, and note the different features, designs, and capacities.
- Ask the dealer to see the warranty before purchasing the appliance. Does the warranty cover the entire product? Only certain parts? Is labor included? How long is the warranty coverage?
- Ask the dealer for the use and care manual. Read it carefully before you purchase the appliance. The dealer should have manuals available from the floor models on display. These manuals will help in asking pertinent questions, and they will explain how the product operates and what special care it needs.
- Decide what special features are essential. Consider the possibility of adding on features at a later date, such as an ice maker for a refrigerator.
- Check the space available for the appliance. Will it fit where it is planned? Is there adequate clearance space in the hallways and doors through which the appliance will have to pass before installation?
- Check the product design carefully prior to purchase. Compare the designs of different brands. If a combination microwave oven/range is being purchased, check the space between the units to be sure that everything will fit.
- Clearly establish the cost of delivery and installation. Are these costs included, or are they extra?
- Ask the dealer if he services the appliances he sells. If not, ask him where to go for authorized factory service.
- Compare price in relation to convenience and service. Both vary according to the model. As more features and conveniences are included, the price increases.
- Be sure the house has adequate electrical service for the appliance in order to avoid overloading circuits. Also, be sure it has adequately grounded, three-hole receptacles.[4]

Appliance Warranties

MACAP urges consumers to compare warranties of different brands the way you would compare price, size, and features—it's a basic consumer responsibility! Yet the panel's experience in working with consumer appliance problems indicates that many consumers are not aware of the importance of warranty buymanship.

Warranty inspection is a legislated consumer right. A federal law, the Magnuson-Moss Warranty/Federal Trade Commission Improvement Act of 1975, requires warranty information to be available at the point of purchase for products costing $15 or more. The law does not, however, require manufacturers to provide warranties on their products.

The store must provide this information in one of the following ways:

- Displayed near the appliance.
- Shown on the package displayed with the appliance.
- In an indexed and updated binder that is prominently displayed, containing all warranties for products sold in the department.
- If ordering the appliance through a catalog, the catalog must include the warranty text or offer it upon request.[5]

Full Versus Limited Warranties

The full warranty offers more protection. Under a full warranty, at a minimum, the warrantor must remedy the problem within a reasonable time and without charge for as long as the warranty is in effect. In addition, the manufacturer might not limit the duration of any implied warranty. Any limitation (or exclusion) of consequential damages (for example, food loss or floor damage) must appear conspicuously on the warranty. If a reasonable number of attempts to correct an in-warranty problem fails, the consumer must be given the choice of a replacement or a refund.

Under a limited warranty, the protection is limited to what is outlined in the terms of the warranty. For example, the warranty might require that the consumer pay for diagnostic costs, labor costs, or other costs of that kind. Also, any implied warranty might be limited to the duration of the written warranty. A limited warranty should be studied carefully to determine exactly what the warrantor will provide and what costs the consumer must pay.[5]

Warranty Time Limitations

Some major appliances are hardly used over a year's time. The window air conditioner in Minnesota, the refrigerator at the lake cabin in Michigan, the range in a Florida vacation condominium—all of these have only seasonal use.

When problems with these appliances arise, some owners feel that four years of seasonal usage should constitute less than one year's worth of warranty coverage. The manufacturers, however, think differently. MACAP knows it is necessary for consumers to be aware of the contents of their appliance warranties, and also of the period covered. The coverage is stated for a specific period of time, not for how often the appliance is used.

The coverage period might vary with the product's components. For example, a window air conditioner warranty might provide free repairs of any part that breaks down in the first year, but only partially cover repair expenses of the sealed system components (compressor, evaporator, etc.) for an extra four years. Manufacturers can offer almost any type of warranty as long as the provisions are clearly stated and the warranty is available for review before buying.

Although seasonal usage might result in less wear and tear on the appliance, MACAP believes that there are at least two valid reasons for not basing warranties on usage:

- Basing a warranty on usage (rather than on appliance age) is unrealistic because it's impossible for a manufacturer to monitor actual consumer usage.

- Some appliances actually suffer more from extended periods of non-use than from continuous daily operation.

For example, refrigerator and air conditioner sealed systems stay vacuum-tight (generally for many years) if the refrigerant gas and oil they contain are kept moving. This way, the various seals are kept lubricated and soft. They might dry out if the mechanisms involved don't move for long periods. The dishwasher provides yet another example. Many consumers don't realize that the seals in the pump area of the dishwasher are meant to be immersed in water at all times. Water keeps the seal soft and pliable, which is necessary for proper operation.

MACAP urges consumers to carefully read the warranty that comes with each major appliance and use the appliance enough during the warranty period so that any defects have time to surface. It is important to use every feature and control on the new appliance soon after it is installed.[6]

Appliance Installation Instructions

The manufacturer supplies the installation instructions with every new appliance or air conditioner that is purchased. These instructions will help the installer to plan, locate, install, and secure the product for safe and proper operation.

Appliance Use and Care Manual

This manual contains information and suggestions to help customers get the best results from their appliances. The manual will disclose to the customer how to start the appliance, maintain it, and use all the features that come with it. Also included in the use and care manual are the following:

- Safety precautions
- What to do before using the appliance
- Maintenance instructions
- Vacation and moving care
- Warranty information

- Parts and features
- How to use the appliance
- Common problems and solutions
- Assistance

Where to Get Help

Keep careful records. Always put complaints in writing, and keep copies of all correspondence and service receipts. Be sure to ask for service receipts, even for no-charge, in-warranty calls. Note details: when the problem was first noticed, when it was reported, and the servicing history (who serviced the appliance or air conditioner, when, what was done, and how often service was required).

If there are complaints about the appliance or air conditioner, there are three steps to follow:

- Read the use and care manual that comes with the product. Also check the plug, as well as fuses, pilots, and controls.

- Call the service company authorized to fix the brand. They have the training and equipment to deal with appliance or air conditioner service problems.

- If not satisfied, contact the manufacturer's main customer relations office. The address and phone number are located in the use and care manual.

If none of these steps solves the problem, write to: Major Appliance Consumer Action Panel, 20 North Wacker Drive, Chicago, IL 60606.

Product Recalls and the Internet

There are times when a product might have an electrical, gas, or mechanical issue that might be hazardous to the consumer. The U.S. Consumer Product Safety Commission (CPSC) was created to protect the consumer from serious injuries, unreasonable risks, death, and property damage.

The CPSC's Web site (http://www.cpsc.gov) includes information on appliance, air conditioner, and other product recalls. It also includes other information beneficial to the consumer. Other Web sites helpful to the consumer and technician include http://www .recalls.gov and http://www.pueblo.gsa.gov. In addition, Table 1-2 provides a listing of Web sites for major appliance and air conditioner manufacturers. You can also look in the use and care manual that comes with the product for the manufacturer's Web site. Remember, safety and education must be considered at all times when operating or repairing any major appliance or air conditioner.

National Appliance Service Technicians Certification (NASTeC)

The roots of NASTeC go back to the Certified Appliance Technician (CAT) certification issued by ISCET (International Society of Certified Electronics Technicians) in the mid-1970s, when the bulk of appliance servicing was done by retail appliance dealers, who serviced the products they sold. In many instances, these same dealers sold items such as televisions, recorders of various types, stereos, radios, etc. Since ISCET was already in contact with these dealers with respect to the other electrical appliances, it was natural that the major appliances known as "white goods" should be served as well.

The North American Retail Dealers Association (NARDA) attracted dealers and thus servicers to join their efforts at promoting the retail dealers' interests through their trade organization. This put NARDA in a position to better offer its members an appliance technician certification. NASTeC was then born as a separate entity from ISCET, but as a partner with equity in NASTeC.

As the "white goods" dealers became more separated from the televisions, recorders, etc., the servicing of these products became independent operations. This diminished the ability of NARDA to offer appliance certifications.

www.amana.com	www.kitchenaid.com
www.americanwaterheater.com	www.kitchenaid.com
www.boschappliance.com	www.kuppersbuschusa.com
www.bradfordwhite.com	www.lge.com
www.broan.com	www.lifeluxurymarvel.com
www.brockwaterheaters.com	www.maytag.com
www.carrier.com	www.miele.com
www.convaircoller.com	www.northlandnka.com
www.dacor.com	www.panasonic.com
www.danby.com	www.rheem.com
www.delonghi.com	www.roperappliances.com
www.dynastyrange.com	www.samsung.com
www.eemaxine.com	www.sears.com
www.electrolux.com	www.sharpusa.com
www.equatoronline.com	www.staber.com
www.fedders.com	www.subzero.com
www.friedrich.com	www.sunfrost.com
www.frigidaire.com	www.thermador.com
www.gaggenau.com	www.thorappliances.com
www.ge.com	www.u-line.com
www.goldstarappliances.com	www.vaughncorp.com
www.haieramerica.com	www.veronaappliances.com
www.hatachi.com	www.vikingrange.com
www.hotwater.com	www.weathercraft.com
www.insinkerator.com	www.whirlpool.com
www.jadeappliances.com	www.wolfappliance.com
www.jennair.com	www.zephyronline.com

TABLE 1-2   Major Appliance and Room Air Conditioner Manufacturers

In 2002, ISCET secured the interests of NARDA in the NASTeC venture and began developing the NASTeC certifications to once again actively work with the appliance-servicing organizations to establish both certification and education in the appliance-servicing segment of the "white goods" industry.

Certification is offered in basic skills, laundry equipment, cooking equipment, and refrigeration equipment. Once a technician acquires all four certificates, he is considered a "universal technician" and is awarded that certificate. Each step is signified by shoulder patches that identify a person as a certified appliance technician. These technicians are known for their abilities and professional approach to both the customers and their work.[7] For more information, visit the NASTeC Web site (http://www.nastec.org).

Endnotes

1. Pyrolytic cleaning is the true self-cleaning system. It uses high heat during a special 1- to 3-hour cycle to decompose food, soil, and grease. During the cycle, which is clock-controlled, the oven door is latched and locked. It cannot be opened until the oven cools down. All of the oven walls, racks, and the door (except for a small area outside the door gasket) are completely cleaned. After cleaning, a small bit of white ash might be found, which can be easily wiped up.

2. Catalytic and continuous-cleaning ovens use a special porous coating on the oven walls that partially absorbs and disperses the soil. This process takes place during normal baking and keeps the oven presentably clean, but the racks and door parts must be cleaned by hand. Some manufacturers recommend occasionally operating an empty oven at 500 degrees Fahrenheit to remove any buildup of soil. This special oven coating cannot be cleaned with soap, detergent, or commercial oven cleaners without causing permanent damage.

3. Eye-level ranges are also referred to as high-low ranges or tri-level ranges.

4. Reprinted from MACAP Consumer Bulletin, Issue no. 8, December 1985.

5. Reprinted from MACAP Consumer Bulletin, Issue no. 1, December 1979.

6. Reprinted from MACAP Consumer Bulletin, Issue no. 5, February 1983.

7. Reprinted with permission from NASTeC, April 2006.

Safety Precautions

S *afety* starts with accident prevention. Injuries are usually caused because learned safety precautions are not practiced. In this chapter are listed some tips to help the technician to correctly and safely install, operate, and repair major appliances.

Any person who cannot use basic tools should *not* attempt to install, maintain, or repair any major appliance or air conditioner. Any improper installation, preventative maintenance, or repairs will create a risk of personal injury, as well as property damage. Call the service manager if installation, preventative maintenance, or the repair procedure is not fully understood.

Every technician should carry a first aid kit and should know how to properly use the contents of that kit and its location. Technicians should carry a fire extinguisher in their service vehicles in case of an emergency. It is also recommended that they take a first aid course, such as those offered by the American Red Cross.

Safety Procedures

Individual, electrical, gas, chemical, appliance and air conditioner, operating, and installation safety precautions are generally the same for all major appliances and air conditioners. Carefully observe all safety cautions and warnings that are posted on the appliance or air conditioner being worked on. Understanding and following these safety tips can prevent accidents.

Individual Safety Precautions

Protecting yourself from injuries is vital. Before installing, maintaining, or servicing any major appliance or air conditioner, do the following:

- Wear gloves. Sharp edges on appliances or air conditioners can hurt hands.
- Wear safety shoes. Accidents are often caused when dropping heavy appliances, especially on feet that are not protected.
- Avoid loose clothing that could get caught in the appliance while it is operating.
- Remove all jewelry when working on appliances.

- Tie long hair back.
- Wear safety glasses to protect your eyes from flying debris.
- Use proper tools that are clean and in good working condition when repairing appliances and air conditioners.
- Have ample light in the work area.
- Be careful when handling access panels or any other components that may have sharp edges.
- Avoid placing hands in any area of the appliance or air conditioner that has not been visually inspected for sharp edges or pointed screws.
- Be sure that the work area is clean and dry from water and oils.
- When working with others, always communicate with each other.
- Always ask for help moving heavy objects.
- When lifting heavy appliances or air conditioners, always use your leg muscles and not your back muscles.

Electrical Safety Precautions

Know where and how to turn off the electricity to the appliance or air conditioner. For example: Know the location of plugs, fuses, circuit breakers, and cartridge fuses in the home. Label them. If a specific diagnostic check requires that voltage be applied, reconnect electricity only for the time required for such a check, and disconnect it immediately thereafter. During any such check, be sure that no other conductive parts come into contact with any exposed current-carrying metal parts. When replacing electrical parts or reassembling the appliance or air conditioner, always reinstall the wires back on their proper terminals according to the wiring diagram. Then check to be sure that the wires are not crossing any sharp areas, are not pinched in some way, are not between panels, and are not between moving parts that may cause an electrical problem.

These additional safety tips are also important to remember:

- Always use a separate, grounded electrical circuit for each major appliance or air conditioner.
- Never use an extension cord for major appliances.
- Be sure that the electricity is off before working on the appliance or air conditioner.
- Never remove the ground wire from a three-prong power cord or any other ground wires from the appliance or air conditioner.
- Never bypass or alter any appliance or air conditioner switch, component, or feature.
- Replace any damaged, pinched, or frayed wiring before repairing the appliance or air conditioner.
- Be sure that all electrical connections within the product are correctly and securely connected.

Gas Safety Precautions

As a technician, I have full respect for electricity and gas fuels. Like electricity, natural and liquefied petroleum (LP or LPG) gas can become dangerous if not handled properly. First, locate the gas shutoff valve by the appliance, and know where the main gas shutoff is located before beginning repairs. The fuel supply (natural or LP gas), supply lines, pressure regulator, and LP storage tank will need servicing at least once a year to make certain that there are no leaks. All gas appliances must be kept clean and free of soot.

The following are some additional safety procedures that a homeowner or technician must know before working on gas appliances:

- Never smoke or light a flame when working on gas appliances.
- Always follow the manufacturer's recommendations as stated in the use and care manual for gas appliances.
- Keep all combustible materials, such as gasoline, liquids, paint, paper materials, and rags, away from gas appliances.
- Do not allow pilot lights to go out. The gas fumes will seep into the room, causing a hazardous condition for anyone who lights a match, turns on a light switch, or uses the telephone.
- When working on gas appliances always have a fire extinguisher nearby.
- When using gas appliances, always have the proper ventilation according to the manufacturer's specifications.
- Keep ventilation systems clean of debris and obstructions.
- If you smell gas where the appliance is located, shut off the gas supply and vent the room. Warning: If you cannot turn the gas supply off, you must leave the area, go to a neighbor, and telephone the gas company and the fire department.
- When the flame turns from a blue color to a yellow color, or if soot forms, you could have a condition known as incomplete combustion. Warning: If not corrected immediately, you could have carbon monoxide buildup, which could be potentially fatal. Carbon monoxide is a poisonous nonirritating gas that is odorless, colorless, and has no taste. If you breathe in carbon monoxide, it will mix with your blood and prevent oxygen from entering into the blood supply, causing illness or death.
- All homes should have smoke and carbon monoxide detectors installed.

Chemical Safety Precautions

Chemicals are also dangerous. Knowledge of chemical safety precautions is essential at all times. The following tips are examples of important practices:

- Remove all hazardous materials from the work area.
- Always store hazardous materials in a safe place and out of the reach of children.
- Before turning on appliances that use water, run all of the hot water taps in the house for approximately five minutes. This clears out the hydrogen gas that can build up in the water heater and pipes if they have not been used for more than two weeks.

Appliance and Air Conditioner Safety

Call the service manager to check out the appliance or air conditioner if the safety of the product is in doubt.

Only use replacement parts of the same specifications, size, and capacity as the original part. If you have any questions, contact your local appliance and air conditioner parts dealer, your service manager, or the manufacturer of the product.

Check water connections for possible water leaks before reconnecting the power supply. Then completely reassemble the appliance, remembering to include all access panels.

Operating Safety

After repairing the appliance or air conditioner, do not attempt to operate it unless it has been properly reinstalled according to the use and care manual and to the installation instructions supplied by the manufacturer. If these instructions are not available, do not operate the appliance or air conditioner. Call the service manager to check out the reinstallation or ask for a copy of the installation instructions from the manufacturer.

Know where the water shutoff valves are located for the washer, dishwasher, ice maker, and water heater, as well as the house's main water shutoff valve. Label them. Following these additional safety tips can also prevent injuries:

- Do not allow children to play on or to operate appliances or air conditioners.

- Never allow anyone to operate an appliance or air conditioner if they are not familiar with its proper operation.

- When discarding an old appliance, remove all doors to prevent accidental entrapment and suffocation.

- The refrigerant in refrigerators, freezers, and air conditioners will have to be reclaimed before disposal of the product can begin.

- Instruct the customer to use the appliance or air conditioner only for the job that it was designed to do.

Installation Safety Precautions

The first step in assuring safety with major appliances and air conditioners is to be sure that they are installed correctly. Read the installation instructions and the use and care manual that comes with the appliance or air conditioner. Observe all local codes and ordinances for electrical, plumbing, and gas connections. Ask your local government agency about these codes. Additional safety tips include the following:

- Carefully observe all safety warnings that are contained in the installation instructions and in the use and care manual.

- The work area should be clear of unnecessary materials so that there is plenty of room to work on the appliance or air conditioner.

- Be sure that the appliances are installed and leveled on a floor that is strong enough to support their weight.

- The appliance or air conditioner should be protected from the weather, as well as from freezing or overheating.

- The appliance or air conditioner should be correctly connected to its electric, water, gas, drain, and/or exhaust system. It should also be electrically grounded.
- Be sure that the appliance has a properly installed anti-tip device, as in the case of kitchen ranges.

Grounding of Appliances

In 1913, the National Electrical Code (NEC) made grounding at the consumer's home mandatory. The NEC required that range frames be grounded to a neutral conductor in 1943. Later, in 1946, it required that receptacles in laundry areas be grounded. Soon after, the NEC required that the frames of automatic dryers be grounded to neutral conductors. Then, in 1959, NEC required that automatic washers, automatic dryers, automatic dishwashers, and certain motor-operated handheld appliances be grounded. All 15-amp and 20-amp branch circuits have grounding type receptacles, as specified in 1962. Finally, in 1968, the code required that refrigerators, freezers, and air conditioners be grounded.

The greatest importance of grounding appliances and air conditioners is that it prevents people from receiving shocks from them. However, the major problem associated with the adequate grounding of appliances and air conditioners is that many homes are not equipped with three-prong or four-prong grounded receptacles. To solve this problem, the consumer must install, or have installed, a properly grounded and polarized three-prong or four-prong receptacle. A qualified electrician should connect the wiring, and properly ground and polarize the receptacle.

Remember that safety is the paramount concern, especially when dealing with electricity. Both the technician and the consumer must be aware that it only takes about 100 milliamperes of current to cause death in one second. Here are some safety tips:

- Do not install or operate an appliance or air conditioner unless it is properly grounded.
- Do not cut off the grounding prong from the appliance or air conditioner plug.
- Where a two-prong wall receptacle is encountered, it must be replaced with a properly grounded and polarized three-prong receptacle.
- Call the service manager if you doubt your abilities. When dealing with electricity, there is no leeway for mistakes.

Checking Appliance and Air Conditioner Voltage

If it becomes necessary to test an appliance or air conditioner with the voltage turned on, observe the following precautions:

- The floor around the appliance or air conditioner must be dry. Water and dampness increase the probability of a shock hazard.
- When using a multimeter, always set the meter correctly for the voltage being checked.
- Handle only the insulated parts of the meter probes.

- Touch components, terminals, or wires with the meter probe tip only.
- Touch the meter probe tips only to the terminals being checked. Touching other components could damage good parts.
- Be sure that the appliances have properly installed anti-tip devices, as found on kitchen ranges.

What to Do If You Smell Gas

If there is a gas smell, the following precautions should be taken:

- Turn off the gas supply.
- Do not light matches.
- Do not turn the lights on or off.
- Do not use the telephone.
- Open the windows to ventilate the room.

Then call the gas company to come out and repair the gas leak.

CHAPTER 3

Tools Needed for Installation and Repair

A basic knowledge of hand tools, electrical tools, and test meters is necessary to effectively complete most installations and repairs. This chapter will cover the basics of each tool. A working knowledge of these tools is a must for the installation and repair of appliances and air conditioners. Always follow safety precautions and manufacturer's recommendations when handling tools.

Before starting on any type of repairs, take the time to put together a toolkit with a selection of good quality hand tools. A partial list of common hand tools includes:

- **Screwdrivers.** A complete set of flat-blade screwdrivers, ranging from 1/8 inch to 5/16 inch. Handle sizes may vary with the blade dimension. Phillips-tip sizes also vary; the two most common are #1 and #2.

- **Nut drivers.** A complete set is recommended. The common sizes are: 3/16, 1/4, 5/16, 11/32, 3/8, and 1/2 inch.

- **Wrenches**

  - Socket wrenches. Either 6-point or 12-point, ranging in size from 5/32 to 1 inch.

  - Box wrenches. Common sizes range from 1/4 to 11/2 inch.

  - Open-end wrenches. Common sizes range from 1/4 to 15/8 inch.

  - Adjustable wrenches. The handle size indicates the general capacity. For example, a 4-inch size will take up to a 1/2-inch nut. A 16-inch handle will take up to a 17/8-inch nut.

  - Allen wrenches. Sizes range from 1/16 to about 1/2 inch.

- **Claw hammer**

- **Adjustable pliers**

- **Flashlight**

- **Drop-cloth**

Safety Precautions

Safety starts with accident prevention. Listed in this chapter are some tips to help the technician when using hand and power tools.

WARNING *Any person who cannot use basic tools should* not *attempt to install, maintain, or repair any major appliance or air conditioner. Any improper installation, preventative maintenance, or repair creates a risk of personal injury and property damage.*

Individual Safety Precautions

Injuries abound when using tools. To be protected from injuries when using hand tools and power tools, do the following:

- Wear gloves.
- Avoid wearing loose clothing when working with power tools.
- Wear safety glasses to protect the eyes from flying debris.
- Use tools according to manufacturer's specifications, and never alter their use.

Safety Precautions When Handling Tools

Regardless of which tool is being used, these same rules of care and safety apply:

- Keep tools clean and in good working order.
- Use the tool only for jobs for which it was designed.
- When using power tools, be certain that the power cord is kept away from the working end of the tool.
- If the tool has a shield or guard, be sure it is working properly and remember to use it.
- If an extension cord is used, be sure it is in good working order. Do not use it if bare wires are showing. Also, use a heavy-gauge wire extension cord to ensure adequate voltage for the tool being used.
- Be sure that the extension cord is properly grounded.
- Grip the tool firmly.
- Never use worn-out tools. A worn-out tool has more potential for causing injuries. For example, with a worn-out screwdriver, there is a greater possibility for slips, which could make medical attention necessary.
- If there is a problem with a power tool, never stick your fingers in the tool. Unplug it first, and then correct the tool's problem.
- Make sure to use insulated hand tools when working with electricity and electrical components.

Screwdrivers

A screwdriver is a hand tool used either to attach or remove screws. The two most common types are the flat-blade and Phillips. The flat-blade screwdriver is used on screws that have a slot in the screw head (Figures 3-1 and Figure 3-3). The flat-blade screwdriver is available in

FIGURE 3-1 Flat-blade screwdriver.

FIGURE 3-2 Phillips screwdriver.

many sizes and shapes. Always use the largest blade size that fits snugly into the slot on the screw head so that it will not slip off the screw. The screwdriver should never be used as a pry bar or a chisel; it was not designed for that purpose.

A Phillips screwdriver is used to attach or remove screws that have two slots crossing at right angles in the center of the screw head (Figures 3-2 and Figure 3-3).

When using a Phillips screwdriver, exert more pressure downward in order to keep the tool in the slots. Always use the largest Phillips size that fits snugly into the slots, just as with the flat-blade screwdriver.

Never use worn-out screwdrivers when working on appliances. A worn screwdriver may damage the head of the screw. It can also damage the product on which you are working.

Nut Drivers

Many manufacturers use metal screws with a hexagonal head. A nut driver is a hand tool similar to a screwdriver, except that the working end of the driver is hexagonal-shaped and fits over a hexagonal nut or a hexagonal bolt head. Each size nut requires a different sized driver (Figure 3-4).

Wrenches

Wrenches are the most frequently used tool. There are many different types and sizes of wrenches. Their purpose is to turn nuts and bolts. Wrenches are generally available in five different types (Figure 3-5):

- Socket wrenches
- Box wrenches

FIGURE 3-3 Combination screwdriver set.

FIGURE 3-4 Hex-nut drivers.

FIGURE 3-5   Wrenches are used to remove and fasten nuts and bolts. They are available in socket wrench, box, open-end, adjustable, and Allen types. (*continued*)

FIGURE 3-5 Wrenches are used to remove and fasten nuts and bolts. They are available in socket wrench, box, open-end, adjustable, and Allen types.

- Open-end wrenches
- Adjustable wrenches
- Allen, or hex, wrenches

Socket wrenches are used to slip over bolt heads, as opposed to other wrenches listed, which are used at right angles to the nut or bolt. This arrangement allows more leverage to be applied to loosen or tighten the nut or bolt (Figure 3-6).

Always choose the correct wrench for the job. Box wrenches should be used for heavy-duty jobs and in certain close-quarter situations. Open-ended wrenches are useful for medium-duty work or situations where it is impossible to fit a socket or box wrench on a nut, bolt, or fitting from the top. Adjustable wrenches help with light-duty jobs, work well with odd-sized nuts and bolts, and are useful where a regular open-end wrench could be used. It is adjustable to fit any size object within its maximum opening. Allen wrenches are used for adjusting and removing fan blades or other components that are held in place by Allen set screws. The Allen wrench has a six-pointed flat face on either end.

FIGURE 3-6
(a) socket wrench,
(b) adjustable wrench,
(c) open-end wrench.

(a)

(b)

(c)

Hammers

A hammer is a hitting tool. There are many sizes and styles of hammers available (Figure 3-7). The most common type used in appliance repairs is the claw hammer. The claw hammer can also be used for prying objects.

Prying Tools

Prying tools are available in many sizes and shapes. The most common are the crowbar, ripping bar, and the claw hammer. The claw hammer is basically used for light-duty work: removing nails and prying small objects. The ripping bar is used for medium-grade work, and the crowbar is used for heavier work.

FIGURE 3-7
While there are many different kinds of hammers, the claw hammer is the one most often used.

Rubber mallet

Claw hammer

Ball-peen hammer

Pliers

Pliers are one of the most frequently used tools. A pliers is a tool for holding or cutting, depending on the type. Generally, they are not made to tighten or unscrew heavy nuts and bolts. They are available in many sizes and shapes (Figure 3-8). Choose a pliers that fills a particular need, being careful that it is the proper pliers for the job. Some of the most common types of pliers include:

- Slip-joint
- Slip-joint adjustable
- Vise grip
- Needle nose
- Diagonal cutting

Slip-joint pliers are pliers for everyday tasks (Figure 3-8a). The jaws can be adjusted into two different positions. Do not use them on nuts, bolts, or fittings. They can easily slip and injure both the technician and the device.

Pump pliers, also known as *slip-joint adjustable pliers,* are also used for general jobs (Figure 3-8i). They would be preferred over slip-joint pliers when working on a larger object. The jaws of slip-joint adjustable pliers can be moved into many different positions.

The *vise grip pliers* is actually four tools in one (Figure 3-8j): a clamp, a pipe wrench, a hand vise, and pliers. The lever holds the jaws in one position, allowing the vise to hold up to one ton of pressure.

The *needle nose pliers* are mostly used with electronic, telephone, and electrical work (Figure 3-8b, Figure 3-8f, and Figure 3-8g). Other uses include in confined areas, to form wire loops, and to grip tiny pieces firmly. (The long nose is particularly useful for this latter task.) They are also available with side cutters.

Diagonal cutting pliers are used in electrical and electronic work (Figure 3-8c, Figure 3-8d, and Figure 3-8e). They are used for cutting wire and rope.

FIGURE **3-8**
Pliers are available in many sizes and shapes.

(a) Heavy duty slip-joint pliers

(b) Long needle nose pliers

(c) Heavy duty diagonal cutter

(d) Diagonal cutter

(e) Heavy duty lineman's side cutter

(f) Long nose pliers

(g) Long reach needle nose pliers

(h) Hose clamp pliers

(i) Slip-joint adjustable pliers

(j) Vise grip pliers

Cutting Tools

Many different types of tools are used for cutting. The key is to know which tool to use in each situation. Chisels are used for cutting metal and wood. They are made of high-carbon steel, which makes them hard enough to carve through metal (Figure 3-9). These should be used when removing rusted bolts and nuts.

Hacksaws are used for cutting metal (Figure 3-10). The hacksaw consists of a handle, frame, and a blade. The frame is adjustable, so it can accept any length of blade. The blades are available with different numbers of teeth per inch.

FIGURE **3-9**
Chisels comes in many different sizes and shapes, such as wood chisels, metal chisels, and concrete chisels.

Pin punches

Solid punches

Center punches

Cold chisels

Star drills

Wood chisel

FIGURE 3-10
Saws are used for cutting metal, wood, etc.

Hacksaw

A file is also a cutting tool. It is used to remove excess material from objects. Files come in a variety of sizes and shapes (Figure 3-11).

Drill bits are also cutting tools. They are designed for cutting holes in metal, wood, and concrete (Figure 3-12).

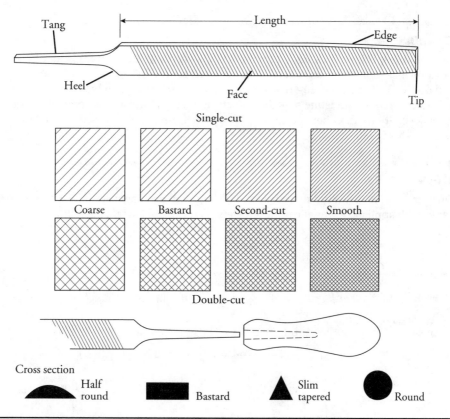

FIGURE 3-11   Files are used to smooth the rough edges off metals, wood, etc.

Figure 3-12
A general-purpose twist drill set will handle most of the technician's needs. The wood bit and masonry drill bit are used in installation work.

Wood bit

Twist drill

Masonry drill
with carbide tip

Power Tools

Power tools do the same job as hand tools. However, they do the job faster (Figure 3-13). The most common power tools can be either electric or battery-powered.

When using power tools, there are some safety precautions that must be followed to prevent accidental injury. Always read the use and care manual that comes with each power tool.

Specialty Tools

These tools are specifically designed for a particular use and are used for in-depth servicing of the appliance. For example, special tools are used for the installation and removal of special screws and nuts; they are required to remove the bearings in washing machines; and they are also used for adjusting switch contacts. Figure 3-14 illustrates some of the many types of specialty tools. Other specialty tools and their uses will be discussed in later chapters of this book.

Test Meters

Test instruments are important tools used in assisting with the diagnosis of the various problems that arise with appliances. Varieties of test meters (Figure 3-15) include the following:

Figure 3-13
Power tools can have either a line- or battery-operated power supply.

Electric drill

Battery drill

"Pinch-off" pliers

Hose clamp pliers

External snap ring pliers

Internal snap ring pliers

Spanner wrench

Wire stripper

FIGURE 3-14 In addition to the basic hand tools, a number of special-purpose tools are used in appliance servicing.

Clamp-on ammeter

Volt-ohm milliammeter

Thermometer

Test leads

FIGURE 3-15 Test meters are available in analog or digital readouts. For example, thermometers measure the temperature.

- The analog volt-ohm-milliammeter is used for testing the resistance, current, and voltage of the appliance or air conditioner. This type of meter moves a needle along a scale. It is also the most important test meter to have in the tool box.
- The digital multimeter displays a readout in numbers, usually on an LCD screen. This type of meter is similar in operation to the analog meter.
- An ammeter is a test instrument that is connected into a circuit to measure the current of the circuit without interrupting the electrical current.
- A wattmeter is a test instrument used to check the total wattage drawn by an appliance.
- A temperature tester is a test instrument used to measure the operating temperatures of the appliance.

Advantages of Digital Meters

- A digital meter offers greater sensitivity, accuracy, and a better resolution in reading measurements.
- Digital meters have a faster readout as compared to analog meters.

Basic Techniques

There are many different ways to diagnose a problem, but all of them basically use the same reasoning of deduction:

- Where does the consumer think the malfunction is located within the appliance or air conditioner?
- Where is the actual problem located within the appliance or air conditioner?
- Are there any related problems with the appliance or air conditioner?
- How can the problem with the appliance or air conditioner be solved?

For example, the consumer states that the dryer does not dry the clothes and believes that the heating element is bad. The actual problem might be a restricted exhaust vent, a clogged lint filter, bad heating elements, faulty operating thermostats or safety thermostat, or improper control settings.

When checking the dryer, you might notice that the control settings are set for air drying instead of heat drying. Thus, the actual problem was that the control settings were not positioned correctly. The related problem is: "How did the control setting move to the air-dry position?" This leads to the question: "Does the consumer know how to operate the dryer?" To solve this problem, you will have to instruct the consumer in the proper operation of the dryer.

All appliances and air conditioners go through a certain sequence of events. Understanding the proper operation and this sequence as indicated in the use and care manual is beneficial when diagnosing the appliance or air conditioner.

Product Failure

Given the information about the product's problem, information and diagnostic charts from this book, and the information you have read in the use and care manual, as the servicer, you will be able to perform the following steps in sequence to diagnose and correct a malfunction. The basic steps to follow when diagnosing an appliance or air conditioner problem are:

1. **Verify the complaint.** Ask the consumer what symptoms were caused by the problem with the appliance or air conditioner.

2. **Check for external factors.** For example: is the appliance or air conditioner installed properly, does the product have the correct voltage, etc.?

3. **Check for physical damage.** Look for internal and external physical damage. Any damage will prevent the appliance or air conditioner from functioning properly. Two examples are broken parts at the base of the washing machine or a damaged cabinet that prevents the doors from closing properly.

4. **Check the controls.** The controls must be set to the proper settings. If the controls are not set correctly, the appliance or air conditioner might not function properly or complete its cycle.

5. **Operate the product.** Operate the appliance or air conditioner, and let it run through its cycle. Check the cycle operation against the operational sequence of events that is listed in the use and care manual.

6. **Is the product operating properly?** If it is, explain to the consumer how to operate the appliance or air conditioner according to the manufacturer's specifications.

7. **The product is not operating properly.** If the appliance or air conditioner is not operating properly, proceed to locate which component has failed. Check the diagnostic charts that are listed in this book to assist you in the correct direction to take.

Diagnosis and Correction Procedure

When diagnosing a problem with an appliance or air conditioner, use your five senses to determine the condition of the product. This will help in analyzing and defining the problem:

- *Example #1:* When turning on the washing machine, there is a smell of something burning. You can track down the location of the burning smell and, therefore, discover which part has failed.

- *Example #2:* When turning on a dishwasher, unusual noises are heard coming from underneath the machine. Stop the dishwasher, and attempt to track down where the noises are coming from.

Along with your hand tools, there are a variety of test meters that can assist you in analyzing and defining the problem. This is the sequence of events to follow when servicing an appliance or air conditioner:

- **Unplug the product.** Change the range setting on the multimeter to voltage. Check the voltage from the appliance or air conditioner's receptacle. If there is an uncertainty, check the name plate rating for the correct voltage rating; this is located on the product. When diagnosing a component failure, there are three types of circuit failure—the open circuit, the grounded circuit, and the short circuit—all of which are thoroughly explained in Chapter 5.

- **Gain access.** Only remove the panels and screws necessary to gain access to where the suspected component failure is located.
- **Isolate and/or remove the defective part.** Using the multimeter, isolate and/or remove the part, set the range to ohms, and check for component failure. This will be further explained in Chapter 5 and then in Chapters 11 through 23.
- **Install the new component.** When you find a defective part, replace it with a new original part. Reconnect all the wires in their original places.
- **Reattach all panels and screws.** Close the appliance or air conditioner, and reattach all panels and screws.
- **Test the product for proper operation.** Plug in the appliance or air conditioner and test it.

Electricity, Electronics, and Gas

Electricity

The technician must be knowledgeable in electrical theory to be able to diagnose and repair major appliances and air conditioners properly. When servicing these products, you will also encounter electronic parts within them. Manufacturers introduced electronic parts for a number of reasons, such as reliability, cost, miniaturization, standardization, and maintainability. Although this chapter cannot cover all there is to know about electricity or electronics, it will provide the basics. In the field of major appliances and air conditioners, the greatest number of potential problems is in the electrical portions of the product.

Electrical Wiring

The flow of electricity from a power source to the home can be made easier to understand by comparing it to a road map. Electricity flows from a power source to a load. This is similar to a major highway that runs from one location to another.

High-voltage transformers are used to increase voltages for transmission over long distances. The power lines that go to different neighborhoods are like the smaller roads that turn off the major highway. The electricity then goes to a transformer that reduces the voltage going into the home. This is the intersection between the small roads and the medium-sized highways. The small road that goes into the neighborhood, and all the local streets, are like the wiring that goes inside the home.

When all the streets, roads, and highways are connected together, the city is accessible. This is similar to having electricity flowing from the power source to all of the outlets in the home.

Imagine driving down a road and coming to a drawbridge (in this case, the switch), and it opens up. This stops the flow of traffic (electricity). In order for traffic (or electricity) to flow again, the drawbridge must close.

What Is a Circuit?

A *circuit* is a complete path through which electricity can flow and then return to the power source. Figure 5-1 is an example of a complete circuit. To have a complete path (or *closed circuit*), the electricity must flow from Point A to Point B without interruption.

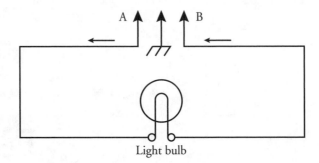

FIGURE 5-1   The complete circuit. Current flows from Point A, through the light bulb, and then back to Point B.

When there is a break in the circuit, the circuit is *open*. For example, a break in a circuit is when a switch is turned to its "off" position. This will interrupt the flow of electricity, or current, as shown in Figure 5-2. When a broken circuit is suspected, it is necessary to discover the location of the opening.

Circuit Components

In an appliance or air conditioner, an electric circuit has four important components:

- **Power source.** This source might be a battery or the electricity coming from the wall outlet. Without the applied voltage, current cannot flow.

- **Conductors.** A conductor will usually be a wire and sometimes the metal chassis (frame). The function of the wire conductor is to connect a voltage source to a load.

- **Loads.** These are the components that do the actual work in the appliance or air conditioner. A load is anything that uses up some of the electricity flowing through the circuit. For example, motors turn the belt, which turns the transmission. That, in turn, turns the agitator in a washing machine. Other examples are heating elements and solenoids.

- **Controls.** These control the flow of electricity to the loads. A control is a switch that is either manually operated by the user of the product or operated by the appliance or air conditioner itself.

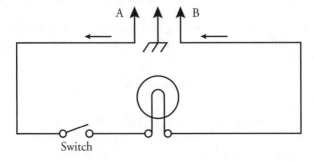

FIGURE 5-2   With the switch open, the current flow is interrupted.

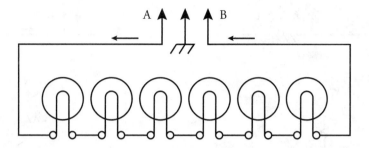

FIGURE 5-3 A series circuit.

Three Kinds of Circuits

You will come across three kinds of circuits: series circuits, parallel circuits, and series-parallel circuits (a combination of series and parallel circuits).

Series Circuit

The components of a series circuit are joined together in successive order, each with an end joined to the end of the next (Figure 5-3). There is only one path that electricity can follow. If a break occurs anywhere in the circuit, the electricity, or current flow, will be interrupted and the circuit will not function (Figure 5-4). Figure 5-5 shows some of the many different shapes of series circuits, all of which are used in wiring diagrams. In each series circuit, there is only one path that electricity can follow. There are no branches in these circuits where current can flow to take another path. Electricity only follows one path in a series circuit.

Parallel Circuits

The components of a parallel circuit are connected across one voltage source (Figure 5-6). The voltage to each of these branches is the same. The current will also flow through all the branches at the same time. The amount of current that will flow through each branch is determined by the load, or resistance, in that branch.

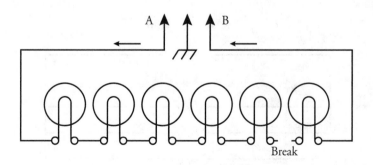

FIGURE 5-4 If there is a break in the wiring, all of the light bulbs will be off.

FIGURE 5-5   When you look at wiring diagrams, you will find series circuits in all sorts of shapes.

Figure 5-7 and Figure 5-8 show examples of parallel circuits. If any branch has a break in it, the current flow will only be interrupted in that branch. The rest of the circuits will continue to function.

Series-Parallel Circuits

A series-parallel circuit is a combination of series circuits and parallel circuits. In many circuits, some components are connected in series to have the same current, but others are

FIGURE 5-6   Parallel circuit.

FIGURE 5-7   Another parallel circuit.

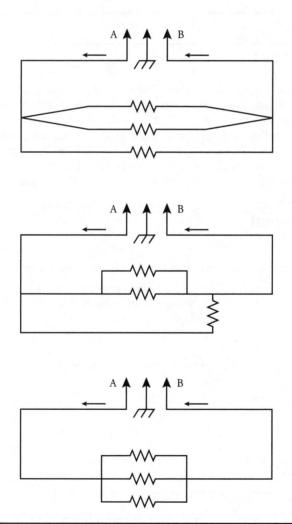

FIGURE 5-8   Notice that as in series circuits, the same parallel circuit can be drawn in many different ways.

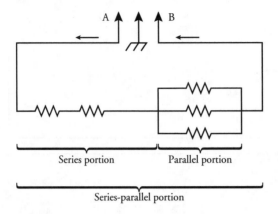

FIGURE 5-9   Series-parallel circuit.

in parallel for the same voltage (Figure 5-9). This type of circuit is used where it is necessary to provide different amounts of current and voltage from the main source of electricity that is supplied to that appliance.

Series and parallel rules apply to this type of circuit. For example, if there is a break in the series portion of the circuit (Figure 5-9), the current flow will be interrupted for the entire circuit. If the break is in the parallel portion of the circuit, the current will be interrupted for only that branch of the circuit. The rest of the circuits will still function.

Types of Electric Current

There are two types of electric current:

- Direct current (DC) flows continuously in the same direction (Figure 5-10).
- Alternating current (AC) flows in one direction and then reverses itself to flow in the opposite direction along the same wire. This change in direction occurs 60 times per second, which equals 60 Hz (Figure 5-11).

FIGURE 5-10   A simple DC electrical circuit. Current flows from the negative side of the battery through the switch and load, and back to the positive side of the battery.

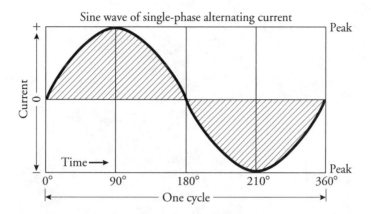

FIGURE 5-11 A waveform of a single-phase alternating current.

Direct current (DC) is used in automobile lighting, flashlights, and cordless electric appliances (such as toothbrushes, shavers, and drills), and in some major appliances and air conditioners.

Alternating current (AC) is used in most homes. This current can be transmitted more economically over long distances than direct current can. Alternating current can also be easily transformed to higher or lower voltages.

Theory of Current Flow

When servicing electronic solid-state circuits in major appliances or air conditioners, the technician should be aware that explanation of the circuit is often given assuming conventional current flow as opposed to electron flow. Conventional current flow theory states that current flow is from positive to negative. Electron current flow theory states that current flows from negative to positive.

Ohm's Law

Ohm's Law states: The current which flows in a circuit is directly proportional to the applied voltage and inversely proportional to the resistance. In other words: the greater the voltage, the greater the current; the greater the resistance, the less the current.

FIGURE 5-12
An ohmmeter connected to read resistance.

Ohms

Resistance is measured in ohms and opposes the flow of electrons (current). An instrument that measures resistance is known as an ohmmeter. Figure 5-12 is a schematic showing an ohmmeter connected to read the resistance of R1. The resistance of any material depends on its type, size, and temperature. Even the

FIGURE 5-13   A digital multimeter for measuring ohms.

best conductor offers some opposition to the flow of electrons. Figure 5-13 shows another type of meter, the digital multimeter, used for measuring ohms. The fundamental law used to find resistance is stated as follows: The resistance (R) in ohms is equal to the potential difference measured in volts (V), divided by the current in amperes (A). The equation is: $R = V \div A$.

Amperes

Current is measured in amperes. The term *ampere* refers to the number of electrons passing a given point in one second. When the electrons are moving, there is current. The ammeter is calibrated in amperes, which we use to check for the amount of current in a circuit.

FIGURE 5-15
In the ammeter, the jaws clamp around a wire to measure the amperage of a circuit.

FIGURE 5-14 An ammeter connected in a circuit, measuring amperes.

An instrument that measures amperes is known as an ammeter. Figure 5-14 is a schematic showing an ammeter connected in a circuit to measure the current in amperes. Figure 5-15 shows an ammeter that is used in diagnosing electrical problems with appliances. Current is the factor that does the work in the circuit (light the light, ring the buzzer). The fundamental law to find current is stated as follows: The current in amperes (A) is equal to the potential difference measured in volts (V), divided by the resistance in ohms (R). The equation is: $A = V \div R$.

Volts

Electromotive force is measured in volts. This is the amount of potential difference between two points in a circuit. It is this difference of potential that forces current to flow in a circuit. One volt (potential difference) is the electromotive force required to force one ampere of current through one ohm of resistance.

An instrument that measures voltage is known as a voltmeter. Figure 5-16 is a schematic showing a voltmeter connected in the circuit to measure the voltage. Voltmeter 1 is connected to read the applied (or source) voltage. Voltmeter 2 is connected to measure the voltage drop,

FIGURE 5-16 A voltmeter connected in a circuit to measure voltage.

Volt-ohm-milliammeter

Test leads

FIGURE 5-17   The volt-ohm-milliammeter with test leads.

or potential difference, across R2. Figure 5-17 shows an actual volt-ohm-milliameter (VOM) that is used in measuring voltage.

The fundamental law to find voltage is stated as follows: The potential difference measured in volts (*V*) is equal to the current in amperes (*A*) multiplied by the resistance in ohms (*R*). The equation is: $V = A \times R$.

FIGURE 5-18
The wattmeter is used to measure watts.

Digital wattmeter

Watts

Power is measured in watts, and an instrument that measures watts is known as a wattmeter (Figure 5-18). One watt of power equals the work done in one second, by one volt of potential difference, in moving one coulomb of charge. One coulomb per second is equal to one ampere. Therefore, the power in watts (*W*) equals the product of amperes (*A*) times volts (*V*). The equation is: $W = A \times V$.

FIGURE 5-19
The Ohm's Law
equation wheel.

Conversion chart for determining amperes, ohms, volts, or watts
(Amperes = A, Ohms = Ω, Volts = V, Watts = W)

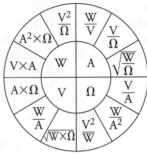

Ohm's Law Equation Wheel

The Ohm's Law equation wheel in Figure 5-19 shows the equations for calculating any one of the basic factors of electricity. Figure 5-20 shows the cross-reference chart of formulas as used in this text. If you know any two of the factors (V = voltage, A = amperage, R = resistance, W = power), you can calculate a third. To obtain any value in the center of the equation wheel for direct or alternating current, perform the operation indicated in one segment of the adjacent outer circle.

Example: A 2400-watt heating element is connected to a 240-volt circuit. How many amps does it draw?

When finding amperage, the formula will be found in the Amperes section of the wheel.

$$\frac{W(\text{watts})}{V(\text{volts})} = A(\text{amps})$$

Then, solving for amperage:

$$\frac{2400(\text{watts})}{240(\text{volts})} = 10\,\text{amps}$$

What is the resistance (ohms)?

$$\frac{V^2(\text{volts squared})}{W(\text{watts})} = \text{ohms}$$

Then, solving for resistance:

$$\frac{240^2(\text{volts squared})}{2400(\text{watts})} = 24\,\text{ohms}$$

FIGURE 5-20
The cross-reference
chart of formulas.

Term	Measured in	Referred to in formulas as	Identification used in this text
Amperage	Amperes (Amps)	I	A for amps
Current	Amperes (Amps)	I	A for amps
Resistance	Ohms	Ω or R	R for resistance
Voltage	Volts	V or E	V for volts
Electromotive force	Volts	V or E	V for volts
Power	Watts	W	W for watts

Resistors

A resistor, when installed into an electrical circuit, will add resistance, which will produce a specific voltage drop, or a reduction in current. Resistors can be either fixed or variable. The most common problem found with resistors when testing them with a milliammeter is that they read an open (infinite ohms). To determine the resistance value in ohms of a resistor, you have to isolate it from the circuit and then use your ohmmeter to read its value. On larger resistors, the resistance value is stamped on the body of the resistor.

A visual inspection of the resistor should also be performed to rule out any physical flaws in the resistor and the surrounding components. If you touch a resistor, and it begins to flake apart, most likely it is bad and it needs to be replaced. If the technician smells a burning smell coming from the resistor, most likely this component is bad and it must be replaced.

Resistance Color Bands

Because of their size, most axial resistors (Figure 5-21a) are color-coded, usually with four bands to indicate their resistance value. Reading the resistor bands from left to right, the first band closest to the end indicates the first numerical digit of resistance. The second band indicates the next numerical digit of resistance, and the third band indicates the numerical multiplier in zeros. The fourth band on a resistor is the resistance tolerance measured in percentage (Figure 5-21b). For example, let's say the color bands are brown, red, orange, and silver:

FIGURE 5-21
(a) A common resistor. (b) Resistor color-code chart.

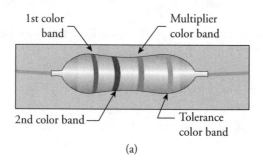

1st color band   Multiplier color band

2nd color band   Tolerance color band

(a)

COLOR	1ST BAND	2ND BAND	MULTIPLIER
Black	0	0	× 1
Brown	1	1	× 10
Red	2	2	× 100
Orange	3	3	× 1000
Yellow	4	4	× 10,000
Green	5	5	× 100,000
Blue	6	6	× 1,000,000
Violet	7	7	× 10,000,000
Gray	8	8	× 100,000,000
White	9	9	× 1,000,000,000

4th BAND = TOLERANCE					
Gold	5%	Silver	10%	No Color	20%

(b)

Brown = 1
Red = 2
Orange = 1000
Silver = + or −10 percent

The resistor is rated at 12,000 ohms with a ±10 percent tolerance. The tolerance is the safe operating percentage difference of the color-coded resistance value. For instance, a 12,000-ohm resistor with a 10 percent tolerance will operate safely at 10 percent below or 10 percent above the 12,000-ohm valve.

Thermistor

Thermistors are constructed from a semiconductor material with a specified temperature range. This electronic device exhibits a change in resistance with a change in temperature. There are two types of thermistors: negative temperature coefficient (NTC) and positive temperature coefficient (PTC), with the most common being NTC. The resistance of the NTC thermistor decreases with an increase of temperature, whereas the resistance of the PTC thermistor increases with an increase of temperature. There are many applications for thermistors, including their use as temperature sensors, resettable fuses, current limiters, and power indicators. They are also used in applications where interchangeability without recalibration is required.

One type of test performed on thermistors to check if they are good or bad is the resistance-versus-temperature test. To perform this test, the technician will have to locate the technical data sheet that comes with the appliance. The data sheet will provide the resistance (ohms) valve at the test temperature listed. Another way to test the thermistor is by testing it with an ohmmeter for an open or a short.

Capacitor

A capacitor is a device that stores electricity. It is rated by its working voltage (WV or WVac) and by its storage capacity in microfarads (μF). Always replace a capacitor with one having the same voltage rating and the same (or up to 10 percent greater) microfarad rating. On larger capacitors, the rating is stamped on the side. Also, watch out for the decimal point on some capacitors. The rating might read .50 μF instead of 50 μF. Small capacitors in electronic circuits are rated by numbering or are color-coded.

Capacitors are used in electrical circuits to perform the following:

- An electrical voltage boost in a circuit
- Control timing in a computerized circuit
- Reduce voltage disruptions and allow voltage to maintain a constant flow
- Block the flow of direct current when fully charged and allow alternating current to pass in a circuit

Testing a Capacitor

Before testing a capacitor, disconnect the electricity. This can be done by pulling the plug from the electrical outlet. Be sure that you only remove the plug for the product you are working on. Or, you can disconnect the electricity at the fuse panel or at the circuit breaker panel.

Figure 5-22   Placing ohmmeter test leads on the capacitor terminals.

Some appliance or air conditioner models have the capacitor mounted on the motor, and some are mounted to the cabinet interior in the rear of the machine. Access might be achieved through the front or rear panel, depending on which model you are working on. Do not touch the capacitor until it's discharged.

Warning: A capacitor will hold a charge indefinitely, even when it is not currently in use. A charged capacitor is extremely dangerous. Discharge all capacitors immediately any time that work is being conducted in their vicinity. Redischarge after repowering the equipment if further work must be done.

Use a screwdriver with an insulated handle to discharge a larger capacitor by shorting it across both terminals. Remove the wire leads one at a time with needle nose pliers. Set the ohmmeter on the highest scale, and place one probe on one terminal and the other probe on the other terminal (Figure 5-22). Observe the meter action. While the capacitor is charging, the ohmmeter will read nearly zero ohms for a short period of time. Then the ohmmeter reading will slowly begin to return toward infinity. If the ohmmeter reading deflects to zero and does not return to infinity, the capacitor is shorted and should be replaced. If the ohmmeter reading remains at infinity and does not dip toward zero, the capacitor is open and should be replaced.

Diodes

A diode is an electrical device that allows current to flow in one direction only. If you connect a positive voltage to the anode (positive side of the diode) and a negative voltage to the cathode side (negative side of the diode), the diode becomes forward-biased (Figure 5-23a). When the diode is forward-biased, current will flow. When a positive voltage is connected to

(a)   (b)

Figure 5-23   (a) When testing the diode with an ohmmeter, the forward bias indicates zero ohms of resistance. (b) When testing the diode with an ohmmeter, the reversed bias indicates infinite resistance.

FIGURE 5-24
The wiring diagram
symbol for a diode.

Cathode

Anode

+

Cathode

Anode

FIGURE 5-25
The band closest to
the end of the diode
is the cathode (+),
and the other end is
known as the anode
(−).

the cathode (negative side of the diode) and a negative voltage is connected to the anode (positive side of the diode), the diode becomes reverse-biased (Figure 5-23b). Then the diode will not allow current to flow. There are many types of uses for diodes besides rectification. These include capacitance that varies with the amount of voltage applied to the diode and photoelectric effects. The wiring diagram symbol for a diode is shown in Figure 5-24.

When a diode is connected in a circuit, there must be a means to identify the cathode and the anode (Figure 5-25). Diodes are manufactured in different case styles. On large diodes, there is a diode symbol stamped on the side to indicate to the technician the cathode and anode. On smaller diodes, there is a band around one end to identify the cathode.

Testing a Diode

A diode can be tested with an ohmmeter. To test a diode, you must first disconnect one side of the diode to isolate it from the remainder of the circuit. Set your ohmmeter to the ohms scale, and connect the leads to the diode. The meter may or may not read continuity. Then reverse the leads on the diode and check for continuity. You should read continuity in one direction only. If continuity is not indicated in either direction, the diode is open (Figure 5-26a). If continuity is indicated in both directions, the diode is shorted (Figure 5-26b).

Transistor

A transistor is a three-element, electronic, solid-state component that is used in a circuit to control the flow of current or voltage. It opens or closes a circuit just like a switch (Figure 5-27a). Other uses for transistors include amplification and oscillation. Transistors are located on circuit boards. Figure 5-27b indicates the wiring diagram symbol for a transistor.

Integrated Circuits and Circuit Boards

An integrated circuit (IC), shown in Figure 5-28, is an electric circuit miniaturized and consisting of transistors, diodes, resistors, capacitors, and all the connecting wiring—all of it manufactured on a single semiconductor chip. Integrated circuits are found in products like computers, televisions, cell phones, and major appliances and air conditioners. IC chips are not repairable.

A printed circuit board (PCB) is a layout of conductive pathways (an electric circuit) secured to a board. Electronic components are soldered on to the board for a particular function within an appliance or air conditioner. Only a small percentage of printed circuit boards are repairable. The remaining percentage of PCBs is replaced with new boards.

Figure 5-26 (a) The ohmmeter indicates no continuity in either direction. The diode is open. (b) The ohmmeter reads continuity in both directions, indicating a shorted diode.

Figure 5-27 (a) A sample of the various types of transistors used in circuits. (b) The wiring symbol for an NPN, PNP, and photosensitive transistor.

Figure 5-28
An integrated circuit
(IC) chip.

Electrostatic Discharge

The electronic parts in an appliance or air conditioner are sensitive to electrostatic discharge (ESD). An example of electrostatic discharge occurs when you rub your feet on a carpet and then touch a metal doorknob. The little spark that you generated by touching the metal doorknob is ESD. This type of electrostatic discharge will damage the electronics in the appliance or air conditioner at the microscopic level. When servicing an appliance or air conditioner or installing a new electronic part, ESD may cause failure in the future due to stress from these occurrences.

To prevent ESD from damaging expensive electronic components, follow these steps:

- Turn off the electricity to the appliance or air conditioner before servicing any electronic component.
- Before servicing the electronics in an appliance or air conditioner, discharge the static electricity from your body by touching your finger repeatedly to an unpainted surface on the appliance or air conditioner. Another way to discharge the static electricity from your body is to touch your finger repeatedly to the green ground connection on that product.
- The safest way to prevent ESD is to wear an anti-static wrist strap.
- When replacing a defective electronic part with a new one, touch the anti-static package that the part comes in to the unpainted surface of the appliance or air conditioner or to the green ground connection of the appliance or air conditioner.
- Always avoid touching the electronic parts or metal contacts on an integrated board.
- Always handle integrated boards by the edges.

Wiring Diagram Symbols

These symbols are commonly used in most wiring diagrams. Study each symbol so that you can identify it by sight (Table 5-1, Table 5-2, Table 5-3, and Table 5-4).

Terminal Codes

Terminal codes are found on all wiring diagrams. To help you identify the color codes, they are listed in Table 5-5.

TABLE 5-1 Wiring Diagram Symbols.

Temperature-actuated components

(Note: Symbols shown to be used for thermostats, bimetal switches, overload protectors, or other similar components, as required.

Temp. actuated
(close on heat rise)

Temp. actuated
(open on heat rise)

S.P.S.T.
(open on heat rise)

S.P.D.T.

S.P.D.T.

S.P.S.T.
(two contacts)

S.P.S.T. (adj.)
(close on heat rise)

S.P.D.T. (adj.)

S.P.S.T. (adj.)
(open on heat rise)

S.P.D.T. (adj.)
(with aux. "off" contacts)
(typical example)

S.P.S.T. (with internal heater)
(close on heat rise)

S.P.S.T. (with internal heater)
(open on heat rise)

Combination devices

Relay-magnetic
(arrangement of
contacts as necessary
to show operation)

Relay-thermal
(arrangement of
contacts as necessary
to show operation)

Timer (defrost)

Manual and mechanical switches

Normally closed (S.P.S.T.)
(single-pole, single-throw)

Normally open (S.P.S.T.)
(single-pole, single-throw)

Transfer (S.P.D.T.)
(single-pole, double-throw)

Multi position

Number of terminals

Timer switch

Automatic switch

N.O.
(normally open)

N.C.
(normally closed)

Integral switch
(timer, clock, etc.)

Push button switch
(momentary or spring return)

Circuit closing
N.O. (normally open)

Circuit opening
N.C. (normally closed)

Two circuit

SPDT
(single-pole, double-throw)

TABLE 5-2   Wiring Diagram Symbols.

Components		Lines and connections	
Buzzers		Integral conductor	
Adjustable		External or harness wire	
Bell		Optional or alternate circuit	
		Crossover	or
Sensor (moisture)		Permanent connection junction	
		Permanent connection	
Thermocouple		Terminal	
		Shield	
Centrifugal switch		Ground (earth)	
		Ground (chassis)	
Pressure switch S.P.D.T.		Grounded service cord (3-prong plug)	
		Service cord (2-prong)	
Humidistat		Mechanical connection	
		Separable connector	
		Motors	
		Timer or clock	TM
Magnetron		Single-speed	
		Two-speed	
Thermistor		Three-speed	
		Compressor	

TABLE 5-3   Wiring Diagram Symbols.

Electronic wiring symbols

TABLE 5-4   Electronic Wiring Diagram Symbols.

Terminal Color Code	Harness Wire Color
BK	Black
BK-Y	Black with yellow tracer
BR	Brown
BR-O or BR-OR	Brown with orange tracer
BR-R	Brown with red tracer
BR-W	Brown with white tracer
BU or BL	Blue
BU-BK or BL-BK	Blue with black tracer
BU-G or BU-GN	Blue with green tracer
BU-O or BU-OR	Blue with orange tracer
BU-Y	Blue with yellow tracer
G or GN	Green
B-Y or GN-Y	Green with yellow tracer
G-BK	Green with black tracer
GY	Gray
GY-P or GY-PK	Gray with pink tracer
LBU	Light blue
O or OR	Orange
O-BK or OR-BK	Orange with black tracer
P or PUR	Purple
P-BK or PUR-BK	Purple with black tracer
P or PK	Pink
R	Red
R-BK	Red with black tracer
R-W	Red with white tracer
T or TN	Tan
T-R	Tan with red tracer
V	Violet
W	White
W-BK	White with black tracer
W-BL or W-BU	White with blue tracer
W-O or W-OR	White with orange tracer
W-R	White with red tracer
W-V	White with violet tracer
W-Y	White with yellow tracer
Y	Yellow
Y-BK	Yellow with black tracer
Y-G or Y-GN	Yellow with green tracer
Y-R	Yellow with red tracer

TABLE 5-5 Terminal Codes.

Timer Sequence (Esterline) Charts

Figure 5-29 represents a sample timer sequence chart. Dishwashers, washers, and dryers have an Esterline Chart, which is also known in the appliance industry as a Timer Sequence Chart, that is part of the wiring schematic. This chart indicates the switch contacts that are open or closed at a specific time (interval) during the operating cycle. It is another diagnostic tool that is used in conjunction with the wiring schematic by technicians to assist in diagnosing problems with the appliance.

Figure 5-29 A sample timer sequence chart.

The timer sequence chart is divided into the following:

- Overview of all of the columns on the chart.

- Correspondence between the timer dial and the intervals on the timer sequence chart.

- The consumer selects the additional options through the push button switch that will effect the switch contacts opening and closing on the wiring schematic.

- At a certain interval in the operating cycle, the timer sequence chart will show the technician the position of the switch contacts that are open or closed on the wiring schematic.

- The timer sequence chart will indicate which components are energized at any time interval.

Wiring Diagrams

In this section, each of the four examples presented will take you through a step-by-step process in how to read wiring diagrams.

Example 1: Take a look at a simple wiring diagram for a refrigerator (Figure 5-30). Note the black wire on the diagram. This is the wire that goes to the temperature control. The circuit is not energized when the temperature control is in the "off" position. When the temperature control knob is turned to the "on" position (switch contacts closed) and the circuit is energized, current will flow through the temperature control, through the red wire, through the overload protector, through the compressor and the relay, and back through the white wire to the line cord.

Example 2: In the wiring diagram in Figure 5-31, trace the active circuits with your finger. Note switch number 1. This is the first switch in the line, and it is the main switch that supplies voltage to the timer. No circuits are energized when the timer dial is in the "off" position (as the diagram indicates). When the user selects the wash cycle with a warm wash

---

FIGURE 5-30
A simple wiring
schematic of a
refrigerator circuit.

Figure 5-31
The wiring schematic
of an automatic
washer circuit.

and turns on the timer, switch contacts 1, 2, 3, 6, and 7 are closed. (Use a pencil to close the switches on the diagram.) Voltage is supplied to the water level switch and to the hot and cold water valves. Warm water is now entering the washing machine tub. When the water level reaches the selected position, the water level switch contacts close from V to number 4, indicating that the water in the tub has reached the selected water level. When the water level switch is in this position, the water is turned off. Voltage is now supplied to the timer motor, drive motor, and the agitate solenoid. The washing machine is now agitating and cleaning the clothes. As the timer advances to the spin cycle, timer switch contacts 2 and 3 open, thus turning off the agitate solenoid and the drive motor. Switch contacts 8 and 2 now close, thus supplying voltage to the drive motor and the spin solenoid. The timer motor advances to the end of the cycle, and timer switch contacts 1, 2, and 8 will open. The washing machine is now off, and no circuits are energized.

Example 3: The wiring diagram in Figure 5-32 is for a refrigerator. Assume that the thermostat is calling for cooling and the compressor is running. With your finger, trace the active circuits. The thermostat in the wiring diagram for the refrigerator is closed. The evaporator fan motor and the condenser fan motor are running. Voltage is supplied

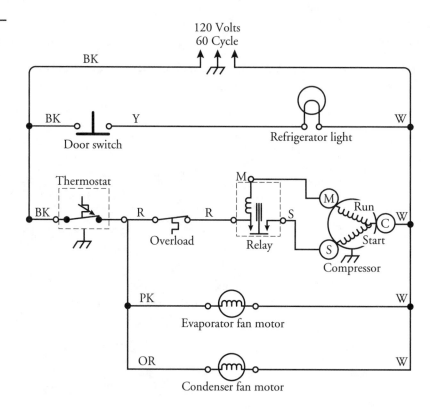

FIGURE 5-32
The wiring schematic of a refrigerator circuit.

through the overload to the relay. Current is flowing through the relay coil to the compressor-run-winding. Also notice that the door switch is open and the refrigerator light is off. When the temperature in the refrigerator satisfies the thermostat, the thermostat switch contacts will open, thus turning off the compressor, the evaporator fan motor, and the condenser fan motor.

Example 4: The wiring diagram in Figure 5-33 is for a no-frost refrigerator. Note the defrost timer in the lower-left part of the diagram. The defrost timer switch contact is closed to contact 4, the thermostat is calling for cooling, and the compressor is running. Trace the active circuits with your finger. Voltage is supplied to the defrost timer terminal 1. Current will flow through the defrost 2 timer motor to the white wire and then back to the line cord. At the same time, current flows through contact 4 in the defrost timer to the thermostat. At this point, the current passes through the thermostat to a junction and splits in two directions. Current will flow through the temperature control, through the overload protector, through the compressor and the relay, and back through the white wire to the line cord. At the same time, the compressor is running and current will flow through the evaporator and condenser fan motors.

When the defrost timer activates the defrost cycle, the defrost timer switch contact is closed to contact 2, and the compressor is not running. The evaporator and condenser fan motors will also stop running. The current will flow from terminal 2, through the defrost bi-metal, through the defrost heater, and back through the white wire to the line cord. With your finger, trace the active circuits.

FIGURE 5-33
The wiring schematic
of a no-frost
refrigerator circuit.

Sample Wiring Diagrams

The diagram in Figure 5-34 is known as a ladder diagram. It is a simplified diagram using symbols for parts and control components attached to wires. The timer is represented by a dotted line running vertically through the switch contacts. To read and understand this type of diagram, assume that the complaint is that the dryer is not heating. Look at the section of the diagram marked HEAT. Starting on the left side of the diagram, trace the circuit with your finger. You will notice that L1 goes to one side of the timer switch. As your finger moves from left to right, you will pass over the cycling thermostat, high-limit thermostat, heater, motor centrifugal switch (CS), and on to L2. These are the components that make up the heating circuit. You must also include the motor and the door switch. If the door switch should fail, the motor will not run, thus opening the motor centrifugal switch (CS). If the motor fails, the high-limit thermostat will open, shutting off the heater element. If any of these components fail, the dryer will not dry the clothing.

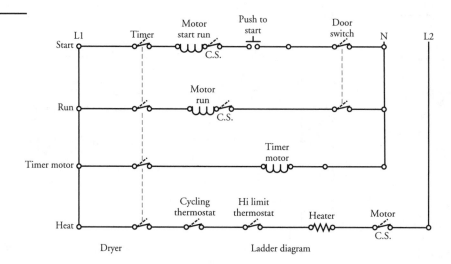

FIGURE 5-34
A simple ladder
diagram.

The diagrams in Figure 5-35 and Figure 5-36 illustrate a pictorial diagram and a schematic diagram. The pictorial diagram shows the actual picture of the components, and the schematic diagram uses symbols for the components. Figure 5-37 shows a pictorial diagram of a refrigerator. In this type of diagram, you can see where the components are actually located.

How to Read the Volt-Ohm-Milliammeter Test Instrument

The volt-ohm-milliammeter (VOM) is sometimes called a multimeter because it can perform more than one function. A typical VOM will allow you to measure voltage, resistance, and current:

- Voltage equals electromotive force.
- Resistance equals the amount of resistance holding back the flow of current (measured in ohms).
- Current equals the amount of electricity flowing through a wire or circuit component (measured in amperes). There are many different types and brands of VOMs (Figure 5-38). However, all VOMs are used in the same way to measure voltage, current, or resistance. Most VOMs will have the following:
  - Test leads. These are the wires coming from the meter to the part being tested.
  - Meter scales and pointer (or a digital display on a digital meter). These show the amount of whatever value you are measuring.
  - Function switch. This allows you to select whether you will be measuring AC or DC voltage (volts), current (amps), or resistance (ohms).
  - Range selector switch. This allows you to select the range of values to be measured. On many meters, you can select functions and ranges with the one switch (as in the meter pictured in Figure 5-38a and Figure 5-38c).

FIGURE 5-35 A pictorial diagram and a schematic diagram. Both show the same components.

Schematic   **Pictorial**

FIGURE 5-36   A pictorial diagram and schematic diagram.

Measuring Voltage

If you don't have the right voltage in an appliance or air conditioner, the product won't function properly. You can find out whether the product is getting the right voltage by measuring the voltage at the wall outlet (receptacle). If an appliance or air conditioner isn't getting the proper voltage, nothing else you do to fix it will help.

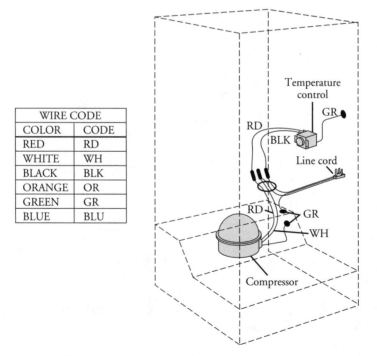

WIRE CODE	
COLOR	CODE
RED	RD
WHITE	WH
BLACK	BLK
ORANGE	OR
GREEN	GR
BLUE	BLU

FIGURE 5-37   A pictorial diagram of a refrigerator, showing where the components are located.

Volt-ohm-milliammeter

Test lead

Test lead

VOM and ammeter multimeter combination

(b)

(c)

FIGURE 5-38 (a) Volt-ohm-milliammeter, (b) VOM and ammeter multimeter combination, (c) a digital multimeter.

For your safety, before using any test instrument, it is your responsibility as a technician to read and understand the manufacturer's instructions on how the test instrument operates.

Making the Measurement

Making voltage measurements is easy once you know how to select and read the scales on your meter. When measuring voltage, you should perform the following steps:

- Attach the probes (another term for *test leads*) to the meter. Plug the black probe into the meter jack marked negative or common. Plug the red probe into the positive outlet.
- Set the function switch to AC VOLTS.
- Select a range that will include the voltage you are about to measure (higher than 125 Vac if you are measuring 120 volts; higher than 230 Vac if you are measuring 220 volts.) If you don't know what voltage to expect, use the highest range, and then switch to a lower range if the voltage is within that amount.
- Touch the tips of the probes to the terminals of the part to be measured.
- Read the scale.
- Decide what the reading means.

Selecting the Scales

To measure the voltage in appliances and air conditioners, use the AC/DC scales (Figure 5-39). These same scales are used for both AC and DC readings. The numbers on the right side of the AC/DC scales tell you what ranges are available to you. The meter face in Figure 5-40 has three scales for measuring voltage: one is marked 10, another 50, and the third is marked 250. Remember—the scale you read is determined by the position of the range switch.

Example: If the range is set on 250 V, as in Figure 5-41, you read the 0- to 250-V scale.

Reading the AC Voltage Scale

When reading the pointer position, be sure to read the line marked AC (Figure 5-40). The spaces on the voltage scales are always equally divided. When the pointer stops between the marks, just read the value of the nearest mark. In Figure 5-42, the pointer is between 115 and 120 volts on the 250 scale. Read it as 120 volts. With an analog meter, you're not gaining anything by trying to read the voltage exactly.

FIGURE 5-39
An example of scales used on some VOMs.

Measuring Line Voltage

Measuring line voltage is the first, and most important, part of checking out an appliance or air conditioner that does not operate. Line voltage is the voltage coming from the wall outlet. There should be approximately 120 volts AC at the outlet under "no-load" conditions.

FIGURE 5-40
The meter face of an
analog meter.

No-load means that no appliance or air conditioner is connected or that an appliance or air conditioner is connected but it is turned off.

To measure line voltage under no-load conditions at 120 volts (Figure 5-43):

1. Set the meter to measure AC volts.

2. Set the range selector to the range nearest to but higher than 120 volts.

3. Insert either test lead into one slot of an empty wall receptacle.

4. Insert the other test lead into the other slot of the same outlet. (Disregard the ground terminal for this test.) *Warning*: Do not touch or handle the test leads by the metal portion of the probe. Hold the probe by the plastic grips that are attached to the test leads to avoid electric shock.

5. Read the meter. The reading should be between 115 and 120 volts.

FIGURE 5-41
The range is set on
the 250-V scale.

FIGURE 5-42
The range is set on
the 250-V scale, and
the pointer reads
120V.

6. When testing for 240 volts, be sure that the range selector is set to the nearest range
higher than 240 volts. *Note:* Most appliances and small BTU-sized air conditioners
are rated at 120 volts, but will work on voltages ranging from 110 to 125 volts. If
the voltage drops more than 10 percent, the appliance or air conditioner will not
operate, and most likely will damage some electrical components if the appliance or
air conditioner keeps running.

FIGURE 5-43
Measuring line voltage
with no load.

To measure line voltage under load at
120 volts (Figure 5-44):

1. Be sure that the appliance or air
   conditioner is plugged into one of the
   receptacles and that the product is
   turned on.

2. Follow steps 1–5 for no-load
   conditions, inserting the test leads
   into the empty receptacle next to the
   one into which the product is
   plugged.

3. Under load conditions (appliance or
   air conditioner is turned on), your
   reading will be slightly less than
   under no-load conditions.

Figure 5-44
Measuring line voltage under load.

4. When testing for 240 volts, be sure that the range selector is set to the nearest range higher than 240 volts. *Note:* Products with drive motors and compressors, such as automatic washers, dishwashers, trash compactors, refrigerators, air conditioners, and microwaves, should also be tested at the moment of start. If the voltage drops more than 10 percent of the supplied voltage when the motor is started, there is a problem with the electrical supply.

Testing for Ground and Polarity

When a component is grounded to a chassis, there is no voltage between the two. When a chassis is grounded to the earth, there is no voltage between the chassis and the earth. If there is voltage between a chassis and the earth, it's dangerous. If you stand on the earth (dirt, concrete slab, etc.) and touch the chassis, electricity will flow through your body. You could injure yourself, or even someone else, and death could occur. It is important to be sure that the electrical outlets in the home from which appliances or air conditioners are powered are correctly wired. That will protect the users from electrical shock.

If an outlet is wired "backward" (that is, the black or red "hot wire" is connected to the long slot of the outlet), the appliance or air conditioner connected to that outlet might be unsafe to operate, blow the fuse, or trip the circuit breaker. Appliances and air conditioners with solid-state controls will not function properly if the outlets are wired backward.

To test for ground:

1. First, test for line voltage. (If there is no voltage, you can't test for ground.)

2. Notice that the receptacle has a longer and a shorter slot (Figure 5-45). If the outlet has been mounted right-side up, the longer slot will be on the left.

3. Test for voltage between the short slot and the ground receptacle (the round hole), as shown in Figure 5-46. If there is line voltage between these two points, the receptacle is grounded.

4. If there is no round hole (ground prong receptacle), touch one of your probes to the screw that fastens the coverplate to the outlet (Figure 5-47).

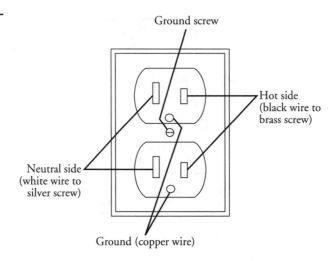

Figure 5-45
Receptacle
identification points.

To test for polarity:

1. Test to be sure that there is line voltage between the longer and shorter slots (Figure 5-48).

2. Test to be sure that there is line voltage between the short slot and the center screw or the round hole (ground prong receptacle) (Figure 5-47 and Figure 5-48).

3. Test to be sure that there is no voltage between the longer slot and the center screw or the round hole (Figure 5-49 to Figure 5-52). *Note:* If these three tests don't test out this way, the outlet is incorrectly wired and should be corrected by a licensed electrician.

Figure 5-46   Testing for ground.

Figure 5-47   Testing for ground again.

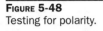
FIGURE 5-48
Testing for polarity.

120 Volts

120 Volts

FIGURE 5-49   Testing the polarity of a receptacle.

Measuring 240-Volt Receptacles

Warning: When testing a three-wire or four-wire receptacle for 240 volts, the technician must be extremely careful not to touch or handle the test leads by the metal portion of the probe. Hold the probe by the plastic grips that are attached to the test leads to avoid electric shock. The multimeter must be set on the AC scale to 240 volts or higher. Do not use the ohms scale—the voltage will damage your meter.

Testing a Three-Wire Receptacle for 240 Volts

To test the voltage on a three-wire receptacle with a no-load condition:

1. Select the correct AC range on your multimeter.

2. Insert the test leads into the receptacle as shown in Figure 5-53. You might have to move the probes around a little to make contact within the receptacle. Check the meter reading—there should be approximately 240 volts. If no voltage is present, check the fuse or circuit breaker.

120 Volts

FIGURE 5-50   Testing the polarity of a receptacle.

Zero Volts

FIGURE 5-51   Testing the polarity of a receptacle.

FIGURE 5-52   Testing the polarity of a receptacle.

FIGURE 5-53   Testing the voltage of a three-wire receptacle for 240 volts.

3. Insert the test leads into the receptacle as shown in Figure 5-54. Check the meter reading—there should be approximately 120 volts. If no voltage is present, check the fuse or circuit breaker.

4. Next, insert the test leads into the receptacle as shown in Figure 5-55. Check the meter reading—there should be approximately 120 volts. If no voltage is present, check the fuse or circuit breaker.

Testing a Four-Wire Receptacle for 240 Volts

The red and black wires attached to the receptacle will each carry 120 volts. The white wire attached to the receptacle is the neutral wire, and the bare wire is the ground wire. This type of receptacle adds additional safety for the consumer and to the product.

To test the voltage on a four-wire receptacle with a no-load condition:

1. Select the correct AC range on your multimeter.

2. Insert the test leads into the receptacle as shown in Figure 5-56. You might have to move the probes around a little to make contact within the receptacle. Check the

FIGURE 5-54
Testing the voltage of a three-wire receptacle for 120 volts.

FIGURE 5-55
Testing the voltage
of a three-wire
receptacle for
120 volts.

120 Volts AC

meter reading—there should be approximately 120 volts. If no voltage is present,
check the fuse or circuit breaker.

3. Insert the test leads into the receptacle as shown in Figure 5-57. Check the meter
   reading—there should be approximately 120 volts. If no voltage is present, check
   the fuse or circuit breaker.

4. Next, insert the test leads into the receptacle as shown in Figure 5-58. Check the
   meter reading—there should be approximately 240 volts. If no voltage is present,
   check the fuse or circuit breaker.

5. Now, insert the test leads as shown in Figure 5-59. The meter should read 0 volts.

Measuring Resistance (OHMS)

Electrical appliances and air conditioners need a complete path around which electricity
can flow. If there is infinite resistance to the flow of electricity, you have an open circuit
infinite resistance between the two points being measured. When there is a complete path,
you have continuity in the circuit. When you test to find out whether there is a break in the
path, you say you are making a continuity check.

FIGURE 5-56
Testing the voltage of
a four-wire receptacle
for 120 volts.

120 Volts AC

Figure 5-57
Testing the voltage of a four-wire receptacle for 120 volts.

120 Volts AC

Figure 5-58
Testing the voltage of a four-wire receptacle for 240 volts.

240 Volts AC

Figure 5-59
Testing the voltage of a four-wire receptacle for 0 volts.

0 Volts

Continuity checks are made by measuring the amount of resistance there is to the flow of electricity. If there is so much resistance that it is too high to measure (called "infinite"), then you say that the circuit is open (there is no complete path for the electricity to follow). If there is some resistance, it means that there is continuity, but that there is also one (or more) load on the line—a light, a motor, etc. *Note:* A load is an electrical component that uses electricity to work (for example, a light bulb, a motor, or a heater coil).

If there is no resistance between the two points, it means that the electricity is flowing directly from one point to the other. If the electricity flowed directly from one point to the other by accident or error, then you say you have a short circuit or a "short."

Setting Up the Meter

Measuring resistance (ohms) is like measuring voltage, except that the measurements are made with the electricity turned off. The steps for measuring ohms are as follows:

1. Attach the leads to the meter. Plug the black lead into the negative outlet and the red lead into the positive outlet.

2. If your meter has a function switch, set the function switch to OHMS.

3. If your meter has a range switch, set the range. The range selector switch will have several ranges of resistance (ohms) measurements (Figure 5-60). The ranges are shown like this:

   - *R* × 1: The actual resistance shown on the meter face times 1.

   - *R* × 10: The resistance reading times 10 (add one zero to the reading).

   - *R* × 100: The resistance reading times 100 (add two zeros to the reading).

   - *R* × 1K: The resistance reading times 1000 (add three zeros to the reading).

   - *R* × 10K: The resistance reading times 10,000 (add four zeros to the reading).

FIGURE 5-60
With the range set on R x 100, the meter reads 400 ohms.

4. Set the range so that it is higher than the resistance you expect. If you don't know what measurement to expect, use the highest setting and adjust downward to a reading of less than 50. The left side of the scale is too crowded for an accurate reading.

5. Zero the meter. You should do this each time you set it. To "zero the meter" means to adjust the pointer so that it reads 0 when the two test leads are touched together. Use the Ohms Adjust knob on the front of the VOM to line up the pointer over the zero on the ohms scale.

6. Attach the test leads to the component you are measuring.

7. Take the measurement. In Figure 5-60, the range selector switch is on $R \times 100$, and the measurement is 400 ohms.

Electrical Safety Precautions

Know where and how to turn off the electricity to your appliance or air conditioner—for example: plugs, fuses, circuit breakers, or cartridge fuses. Know their location in the home and label them. When replacing parts or reassembling the appliance or air conditioner, you should always install the wires on their proper terminals according to the wiring diagram. Then check to be sure that the wires are not crossing any sharp areas, are pinched in some way, or are between panels or moving parts that might cause an electrical problem.

These additional safety tips can help you and your family:

- Always use a separate, grounded electrical circuit for each major appliance or air conditioner.

- Never use an extension cord for major appliances.

- If you have to use an extension cord for a room air conditioner, make sure to use one that is properly rated for the size air conditioner installed.

- Be sure that the electricity is off before working on the appliance or air conditioner.

- Never remove the ground wire of a three-prong power cord, or any other ground wires from the appliance or air conditioner.

- Never bypass or alter any appliance or air conditioner switch, component, or feature.

- Replace any damaged, pinched, or frayed wiring that might be discovered when repairing the appliance or air conditioner.

Operating Appliances and Room Air Conditioners on a Generator

There are times when we have to seek alternate sources of electricity to run appliances and room air conditioners. Many people are installing auxiliary power generators as an alternative. In the 2004–2005 hurricane season, my family and neighbors were without electricity for up to three weeks per each hurricane. Luckily, we had a generator to power our home. When I purchased the generator, I had to take into account what appliances I intended to run off it.

Product	Starting Wattage	Running Wattage
Dishwasher (cool dry only)	540	216
Electric range (8-inch element only)	2100	2100
Electric range (all functions)	10,000 to 12,000	10,000 to 12,000
Microwave	1000 to 14,000	1000 to 14,000
Refrigerator or freezer	1200	132 to 192
Automatic washer	1200	1200
Electric dryer	6750	5400
Room air conditioner (10,000 BTU)	2200	1500

Formula's Watt = 3.1412 BTU/hour, BTU per hour × 0.293 = Watts, Volt × Amps = Watts

Table 5-6   Wattage Requirement Chart.

When you consider running lights, appliances, and room air conditioners on a generator, you must calculate how much wattage is needed for each product, including the start-up wattage. In addition, the voltage (120 or 240) and cycles (60) are critical for proper operation of appliances and air conditioners. If the voltage is less than 10 percent of the operating voltage, refrigerators, freezers, and room air conditioners will not operate properly, which might cause damage to the components in the product. Any variation in the cycles could speed up the clocks or slow them down. It is also important when searching for a generator that it has the ability to regulate the voltage and has surge protection for use with appliances and room air conditioners. Table 5-6 provides a guide for the average wattage requirements for some products when choosing the appropriate size generator to purchase.

Gas

As a technician or novice servicer, this chapter on gas is very important to the safety and technical knowledge that is needed to repair major gas appliances. In addition, consumers should be knowledgeable of safety procedures and basic characteristics of natural and liquefied petroleum gas (LP or LPG). Any person who cannot use basic tools or follow written instructions should not attempt to install, maintain, or repair gas appliances.

If you do not fully understand the procedures in this chapter, or if you doubt your ability to complete the task on your gas appliance, please call your service manager.

Currently, gas is used in millions of homes for heating, cooling, cooking, drying laundry, and water heating. Understanding gas theory, gas conversion, ignition systems, and the different types of gases used is something that every technician needs in his arsenal to service gas appliances. In addition, the technician needs to understand two other important elements when servicing gas appliances: combustion and ventilation. Some appliances use only gas as the main fuel, while others use gas and electricity. Although this chapter cannot cover all there is to know about gas, it will provide the basics.

Gas Safety

The following are a few safety tips to help you in handling major gas appliances in your home:

- Always follow the manufacturer's use and care manual for the gas appliance.
- Always keep combustible products away from gas appliances.
- Keep your gas appliance clean from soot, grease, and food spillages.
- Teach your children not to play near or with gas appliances.
- Always have a fire extinguisher nearby just in case of mishaps that might lead to a fire.
- Have a smoke detector and a carbon monoxide detector installed in the home, and check the batteries yearly.
- Never use a gas range to heat the home.
- Make sure that gas appliances have proper venting according to the manufacturers' recommendations.

Types of Gas

The two most common gases that are used in homes today are natural gas and liquefied petroleum gas (LP or LPG). Gas is a form of chemical energy, and when it is converted by combustion, it becomes heat energy. This type of heat energy is used for cooking, drying, heating, cooling, and lighting.

Natural Gas

Natural gas is a naturally occurring product made up of hydrocarbon and non-hydrocarbon gases. The main ingredient found in natural gas is methane (70 to 90 percent), with the remainder of the ingredients being nitrogen, ethane, butane, carbon dioxide, oxygen, hydrogen sulphide, and propane. These gases are located beneath the earth and can be removed through constructed wells. Another method for producing and harvesting natural gas is through landfills. The methane gas that is produced by the decomposition of materials can be harvested and added to the natural gas supply.

The heating value of natural gas is between 900 and 1200 BTUs per cubic foot. The air we breathe has a specific gravity of 1.00, and natural gas is lighter than air, with a specific gravity varying from 0.58 to 0.79. Natural gases are odorless and colorless—gas companies add an odor agent to warn for leaks.

The pressure of natural gas that is supplied to a residence will vary, between a 5- and 9-inch water column. A gas pressure regulator that is connected to the appliance will further reduce the pressure. For example, on some gas ranges, the pressure will be reduced to a 4-inch water column, while on other models, the pressure might be a 6-inch water column. On some water heaters manufactured today, the pressure is between a 4- and 7-inch water column. To determine the correct pressure rating, the technician must refer to the manufacturer's specifications or the installation instructions for that product.

Liquefied Petroleum Gas

Liquefied petroleum gas (LP or LPG) is obtained from natural gas sources or as a by-product of refining oil. LP gas for domestic use is usually propane, butane, or a mixture of the two. This type of gas is compressed and stored in storage tanks under pressure in a liquid state at approximately 250 pounds per square inch. The pressure in an LP tank will vary according to the surrounding temperatures and altitude. LP gas tanks can be transported to areas that are not supplied by natural gas supply lines.

The heating value of propane gas is 2500 BTUs per cubic foot, with a specific gravity of 1.53. The heating value of butane is much higher—about 3200 BTUs per cubic foot, with a specific gravity of 2.0. Liquefied petroleum gas is heavier than air and will accumulate in low-lying areas on the floor, in an enclosure, or in pockets beneath the ground, creating a hazard if it encounters an open flame. As mentioned, these gases are odorless and colorless, and gas companies add an odor agent to warn of leaks.

LP or LPG gas pressure for residential appliances, as established by the gas industry, will be between a 9- and 11-inch water column. To determine the correct pressure rating, the technician must refer to the manufacturer's specifications or the installation instructions for that product.

Combustion

Combustion is a rapid chemical reaction (burning) of LP or natural gas and air to produce heat energy and light. To sustain combustion in a gas appliance, an ignition source—such as that produced by a flame or by electrical means—is used to ignite the gas vapors. Three elements are needed to produce a flame when burning gas vapors: fuel, heat, and oxygen. If any one element is missing, flame or burning of gas vapors will not exist (Figure 6-1).

It takes 1 cubic foot of gas mixed with 10 cubic feet of air to have complete combustion of natural gas. The process produces approximately 11 cubic feet of combustion product, consisting of approximately 2 cubic feet of water vapor, 1 cubic foot of carbon dioxide, 8 cubic feet of nitrogen, and the excess air from the gas appliance. These combustion products must be properly vented or discharged safely from the gas appliance.

It takes 1 cubic foot of gas mixed with 24 cubic feet of air for complete combustion to happen with propane gas. The process produces approximately 25 cubic feet of combustion product. With butane gas, it takes 1 cubic foot of gas mixed with 31 cubic feet of air and produces 32 cubic feet of combustion product (Figure 6-2). With the proper mixture of air, gas, and flame, complete combustion will take place. The combustion product that is produced from complete combustion will be carbon dioxide and water vapor.

Inadequate venting of a gas appliance will restrict the flow of air into the gas appliance. This lack of proper ventilation will reduce the amount of oxygen within the air for complete combustion to occur. The air that we breathe is a mixture of gases containing nitrogen, oxygen, argon, carbon dioxide, water vapor, and other trace gases (Figure 6-3). Incomplete combustion will cause the reduction of oxygen levels in the air supply within a room. Proper ventilation—adequate fresh air into the room—is important and cannot be stressed enough for the proper operation of a gas appliance, as well as for the safety of human life within the home.

Carbon Monoxide

Carbon monoxide (CO) is a toxic gas that can cause death if inhaled in copious amounts. It is odorless, colorless, and has no taste. The human body cannot detect carbon monoxide with its senses. When carbon monoxide is inhaled, it is absorbed into the bloodstream and stays there longer, preventing oxygenated blood from performing its job in the body. Two factors affect the amount of carbon monoxide absorbed into the bloodstream: the amount of

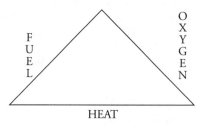

FIGURE 6-1   The combustion triangle.

Properties of		Natural gas	Propane	Butane
Chemical formula		CH_4	C_3H_8	C_4H_{10}
Boiling point of liquid at atmospheric pressure	°F	−258.7	−44	32
Specific gravity of vapor (air = 1)		0.6	1.53	2.00
Specific gravity of liquid (water = 1)		0.6	0.51	0.58
Calorific value @ 60°F	BTU/cubic foot	1012	2516	3280
	BTU/gallon		91,690	102,032
	BTU/lb		21,591	21,221
Latent heat of vaporization	BTU/gallon	712	785.0	808.0
Liquid weight	lbs/gallon	2.5	4.24	4.81
Vapor volume from 1 gallon of liquid at 60°F	Cubic foot		36.39	31.26
Vapor volume from 1 lb of liquid at 60°F	Cubic foot		8.547	6.506
Combustible limits	% of gas in air	5–15	2.4–9.6	1.9–8.6
Amount of air required to burn 1 cubic foot of gas	Cubic foot	9.53	23.86	31.02
Ignition temperature in air	°F	1200	920–1020	900–1000
Maximum flame temperature in air	°F	3568	3595	3615
Octane number		100	Over 100	92
All data are approximate.				

FIGURE 6-2   Properties of utility gases.

carbon monoxide in a room and the length of exposure. Lower levels of CO inhalation can cause flu-like symptoms, including headaches, dizziness, disorientation, fatigue, and nausea. Other exposure effects can vary depending on the age and health of the individual.

Carbon monoxide in homes without gas appliances varies between 0.5 to 5 parts per million (ppm). CO levels in homes with properly maintained gas appliances will vary from 5 to 15 parts per million. For those gas appliances that are not maintained properly, carbon monoxide levels may be 30 ppm or even higher.

Testing for Carbon Monoxide

Consumer products for detecting carbon monoxide in a home have been on the market for years. Carbon monoxide detectors should have an alarm that alerts consumers before they are exposed to hazardous levels of carbon monoxide. In order to prevent false alarms, CO detectors must be able to distinguish carbon monoxide gases from other types of gases, such as butane, heptane, alcohol, methane, and ethyl acetate. Two manufacturers that you can visit on the Internet to view the different types of carbon monoxide detectors available are www.kidde.com and www.firstalert.com.

Technicians who test for carbon monoxide use a special handheld meter to check the levels of carbon monoxide in a room or home. For a handheld carbon monoxide test meter, visit www.fluke.com.

When to check for carbon monoxide in a home:

- When the consumer complains of headaches or nausea
- Houseplants are dying

Component	Volume	Symbol
Ammonia	Trace	NH_3
Argon	0.93%	Ar
Carbon dioxide	0.03%	CO_2
Carbon monoxide	Trace	CO
Helium	5.2 ppm	He
Hydrogen	0.5 ppm	H_2
Iodine	0.01 ppm	I_2
Krypton	1.1 ppm	Kr
Methane	2.0 ppm	CH_4
Neon	18.2 ppm	Ne
Nitrogen	78.08%	N_2
Nitrous oxide	0.5 ppm	N_2O
Oxygen	20.95%	O_2
Sulfur dioxide	1.0 ppm	SO_2
Water vapor	0 to 5%	H_2O

FIGURE 6-3   Composition of air. The air we breathe is made up of 99.99% of nitrogen, oxygen, carbon dioxide, and argon.

- Unknown chronic odors from unknown sources
- Condensation on cool surfaces that might lead to flue gas products in the home

Figure 6-4 illustrates the locations in a home to test for carbon monoxide gas. These tests should be conducted near gas appliances, gas heating systems, heating ducts, and the atmosphere in a room, approximately 6 feet above the floor.

When testing for CO in a gas appliance that has not been running for a while, you should follow these steps:

1. Test the air near the appliance and the surrounding air in the room before you turn on the appliance.

2. Test the air after you turn on the appliance.

3. Test the air near the appliance after the appliance has been running for 15 minutes.

(a) Test in the atmosphere about 5 to 6 feet off the ground
(b) Test near gas appliances
(c) Test near heating ducts
(d) Test near appliance diverters and fire-doors on appliances in the basement or
    utility room

FIGURE 6-4   Typical test locations for carbon monoxide.

Flue Gases

Flue gases are the by-products of combustion that are exhausted through a chimney or flue to the outside of the home. These gases consist of the following:

- Nitrogen (colorless, odorless, and tasteless). The air we breathe is made up of 79 percent nitrogen.

- Carbon dioxide (colorless and odorless). The human respiration process produces carbon dioxide.

- Oxygen (colorless, odorless, and tasteless). Oxygen is the main ingredient for combustion.

- Carbon monoxide (colorless, odorless, and tasteless).

- Nitrogen oxide (colorless and odorless). Nitrogen oxide forms in the combustion process of gas fuels.

- Sulphur dioxide (colorless and smells like burnt matches). Sulphur dioxide is irritating to the lungs.

- Hydrocarbons (colorless, odorless, and tasteless). Hydrocarbons are found in natural and liquefied petroleum gases.

- Water vapor. Approximately 10 percent will be vented through the flue or chimney.

- Soot. The remains of incomplete combustion.

The "Flame"

Prior to the invention of the Bunsen burner in 1842, gas burners would produce a yellow flame for light and heat. The process allowed the gas to enter into a tube, expel through a port, and, when lit with an ignition source, the gas would burn without pre-mixing the air before it left the burner (Figure 6-5). As the flame burned, the gas temperature began to rise within the flame. Without the presence of pre-mixed air and gas, carbon particles began to pass through the flame, causing it to turn yellow. As more air was introduced into the combustion process, the flame would turn blue or blue with yellow tips.

FIGURE 6-5 Gas lamp.

Appliance manufacturers had the freedom to tailor the flame patterns from the Bunsen burner design to design gas appliances. The Bunsen burner worked on the principle of introducing air into the mixture of the gas before it left the burner port. An orifice was used to regulate the gas flow within the burner body (Figure 6-6). Attached to the burner body is an adjustable shutter to control the primary air mixture that enters into it. The flame that is produced by the Bunsen burner has multiple colors within it. Each color within the flame marks a stage of the burning process of gas (Figure 6-7). The stages are as follows:

- The inner cone is the first stage of the burning process. Within this cone, the gas is burned, which forms by-products, such as aldehydes, alcohols, carbon monoxide, and hydrogen.

FIGURE 6-6   Bunsen burner.

FIGURE 6-7   Different flame types of a Bunsen burner, depending on the amount of oxygen supplied.

- The outer cone surrounds the inner cone, and as air diffuses into the flame, it continues the burning process. If enough air is present during the burning process, the by-products from the inner cone will burn up in the outer cone. The by-products that are produced in the outer cone are carbon dioxide and water vapor.

- The outer mantle that is produced by the Bunsen burner is nearly invisible, with all the gas burned up. The reason it glows is due to the high temperature produced by the complete combustion of the gas products from the outer cone.

The temperature of a flame can vary, depending on what type of gas is used and how much air is pre-mixed with the gas. Temperatures within the flame will vary; the inner cone is cooler than the outer cone, and the hottest flame temperature is just above the outer cone.

Air Supply

When air is mixed with gas before it leaves the burner port, it is known as the primary air supply. Under ideal conditions, most burners only use 50 percent of the primary air supply to burn the gas, and the remaining 50 percent of the air is supplied by the secondary air supply from around the flames. In reality, additional excess air is needed to ensure that enough air is present to completely burn off the gas. The total air supply is accomplished by adding the sum of the primary, secondary, and excess air supplies. The total air supply needed for combustion is measured in percentages of air needed for complete combustion. Every manufacturer has its own specifications in the installation instructions or service manual for the total air percentages needed for the appliance model that you are servicing.

Flame Appearance

The appearance and stability of a flame is influenced by the amount of the primary air supply, the gas-burning speed, and the burner port. When the primary air supply is at 100 percent, under ideal conditions, the speed of burning is at its maximum. The speed of burning is also affected by the type of gas being burned. The flow velocity of the gas depends on the size of the orifice—the smaller the orifice size, the greater the flow velocity. Flow velocity from a burner port will also vary, depending on the size of the opening, with the greatest flow from the center of the port (Figure 6-8).

After the air-gas mixture leaves the port, the flow velocity slows down, and the flame begins to stabilize when the flow velocity equals the burning speed. At the same time, the flame cone begins to take shape when the flow velocity of the air-gas mixture levels off when it leaves the burner port. The burning speed increases and the flame temperature increases near the top of the flame cone. Therefore, the shape of the flame cone is rounded off at its tip. The inner cone is determined by the effects of the velocity away from the centerline of the port for each layer of the flame burning.

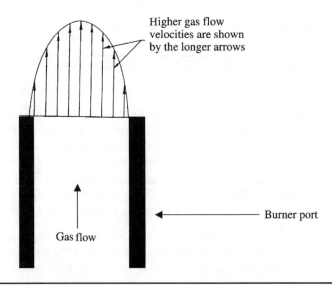

FIGURE 6-8   Flow velocities of gas.

Appliance manufacturers design gas appliances to have a stable burner flame by port loading. Port loading is expressed as BTU per hour per square inch of open burner port area. It is obtained by dividing the gas input rate by the total area of the port opening.

Flame Lifting

When enough primary air is introduced into a burner, a stabilized flame begins to form, and the flame burns quietly. If too much air is introduced, or if the port loadings are increased, the flame on the burner will have the tendency to lift (Figure 6-9). When the flames begin to lift (blowing) off the burner, they become very noisy. If this condition persists, the flame cones begin to rupture and complete combustion will not take place. In addition, the efficiency of the gas burner begins to drop along with the BTU's and the appliance begins to lose its heat content. If this condition is not corrected, aldehydes and carbon monoxide will begin filling the room.

Additional flame-lifting possibilities include:

- Flame lifting is more likely to occur with natural gas appliances.
- Flame lifting will also occur if any operating or design factors cause the burner to decrease the burning speed.
- If any operating or design factors increase the flow velocity from the ports, flame lifting will occur.
- The port size and depth will affect flame lifting.
- Overrating the burners is a primary cause of flame lifting.
- A cold burner will cause the flame to lift, but it should settle down after a while.

Flashback

When the flame begins to ignite from within the burner head, a condition known as flashback occurs. This is caused when the reduced flow velocity of the air-gas mixture is less than the

Figure 6-9 Too much air supply will cause the flames to lift off the burner.

burning speed near the burner port. In other words, the inner flame cone becomes inverted and ignites the gas from within the burner port. If the burner is properly adjusted, flashback will not occur during normal operation of the gas appliance.

Additional flashback possibilities include:

- Flashback occurs in faster-burning gases.
- Flashback could occur when primary air is increased, which will increase the burning speed.
- Flashback occurs with underrated burners with incorrect orifice size.
- Incorrect gas pressure will cause flashback.
- Decreased flow velocities from a burner port will cause flashback.
- Leaking burner valves cause flashback.
- Flashback can occur when a burner is first turned on if the air-gas mixture is too rich or too lean.

When a burner is turned off, a condition known as "extinction pop," or flashback on extinction, could occur. This could happen immediately, or it may take a few seconds. The extinction pop is caused by excessive air entering the burner when the gas burner valve is turned off. The reduced flow rate of the normal air-gas mixture in a burner is replaced with all air. In addition, it is possible for the flame speed to exceed the flow velocity of the gas, causing a flashback. This condition is not hazardous—it is just annoying to the appliance owner. With a burner that is properly designed, maintained, and adjusted, flashback will not occur under normal operating conditions.

Yellow Flame Tips

In gas appliances, if the primary air supply is reduced, the inner cones of the flames will begin to lengthen, and eventually they will disappear, turning the blue flame tips to yellow. If the primary air supply is cut off completely, the flames will turn yellow (Figure 6-10). When these yellow flames and glowing carbon particles impinge on a cool surface, they begin to quench the complete combustion process and the carbon particles will not burn off. In addition, carbon monoxide, soot, or both will begin to form when the yellow flames impinge on a cool surface.

Dust particles in the air will begin to glow as they pass through a flame (Figure 6-11); the flame's appearance begins to have color streaks. These streaks will appear orange-yellow and will not have an effect on complete combustion. The technician should not confuse dust particles with yellow flame tips. True yellow flame tips will be a pale yellow color and they can be eliminated by increasing the primary air supply. Depending on the air conditions of the surrounding area near the gas appliance, when you increase the primary air supply, you might also increase the amount of dust particles entering into the flame. Another way to distinguish yellow flame tips from dust particles is to put on a pair of brazing goggles and look into the flame. True yellow flame tips will not disappear.

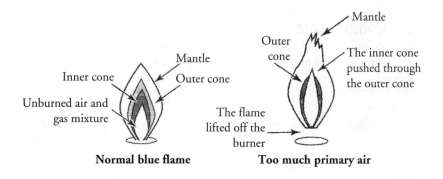

Mantle

Inner cone

Outer cone

Unburned air and
gas mixture

Normal blue flame

Mantle

Outer
cone

The inner cone
pushed through
the outer cone

The flame
lifted off the
burner

Too much primary air

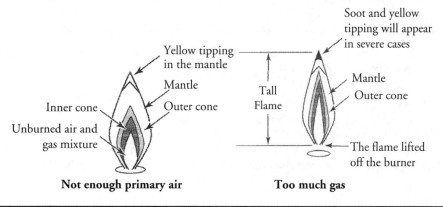

Yellow tipping
in the mantle

Mantle

Inner cone

Outer cone

Unburned air and
gas mixture

Not enough primary air

Soot and yellow
tipping will appear
in severe cases

Mantle

Outer cone

Tall
Flame

The flame lifted
off the burner

Too much gas

FIGURE 6-10   Characteristics of gas flames.

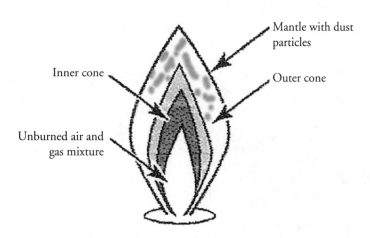

Mantle with dust
particles

Inner cone

Outer cone

Unburned air and
gas mixture

FIGURE 6-11   Dust particles passing through a normal flame.

Appliance Burner Components

Over the years, appliance manufacturers have come out with many types of burner designs (see Figure 1-1 in chapter 1 and Figure 6-12). In fact, appliance burners must have the following features:

- The burner design must be able to provide complete combustion of the gas.
- The burner design must provide for rapid ignition of the gas and be able to carry over the flame across the entire burner.
- The burner must operate reasonably quietly during ignition, burning, and extinction of the flame.
- The burner design must not allow for excessive flame lifting, flashback, or flame outage.
- The burner must provide uniform heat over the heated area.
- Most important to the consumer and manufacturer, the burner must have a long service life.

Gas Orifice

The gas orifice is not part of the burner, but it plays an important role in its operation (Figure 6-13). In the service field, the gas orifice goes by many names, such as orifice hood, spud, hood, or cap. The purpose of the gas orifice is to regulate or limit the flow of gas into the burner. The size of the hole in the orifice will depend on what type of gas is in

Figure 6-12   A double gas burner.

FIGURE 6-13
Gas orifices are used to limit gas flow to burners.

use and what constant pressure is needed to light the burner evenly. Another reason for use of an orifice is to force primary air into the burner. Three types of orifices that are in use are:

- The fixed orifice has a predetermined opening in the orifice hood to allow a certain rate of gas to flow (Figure 6-14).

- The adjustable orifice is mainly used in ranges. This type of orifice will allow you to control the gas rate, from zero to its maximum rated flow (Figure 6-15).

- The universal orifice is also mainly used on ranges. This type of orifice was designed to allow the range to be operated on natural or liquefied petroleum gas (Figure 6-16).

Most gas appliances that operate below the elevation of 2000 feet may not need to have the orifices replaced. If you have to replace the orifice, or if you have to check to see if the correct orifice size is installed, refer to Tables 6-1, 6-2, 6-3, and 6-4. *Note:* Gas appliances shipped from the manufacturer are set up for natural gas. If the home has LP gas, then the appliance will have to be converted for LP use.

Air Shutter

Most gas appliances have an adjustable primary air shutter to adjust the air intake, and they come in many designs. In addition, manufacturers have designed the air shutter to withstand rust and corrosion and to be secured in any position (Figure 6-17).

FIGURE 6-14   Different types of fixed orifices.

Adjustable orifice hood

Fixed orifice hood

Gas

Gas

Gas

Gas

Gas

Fixed needle

Movable needle

FIGURE 6-15 Adjustable orifices.

Venturi Throat

When gas passes through an orifice, it enters the throat of the mixing tube (Figure 6-18). Called the Venturi throat, this was designed to provide a more constant flow of primary air into the mixing tube of the burner.

Mixing Tube

The mixing tube (see Figure 6-17) is designed to mix the air and gas uniformly after it leaves the Venturi throat but before entering the burner head.

Gas flow

Gas flow

Natural gas position

LP gas position

FIGURE 6-16 Universal orifices.

Orifice Size Drill Size (Decimal or DMS)	Gas Pressure at Orifice—Inches Water Column								
	3	**3.5**	**4**	**5**	**6**	**7**	**8**	**9**	**10**
0.008	0.17	0.18	0.19	0.23	0.24	0.26	0.28	0.29	0.30
0.009	0.21	0.23	0.25	0.28	0.30	0.33	0.35	0.37	0.39
0.010	0.27	0.29	0.30	0.35	0.37	0.41	0.43	0.46	0.48
0.011	0.33	0.35	0.37	0.42	0.45	0.48	0.52	0.55	0.59
0.012	0.38	0.41	0.44	0.50	0.54	0.57	0.62	0.65	0.70
80	0.48	0.52	0.55	0.63	0.69	0.73	0.79	0.83	0.88
79	0.55	0.59	0.64	0.72	0.80	0.84	0.90	0.97	1.01
78	0.70	0.76	0.78	0.88	0.97	1.04	1.10	1.17	1.24
77	0.88	0.95	0.99	1.11	1.23	1.31	1.38	1.47	1.55
76	1.05	1.13	1.21	1.37	1.52	1.61	1.72	1.83	1.92
75	1.16	1.25	1.34	1.52	1.64	1.79	1.91	2.04	2.14
74	1.33	1.44	1.55	1.74	1.91	2.05	2.18	2.32	2.44
73	1.51	1.63	1.76	1.99	2.17	2.32	2.48	2.64	2.78
72	1.64	1.77	1.90	2.15	2.40	2.52	2.69	2.86	3.00
71	1.82	1.97	2.06	2.33	2.54	2.73	2.91	3.11	3.26
70	2.06	2.22	2.39	2.70	2.97	3.16	3.38	3.59	3.78
69	2.25	2.43	2.61	2.96	3.23	3.47	3.68	3.94	4.14
68	2.52	2.72	2.93	3.26	3.58	3.88	4.14	4.41	4.64
67	2.69	2.91	3.12	3.52	3.87	4.13	4.41	4.69	4.94
66	2.86	3.09	3.32	3.75	4.11	4.39	4.68	4.98	5.24
65	3.14	3.39	3.72	4.28	4.62	4.84	5.16	5.50	5.78
64	3.41	3.68	4.14	4.48	4.91	5.23	5.59	5.95	6.26
63	3.63	3.92	4.19	4.75	5.19	5.55	5.92	6.30	6.63
62	3.78	4.08	4.39	4.96	5.42	5.81	6.20	6.59	6.94
61	4.02	4.34	4.66	5.27	5.77	6.15	6.57	7.00	7.37
60	4.21	4.55	4.89	5.52	5.95	6.47	6.91	7.35	7.74
59	4.41	4.76	5.11	5.78	6.35	6.78	7.25	7.71	8.11
58	4.66	5.03	5.39	6.10	6.68	7.13	7.62	8.11	8.53
57	4.84	5.23	5.63	6.36	6.96	7.44	7.94	8.46	8.90
56	5.68	6.13	6.58	7.35	8.03	8.73	9.32	9.92	10.44
55	7.11	7.68	8.22	9.30	10.18	10.85	11.59	12.34	12.98
54	7.95	8.59	9.23	10.45	11.39	12.25	13.08	13.93	14.65
53	9.30	10.04	10.80	12.20	13.32	14.29	15.27	16.25	17.09
52	10.61	11.46	12.31	13.86	15.26	16.34	17.44	18.57	19.53
51	11.82	12.77	13.69	15.47	16.97	18.16	19.40	20.64	21.71
50	12.89	13.92	14.94	16.86	18.48	19.77	21.12	22.48	23.65
49	14.07	15.20	16.28	18.37	20.20	21.60	23.06	24.56	25.83
48	15.15	16.36	17.62	19.88	21.81	23.31	24.90	26.51	27.89
47	16.22	17.52	18.80	21.27	23.21	24.93	26.62	28.34	29.81

TABLE 6-1   Orifice Gas Pressure. *(continued)*

Orifice Size Drill Size (Decimal or DMS)	Gas Pressure at Orifice—Inches Water Column								
	3	**3.5**	**4**	**5**	**6**	**7**	**8**	**9**	**10**
46	17.19	18.57	19.98	22.57	24.72	26.43	28.23	30.05	31.61
45	17.73	19.15	20.52	23.10	25.36	27.18	29.03	30.90	32.51
44	18.45	21.01	22.57	25.57	27.93	29.87	31.89	33.96	35.72
43	20.73	22.39	24.18	27.29	29.87	32.02	34.19	36.41	38.30
42	23.10	24.95	26.50	29.50	32.50	35.24	37.63	40.07	42.14
41	24.06	25.98	28.15	31.69	34.81	37.17	39.70	42.27	44.46
40	25.03	27.03	29.23	33.09	36.20	38.79	41.42	44.10	46.38
39	26.11	28.20	30.20	34.05	37.38	39.97	42.68	45.44	47.80
38	27.08	29.25	31.38	35.46	38.89	41.58	44.40	47.27	49.73
37	28.36	30.63	32.99	37.07	40.83	43.62	46.59	49.60	52.17
36	29.76	32.14	34.59	39.11	42.76	45.77	48.88	52.04	54.74
35	32.36	34.95	36.86	41.68	45.66	48.78	52.10	55.46	58.34
34	32.45	35.05	37.50	42.44	46.52	49.75	53.12	56.55	59.49
33	33.41	36.08	38.79	43.83	48.03	51.46	54.96	58.62	61.55
32	35.46	38.30	40.94	46.52	50.82	54.26	57.95	61.70	64.89
31	37.82	40.85	43.83	49.64	54.36	58.01	61.96	65.97	69.39
30	43.40	46.87	50.39	57.05	62.09	66.72	71.22	75.86	79.80
29	48.45	52.33	56.19	63.61	69.62	74.45	79.52	84.66	89.04
28	51.78	55.92	59.50	67.00	73.50	79.50	84.92	90.39	95.09
27	54.47	58.83	63.17	71.55	78.32	83.59	89.27	95.04	99.97
26	56.73	61.27	65.86	74.57	81.65	87.24	93.17	99.19	104.57
25	58.87	63.58	68.22	77.14	84.67	90.36	96.50	102.74	108.07
24	60.81	65.67	70.58	79.83	87.56	93.47	99.83	106.28	111.79
23	62.10	67.07	72.20	81.65	89.39	94.55	100.98	107.49	113.07
22	64.89	70.08	75.21	85.10	93.25	99.60	106.39	113.24	119.12
21	66.51	71.83	77.14	87.35	95.63	102.29	109.24	116.29	122.33
20	68.22	73.68	79.08	89.49	97.99	104.75	111.87	119.10	125.28
19	72.20	77.98	83.69	94.76	103.89	110.67	118.55	125.82	132.36
18	75.53	81.57	87.56	97.50	108.52	116.03	123.92	131.93	138.78
17	78.54	84.82	91.10	103.14	112.81	120.33	128.52	136.82	143.91
16	82.19	88.77	95.40	107.98	118.18	126.78	135.39	144.15	151.63
15	85.20	92.02	98.84	111.74	122.48	131.07	139.98	149.03	156.77
14	87.10	94.40	100.78	114.21	124.44	133.22	142.28	151.47	159.33
13	89.92	97.11	104.32	118.18	128.93	138.60	148.02	157.58	165.76
12	93.90	101.41	108.52	123.56	135.37	143.97	153.75	163.69	172.13
11	95.94	103.62	111.31	126.02	137.52	147.20	157.20	167.36	176.03
10	98.30	106.16	114.21	129.25	141.82	151.50	161.81	172.26	181.13
9	100.99	109.07	117.11	132.58	145.05	154.71	165.23	175.91	185.03
8	103.89	112.20	120.65	136.44	149.33	160.08	170.96	182.00	191.44

TABLE 6-1   Orifice Gas Pressure. *(continued)*

Orifice Size Drill Size (Decimal or DMS)	Gas Pressure at Orifice—Inches Water Column								
	3	**3.5**	**4**	**5**	**6**	**7**	**8**	**9**	**10**
7	105.93	114.40	123.01	139.23	152.56	163.31	174.38	185.68	195.30
6	109.15	117.88	126.78	142.88	156.83	167.51	178.88	190.46	200.36
5	111.08	119.97	128.93	145.79	160.08	170.82	182.48	194.22	204.30
4	114.75	123.93	133.22	150.41	164.36	176.18	188.16	200.23	210.71
3	119.25	128.79	137.52	156.26	170.78	182.64	195.08	207.66	218.44
2	128.48	138.76	148.61	168.64	184.79	197.66	211.05	224.74	235.58
1	136.35	147.26	158.25	179.33	194.63	209.48	223.65	238.16	250.54
A	145.34	155.48	165.62	189.28	206.18	219.70	236.60	250.12	263.63
B	150.36	160.85	171.33	195.81	213.29	227.25	244.76	258.74	272.73
C	155.45	166.30	177.14	202.44	220.52	234.98	253.06	267.51	281.97
D	160.63	171.84	183.04	209.19	227.87	242.81	261.48	276.44	291.38
E	165.89	177.47	189.05	216.05	235.34	250.77	270.06	285.49	300.93
F	175.43	187.61	199.78	227.75	248.70	265.02	285.40	301.70	318.02
G	180.81	193.43	206.04	235.49	256.50	273.33	294.35	311.18	327.99
H	187.81	200.91	214.01	244.59	266.42	283.89	305.74	323.20	340.67
I	196.38	210.08	223.77	255.75	278.58	296.85	319.68	337.95	356.20
J	203.66	217.87	232.08	265.24	288.92	307.87	331.55	350.49	396.44
K	209.59	224.22	238.84	272.95	297.33	316.82	341.19	360.69	380.18
L	223.23	240.49	257.75	290.71	316.68	337.44	363.40	384.17	404.92
M	231.00	247.12	263.23	300.83	327.69	349.18	376.03	397.52	419.01
N	242.09	258.98	275.06	315.27	343.42	365.94	394.09	416.61	439.13
O	265.05	283.54	302.03	345.18	376.00	400.66	431.47	456.13	480.79
P	276.92	296.24	315.56	360.64	392.84	418.60	450.80	476.56	502.32
Q	292.57	312.98	333.39	381.03	415.05	442.26	476.28	503.49	530.71
R	305.03	326.33	347.62	397.26	432.73	461.10	496.59	524.95	553.32
S	321.45	344.01	366.57	418.62	456.01	485.91	523.28	553.19	583.09
T	340.19	363.92	387.65	443.04	482.59	514.24	553.79	585.44	617.09
U	359.46	383.54	409.61	468.14	509.93	543.36	585.17	618.60	652.04
V	377.26	403.58	429.90	491.31	535.17	570.27	614.14	649.23	684.33
W	395.48	423.07	450.66	515.05	561.04	597.83	643.82	680.60	717.39
X	418.34	447.53	476.72	544.82	593.47	632.39	681.03	719.94	758.86
Y	433.23	463.45	493.67	564.20	614.58	654.87	705.25	745.56	785.86
Z	452.75	484.33	515.91	589.64	642.26	684.38	737.02	779.14	821.26

UTILITY GASES

(Cubic feet per hour at sea level)

Specific gravity = 0.60

Orifice coefficient = 0.9

For utility gases of another specific gravity, select factor from Table 6-3

For altitudes above 2,000 feet, first select the equivalent orifice size at sea level from Table 6-4

TABLE 6-1   Orifice Gas Pressure.

Orifice Drill Size (Decimal or DMS)	Gas Input, BTU Per Hour For:	
	Propane	Butane or Butane-Propane Mixture
0.008	500	554
0.009	641	709
0.01	791	875
0.011	951	1053
0.012	1130	1250
80	1430	1590
79	1655	1830
78	2015	2230
77	2545	2815
76	3140	3480
75	3465	3840
74	3985	4410
73	4525	5010
72	4920	5450
71	5320	5900
70	6180	6830
69	6710	7430
68	7560	8370
67	8040	8910
66	8550	9470
65	9630	10,670
64	10,200	11,300
63	10,800	11,900
62	11,360	12,530
61	11,930	13,280
60	12,570	13,840
59	13,220	14,630
58	13,840	15,300
57	14,550	16,090
56	16,990	18,790
55	21,200	23,510
54	23,850	26,300
53	27,790	30,830
52	31,730	35,100
51	35,330	39,400
50	38,500	42,800
49	41,850	45,350
48	45,450	50,300
47	48,400	53,550

TABLE 6-2 LP Orifice Drill Size.

Orifice Drill Size (Decimal or DMS)	Gas Input, BTU Per Hour For:	
	Propane	Butane or Butane-Propane Mixture
46	51,500	57,000
45	52,900	58,500
44	58,050	64,350
43	62,200	69,000
42	68,700	76,200
41	72,450	80,200
40	75,400	83,500
39	77,850	86,200
38	81,000	89,550
37	85,000	94,000
36	89,200	98,800
35	95,000	105,300
34	97,000	107,200
33	101,000	111,900
32	105,800	117,000
31	113,200	125,400
30	129,700	143,600
29	145,700	163,400
28	154,700	171,600
27	163,100	180,000
26	169,900	187,900
25	175,500	194,600
24	181,700	201,600
23	186,800	206,400
22	193,500	214,500
21	198,600	220,200
20	203,700	225,000
19	217,100	241,900
18	225,600	249,800

Liquefied Petroleum Gas (LP or LPG)

(BTU per hour at sea level)

	Propane	Butane
BTU per cubic foot	2500	3175
Specific gravity	1.53	2.00
Pressure at orifice measured in inches water column	11	11
Orifice coefficient	0.9	0.9

For altitudes above 2000 feet, first select the equivalent orifice size at sea level from Table 6-4

TABLE 6-2 LP Orifice Drill Size. (*continued*)

Specific Gravity	Factor	Specific Gravity	Factor
0.45	1.155	0.95	0.795
0.50	1.095	1.00	0.775
0.55	1.045	1.05	0.756
0.60	1.000	1.10	0.739
0.65	0.961	1.15	0.722
0.70	0.926	1.20	0.707
0.75	0.894	1.25	0.693
0.80	0.866	1.30	0.679
0.85	0.840	1.35	0.667
0.90	0.817	1.40	0.655

To select an orifice size to provide a desired gas flow rate:

1) Obtain the factor for the known specific gravity of the gas from above.
2) Divide the desired gas flow rate by this factor, which gives the equivalent flow rate with a 0.60 specific gravity gas.
3) Calculate the gas flow from a gas meter using the following formula:

$$\text{Gas flow measured in cubic feet per hour} = \frac{(\text{cubic feet per revolution of gas meter dial}) \times 3600}{\text{Time one revolution of the gas meter dial in seconds}}$$

The number 3600 converts seconds to hours.

Example: If the gas meter measures ½ cubic foot per revolution and the amount of time for that revolution is found to be 15 seconds, then the gas flow will calculate as follows:

$$\text{Gas flow rate} = \frac{0.5 \times 3600}{15} = 120 \text{ cubic feet per hour}$$

4) Use Table 6-1 to select orifice size.

To estimate gas flow rate for a given orifice size and gas pressure when it is not possible to meter gas flow:

1) Use Table 6-1 to obtain equivalent gas flow rate with a 0.60 specific gravity gas.
2) Obtain the factor from Table 6-3 for the specific gravity of the gas being burned.
3) Multiply the rate obtained in (1) by the factor from (2) to estimate gas flow rate to the burner.

TABLE 6-3 Factors for Utility Gases of Another Specific Gravity.

Orifice Size at Sea Level	Orifice Size Required at Other Elevations								
	2000	3000	4000	5000	6000	7000	8000	9000	10,000
1	2	2	3	3	4	5	7	8	10
2	3	3	4	5	6	7	9	10	12
3	4	5	7	8	9	10	12	13	15
4	6	7	8	9	11	12	13	14	16
5	7	8	9	10	12	13	14	15	17
6	8	9	10	11	12	13	14	16	17
7	9	10	11	12	13	14	15	16	18
8	10	11	12	13	13	15	16	17	18
9	11	12	12	13	14	16	17	18	19
10	12	13	13	14	15	16	17	18	19
11	13	13	14	15	16	17	18	19	20
12	13	14	15	16	17	17	18	19	20
13	15	15	16	17	18	18	19	20	22
14	16	16	17	18	18	19	20	21	23
15	16	17	17	18	19	20	20	22	24
16	17	18	18	19	19	20	22	23	25
17	18	19	19	20	21	22	23	24	26
18	19	19	20	21	22	23	24	26	27
19	20	20	21	22	23	25	26	27	28
20	22	22	23	24	25	26	27	28	29
21	23	23	24	25	26	27	28	28	29
22	23	24	25	26	27	27	28	29	29
23	25	25	26	27	27	28	29	29	30
24	25	26	27	27	28	28	29	29	30
25	26	27	27	28	28	29	29	30	30
26	27	28	28	28	29	29	30	30	30
27	28	28	29	29	29	30	30	30	31
28	29	29	29	30	30	30	30	31	31
29	29	30	30	30	30	31	31	31	32
30	30	31	31	31	31	32	32	33	35
31	32	32	32	33	34	35	36	37	38
32	33	34	35	35	36	36	37	38	40
33	35	35	36	36	37	38	38	40	41
34	35	36	36	37	37	38	39	40	42
35	36	36	37	37	38	39	40	41	42
36	37	38	38	39	40	41	41	42	43
37	38	39	39	40	41	42	42	43	43
38	39	40	41	41	42	42	43	43	44
39	40	41	41	42	42	43	43	44	44
40	41	42	42	42	43	43	44	44	45
41	42	42	42	43	43	44	44	45	46

TABLE 6-4   Equivalent Orifice Sizes at High Altitudes. (*continued*)

Orifice Size at Sea Level	Orifice Size Required at Other Elevations								
	2000	3000	4000	5000	6000	7000	8000	9000	10,000
42	42	43	43	43	44	44	45	46	47
43	44	44	44	45	45	46	47	47	48
44	45	45	45	46	47	47	48	48	49
45	46	47	47	47	48	48	49	49	50
46	47	47	47	48	48	49	49	50	50
47	48	48	49	49	49	50	50	51	51
48	49	49	49	50	50	50	51	51	52
49	50	50	50	51	51	51	52	52	52
50	51	51	51	51	52	52	52	53	53
51	51	52	52	52	52	53	53	53	54
52	52	53	53	53	53	53	54	54	54
53	54	54	54	54	54	54	55	55	55
54	54	55	55	55	55	55	56	56	56
55	55	55	55	56	56	56	56	56	57
56	56	56	57	57	57	58	59	59	60
57	58	59	59	60	60	61	62	63	63
58	59	60	60	61	62	62	63	63	64
59	60	61	61	62	62	63	64	64	65
60	61	61	62	63	63	64	64	65	65
61	62	62	63	63	64	65	65	66	66
62	63	63	64	64	65	65	66	66	67
63	64	64	65	65	65	66	66	67	68
64	65	65	65	66	66	66	67	67	68
65	65	66	66	66	67	67	68	68	69
66	67	67	68	68	68	69	69	69	70
67	68	68	68	69	69	69	70	70	70
68	68	69	69	69	70	70	70	71	71
69	70	70	70	70	71	71	71	72	72
70	70	71	71	71	71	72	72	73	73
71	72	72	72	73	73	73	74	74	74
72	73	73	73	73	74	74	74	74	75
73	73	74	74	74	74	75	75	75	76
74	74	75	75	75	75	76	76	76	76
75	75	76	76	76	76	77	77	77	77
76	76	76	77	77	77	77	77	77	77
77	77	777	77	78	78	78	78	78	78
78	78	78	78	79	79	79	79	80	80
79	79	80	80	80	80	0.013	0.012	0.012	0.012
80	80	0.013	0.013	0.013	0.012	0.012	0.012	0.012	0.011

Equivalent orifice sizes at high altitudes
(Includes 4% input reduction for each 1000 feet)

TABLE 6-4 Equivalent Orifice Sizes at High Altitudes.

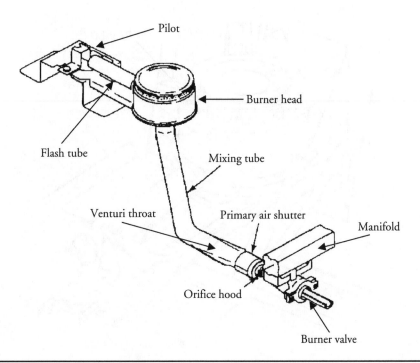

FIGURE 6-17   The primary air shutter is located at the end of the mixing tube. The burner head is designed with ports around the burner to allow for even burning and a blue flame.

Burner Head

When gas leaves the mixing tube, it enters the burner head, which contains the burner ports. The air-gas mixture is then distributed evenly among the burner ports. As the air-gas mixture emerges from the burner ports, it is lit with an ignition source. The size and shape of the burner head will allow even heat transfer to occur (see Figure 6-17).

FIGURE 6-18   The Venturi throat is incorporated into the mixing tube of the burner.

Top burner ports

Climber ports

Charge ports

Flash tube

Pilot flame

Figure 6-19   The design of the burner ports and the location of the burner portholes.

Burner Ports

The burner ports (Figure 6-19) serve to distribute the flame evenly, provide even heat transfer, and provide a stable blue flame. In addition, when the flames are distributed throughout the burner ports, they will then pick up secondary air to assist in complete combustion.

Burner Operation

When the gas is turned on, whether by a knob or automatically by a control device, gas will flow from the supply line through a gas pressure regulator, gas manifold, burner valves, and orifice into the Venturi throat and mixing tube. The-air gas mixture then enters the burner head to be ignited. As the air-gas mixture leaves the burner head, when ignited, the flame begins to burn evenly around the burner head. Figure 6-20 illustrates an exploded view of the gas range's components.

Ventilation

When you properly install a ventilation system, it will serve the following important purposes:

- The by-products of hot combustion gases will be transported outside the home.
- The home will be protected from fire hazards.

FIGURE 6-20   Exploded view illustrating the various components used in a gas range and their location.

- A ventilation system will provide good air circulation and adequate oxygen supply for the gas appliance and the occupants of the residence.
- It will remove the water vapors produced when burning gas.

A wide variety of venting options are available from vent manufacturers, and every gas appliance manufacturer will provide venting instructions for properly installing their products. It is strongly recommended that you follow these instructions. You must use the proper venting materials as described in the installation instructions for the model being installed, and install the venting system according to your local building codes. For additional information on proper ventilation, visit www.epa.gov/iaq/homes/hip-combustion.html.

Troubleshooting Ventilation Problems

The most common problems with ventilation systems are:

- Improper maintenance of the ventilation system
- Incorrect sizing and installation of the ventilation system
- Inadequate air supply

If you suspect a problem with the ventilation system, check the following:

- Make sure that you have the correct vent sizing.
- Check if there are too many elbows or if the length of the venting system exceeds the manufacturer's recommendations or local building codes.
- Inspect the entire ventilation system for faults, such as a disconnected or crushed pipe.
- Inspect for an obstruction, such as a clogged vent cap.
- Inspect all air openings for obstructions.

Measuring Gas Pressure

When installing or repairing gas appliances, it is often necessary to measure the pressure of the gas supply to the appliance. There also may be times when the technician will have to test the gas pressures at the manifold or orifices. The two types of test instruments used today to test the gas pressures are the manometer and the magnehelic gauge.

The manometer is a U-shaped tube equipped with a scale to measure gas pressure in water column inches (Figure 6-21). Before you check the gas pressure on an appliance, locate the model and serial number nameplate. This nameplate will have the gas rating stamped on it. Now turn off the gas supply to the appliance. Before you can begin to test the gas pressure, you must first set up the manometer by adding water to the U-shaped tube until both columns read zero on the scale (Figure 6-22a). If reading the scale is difficult, you can add a food coloring to the water to enhance the color so that you can read the scale better. When measuring gas pressure in an appliance, the tubing from the manometer is attached to the orifice, manifold, or gas supply line, and the other end of the tube on the manometer is left open to the room atmosphere. To test the gas pressure at the burner orifice, remove the burner (Figure 6-23). Next, attach the long hose from the manometer to the burner orifice. Now turn on the gas supply to the appliance. In addition, turn on the burner valve being tested, and turn on one other burner to serve as the load. When the gas pressure is applied to the manometer, it pushes down on the water column and the water column rises on the other side of the manometer (Figure 6-22b). To read the manometer scale, observe where the water column stops. In Figure 6-22b, the incoming gas has pushed the water in the manometer down 2 inches below zero, and the water column on the other side is pushed up 2 inches above zero. The pressure reading is obtained by adding the two readings together. Figure 6-22b shows that the total change equals a 4-inch water column. The reading should be within the rating on the model and serial nameplate.

The magnehelic gauge shown in Figure 6-24 also measures gas pressure. This gauge can measure gas pressure faster than the manometer can. To use this type of gauge, just follow the same procedures for the manometer. The only difference is that you read the gas pressure directly from the gauge scale. Some models have different scales on the gauge dial—just read the gauge that indicates water column pressure.

Many companies manufacture test instruments for checking gas pressures. Web sites to check out include:

- http://www.uniweld.com/catalog/gauges_thermometers/gas_pressure_test_kit.htm
- http://www.dwyer-inst.com
- http://www.marshbellofram.com

FIGURE 6-21
A manometer. This instrument is used to test gas pressure.

Testing for Gas Leaks

When testing for a gas leak, do not use a lighter or a lit match. That is not the safe or correct way to test for gas leaks. All gas appliances, when they are installed or repaired, must be tested for leaks before placing the appliance in operation. Remember, LP or LPG gas is heavier than air and settles in low-lying areas. If you do not have a combustible gas detector (and I recommend that you purchase one), use your nose to smell for a gas leak all around the appliance at floor level. A safe and effective way to test for gas leaks is to use a chloride-free soap-and-water solution to check all connections and fittings. The solution can be sprayed, applied directly, or an applicator can be used to spread the solution. If a gas leak is detected by the soap solution it will begin to form bubbles.

Sometimes, air might be present in the gas lines from installation or repairs, and it could prevent the pilot or burner from lighting on initial startup. The gas lines should be purged of air in a well-vented area. Before proceeding with purging the gas lines, inspect the work area for any other sources of ignition, such as flames, burning candles, running electrical appliances, etc. With a safe working environment, you can begin to purge the gas lines. On gas ranges/ovens, just turn on the burners until you smell the gas or until the burners are lit properly. With gas dryers and gas water heaters, turn off the gas supply valve. Remove the gas line from the appliance, and slowly begin to turn on the gas

(a)

FIGURE 6-22a Before using a manometer, check to see if the columns have equal amounts of water.

(b)

FIGURE 6-22b When reading a manometer, you add the change in both columns; the sum equals the pressure of the gas.

Remove the screw
to remove burner

Orifice

FIGURE 6-23   Removing the burner to test the gas pressure at the gas valve.

supply valve. When you begin to smell gas, turn off the main supply valve. Next, reconnect the gas line to the appliance, open the gas supply valve, and test for gas leaks. If there are no gas leaks, run the appliance.

Use an approved pipe joint compound for connecting gas piping and fittings to the appliance. In addition, never use white Teflon tape on gas lines. White Teflon tape does not guarantee a leak-free connection. You can use a PTFE yellow Teflon tape on gas connections

FIGURE 6-24   A magnehelic gauge. This instrument is used to test gas pressure.

FIGURE **6-25**   Wrap tape in the thread direction on a pipe fitting.

and fittings. This type of approved Teflon tape is a specially designed thread-sealing tape for use on gas appliances. To attach PTFE yellow Teflon tape to pipe thread, wrap the tape in the direction with the thread, as shown in Figure 6-25.

Combustible Gas Detector

Combustible gas detectors are another general-purpose tool that the service technician could use to detect gas leaks (Figure 6-26). This instrument uses an audible alarm that gets louder when you approach a gas leak. This device has the capabilities of detecting LP (LPG) or natural gas and other hydrocarbons and halogenated hydrocarbons, depending on the manufacturer. The sensitivity of a combustible gas detector can be between 50 and 1500 ppm of gas detection, with an instantaneous response.

One manufacturer of combustible gas detectors can be viewed on the Web at www. aprobe.com.

Gas Conversion Procedures

Before installing a new gas appliance, look at the model and serial dataplate to determine that the appliance is made for the type of gas supplied in the home. Do not install an LP or LPG gas appliance when natural gas is supplied to the home. Conversely, do not install a natural gas appliance when LP or LPG gas is supplied to the home.

Appliance manufacturers do make gas appliances that can be installed with either LP or natural gas. All you have to do is follow the manufacturer's recommendations for converting the appliance from one type of gas to the other.

Gas Pressure Regulator

To convert a gas pressure regulator from one type of gas to the other, you must first shut off the main gas supply to the appliance (Figure 6-27). Also, if the gas range has an electrical

FIGURE 6-26   A combustible gas detector. This meter is used for detecting gas leaks.

connection, turn off the electricity. Locate the gas pressure regulator valve. On some models, the gas pressure regulator valve is located in the lower rear of the range (Figure 6-28a) or under the cooktop (Figure 6-28b). To convert the gas pressure regulator valve, you do not have to remove it from the range. Just gain access to it. As you can see in Figure 6-29, remove the

a. The gas supply line from the shut-off valve to the range
b. Gas supply-line shut-off valve
c. The main gas supply entering the home to the shut-off valve

FIGURE 6-27   The main gas supply shut-off valve.

FIGURE 6-28a   On some models, the gas pressure regulator is located in the rear of the range behind the lower draw.

FIGURE 6-28b   The location of gas components under the range cooktop.

a. Plastic cover
b. Pressure regulator cap with the solid end facing toward you
c. Pressure regulator cap with the hollow end facing toward you
d. Washer
e. Pressure regulator cap

FIGURE 6-29 a. Plastic cover. b. Pressure regulator cap with the solid end facing outwards for natural gas. c. Pressure regulator cap with the hollow end facing outwards for LPG gas. d. washer. e. Gas pressure regulator cap.

plastic cover from the regulator cap. Next, unscrew the gas pressure regulator cap with a wrench; turn it counterclockwise to remove the cap and leave the spring in the regulator. *Note:* In Figure 6-29, the regulator cap has a gas designation stamped on it. For LP gas operation, the "<-LP" indicator must face in the direction of the gas pressure regulator, and the hollow end faces away from the regulator. For natural gas operation, the "<-NG" indicator must face in the direction of the regulator, with the dimple end facing away from the valve. Reinstall the cap and tighten with a wrench in the clockwise direction. Do not overtighten the cap, as you might damage it. Finally, replace the plastic cover on the regulator cap.

Converting Surface Burners

To convert the surface burners from one type of gas to the other, you must first shut off the main gas supply to the appliance (see Figure 6-27). Also, if the gas range has an electrical connection, turn off the electricity. Remove the burner grate and burner cap (Figure 6-30). Remove the screws from the burner base, and reinstall one screw to hold the orifice holder

Figure 6-30
The gas range, burner, and grate location.

Burner grate

Burner cap

(Figure 6-31) secured to the cooktop. This will steady the orifice holder so that you can remove the orifice without damaging the orifice holder. To remove the orifice, use a 5/16 nut driver with masking tape applied to the end of it. The masking tape will hold the orifice in the nut driver when you remove it. Turn the nut driver counterclockwise to remove the orifice. Locate the extra orifices that came with the range. Most likely, they will be attached somewhere near the main gas inlet in a holder or small bag. To choose the correct size orifice for the burner that you are converting, refer to the model and serial dataplate. Orifices are stamped with a number and are color-coded, with a groove in the center side of the hex head (Figure 6-32). Use Table 6-5 for the proper sizing of the orifices. With the correct orifice size in hand, just reverse the order of disassembly, and reassemble the burner.

a. Orifice
b. Orifice holder
c. Screw
d. Spark electrode

Figure 6-31   Location of orifice on a drip-proof burner after the burner cap is removed.

Figure 6-32
The burner orifice.

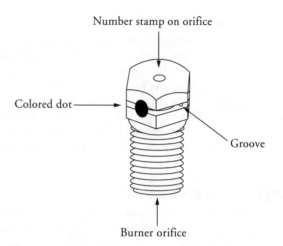

Number stamp on orifice

Colored dot

Groove

Burner orifice

If the range you are servicing resembles the one in Figure 6-28, convert the burner by first removing the screw that secures it to the burner bar. Remove the burner by lifting it up and out of the range. The orifice is now ready to be converted. To convert the orifice from natural to LP gas, use a wrench around the orifice head and turn the orifice clockwise to seat the orifice. Do not overtighten; you might damage the orifice opening. On some models, the orifice hood will be replaced entirely with the correct orifice size for the type of gas used. With the correct orifice adjustment completed, just reverse the order of disassembly, and reassemble the burner.

LP or LPG Gas Orifice Sizing Chart		
Burner BTU Rating	**Color**	**Orifice Size**
12,000	Green/Magenta	1.03 mm
11,000	Green/Yellow	0.99 mm
8000	Green/Light blue	0.80 mm
5000	Green/Brown	0.65 mm
Natural Gas Orifice Sizing Chart		
Burner BTU Rating	**Color**	**Orifice Size**
13,500	Blue/Orange	1.86 mm
12,000/12,500	Blue/Red	1.76 mm
9500	Blue/Brass	1.51 mm
8000	Blue/Black	1.39 mm
5000	Blue/White	1.12 mm
Note: To choose the correct size orifice for the burner that you are converting, refer to the model and serial dataplate for the proper sizing of the orifices for each burner location.		

Table 6-5 LP or LPG Gas Orifice Sizing Chart.

Converting the Oven Bake Burner

To convert the oven bake burner from one type of gas to the other, you must first shut off the main gas supply to the appliance (see Figure 6-27). Also, if the gas range has an electrical connection, turn off the electricity. When converting the oven bake burner, remove the racks and gain access to the oven burner (Figure 6-33). Place a wrench on the orifice, and turn the wrench clockwise for the LP gas setting. Turn the wrench in a counterclockwise direction for natural gas. Only turn the wrench about two or two and a half turns. Do not overtighten the orifice hood; you might damage the orifice pin. With the correct orifice adjustment completed, just reverse the order of disassembly, and reassemble the oven.

Converting the Oven Broil Burner

To convert the oven broil burner from one type of gas to the other, you must first shut off the main gas supply to the appliance (see Figure 6-27). Also, if the gas range has an electrical connection, turn off the electricity. When converting the oven broil burner, gain access to the burner (Figure 6-34). Use a wrench to turn the orifice hood counterclockwise for natural gas or clockwise for LP or LPG gas. Only turn the wrench about two or two and a half turns. Do not overtighten the orifice hood; you might damage the orifice pin. With the correct orifice adjustment completed, just reverse the order of disassembly, and reassemble the oven.

a. Orifice
b. Orifice pin
c. Turn in this direction for natural gas
d. Turn in this direction for LP or LPG gas

FIGURE 6-33   The oven burner is located underneath the bottom oven cover. Remove the racks and lift out the bottom cover for access.

a. Orifice

FIGURE 6-34 The broiler burner location.

Converting the Bake/Broil Thermostat

Gas ranges with oven pilot lights will have to be converted from one type of gas to the other. Remove the oven thermostat knob to expose the oven thermostat selector. Insert a small screwdriver into the pilot control screw, turn the screw clockwise for LP or LPG gas or counterclockwise for natural gas (Figure 6-35). Replace the oven thermostat knob. Gas ranges with electronic spark ignition for pilot control will be discussed later in the book.

FIGURE 6-35 Oven pilot conversion. The pilot control screw is located behind the thermostat knob.

Gas Appliance Maintenance

To maintain gas appliances, always follow the manufacturer's recommendations for periodic maintenance as stated in the use and care manual. The range, oven, or cooktop can be cleaned with warm water, mild detergent, and a soft cloth on all cleanable parts, as recommended in the use and care manual. Also, never use abrasive cleaners that are not recommended by the manufacturer.

Do not allow grease spillovers to accumulate after cooking on top of the range; it will become a fire hazard. When cleaning the burners, always make sure that all of the portholes are free of debris. If, for any reason, the burner portholes are blocked, the flame appearance will be different. Blocked portholes will reduce gas flow, and the heating value of the burner will be reduced. Maintenance procedures for gas water heaters and gas dryers will be discussed in later chapters.

PART III

Refrigeration
and Air Conditioning

Principles of Air Conditioning and Refrigeration

Many excellent books have been written on the subject of air conditioning and refrigeration, and this chapter will only cover the basics needed to diagnose and repair the electrical portion of domestic refrigerators and room air conditioners (RACs). The chapter will cover the refrigeration process, but will not cover the replacement procedures of any sealed system components. This is a specialized area, and only an EPA-certified technician can repair the sealed system of a refrigerator or air conditioner (AC). The HVAC/R-certified technician has the proper training, tools, and equipment required to make the necessary repairs to the sealed system. An individual who is not certified to service the sealed system could raise the risk of personal injury, as well as property damage. The uncertified individual could void the manufacturer's warranty on the product if he or she attempts entry into the sealed system.

Introduction to Air Conditioning and Refrigeration

Before the 1900s and mechanical refrigeration, people cooled their foods with ice and snow. The ice and snow were transported long distances and stored in cellars below ground or in icehouses. Refrigeration is the process of removing heat from an enclosed area, thus lowering the temperature in it. Air conditioning and refrigeration began to come to life in the early 1900s. The first practical refrigerating machine was invented in 1834, and it was improved a decade later.

The two most important figures involved with refrigeration and air conditioning were John Gorrie and Willis Carrier. John Gorrie, an early pioneer of refrigeration and air conditioning, was granted a patent in 1851 for his invention of the first commercial refrigeration machine to produce ice. Improvements were made to Gorrie's invention in the late 1880s, and the reciprocating compressor was used commercially in meat and fish plants and for ice production. In 1902, Willis Carrier designed and invented the first modern air conditioning system, and in 1922, Carrier invented the first centrifugal refrigeration machine. Willis Carrier went on to found the company we know today as Carrier Corporation. Toward the end of the 1920s, the first self-contained room air conditioner was introduced.

Refrigerators from the late 1880s until 1929 used ammonia, methyl chloride, or sulfur dioxide as refrigerants. These refrigerants were highly toxic and deadly if a leak occurred. In 1931, the DuPont Company began producing commercial amounts of "R-12," also known by its trade name "Freon." With safer refrigerants being introduced, the refrigerator and air conditioner market began to grow. Only half a century later did people realize that these chlorofluorocarbons endangered the earth's ozone layer. Today, refrigerant companies are working harder to produce even safer refrigerants that will not endanger the earth's ozone layer.

Saving the Ozone Layer

High above the earth is a layer of ozone gas that encircles the planet. The purpose of the gas is to block out most of the damaging ultraviolet rays from the sun. Such compounds as CFCs, HCFCs, and halons have depleted the ozone layer, allowing more UV (ultraviolet) radiation to penetrate to the earth's surface.

In 1987, the United States, the European Economic Community, and 23 other nations signed the Montreal Protocol on Substances that Deplete the Ozone Layer. The purpose of this agreement was to reduce the use of CFCs throughout the world. To strengthen the original provisions of this protocol, 55 nations signed an agreement in London on June 29, 1990. At this second meeting, they passed amendments that called for a full phaseout of CFCs and halons by the year 2000. Also passed at that meeting was the phaseout of HCFCs by the year 2020, if feasible, and no later than the year 2040, in any case.

On November 15, 1990, President George H.W. Bush signed the 1990 Amendment to the Clean Air Act, which established the National Recycling and Emissions Reduction Program. This program minimizes the use of CFCs and other substances harmful to the environment, while calling for the capture and recycling of these substances. The provisions of the Clean Air Act are more stringent than those contained in the Montreal Protocol as revised in 1990.

Beginning on July 1, 1992, the Environmental Protection Agency (EPA) developed regulations under Section 608 of the Clean Air Act (the Act) that limit emissions of ozone-depleting compounds. Some of these compounds are known as *chlorofluorocarbons* (CFCs) and *hydrochlorocarbons* (HCFCs). The Act also prohibits releasing refrigerant into the atmosphere while maintaining, servicing, repairing, or disposing of refrigeration and air conditioning equipment. These regulations also require technician certification programs. A sales restriction on refrigerant is also included, whereby only certified technicians will legally be authorized to purchase such refrigerant. In addition, the penalties and fines for violating these regulations can be rather severe.

Matter

In order to understand how refrigeration and air conditioners work, it is necessary to understand several basic laws of matter. There are currently five states of matter: solid, liquid, gas, plasma, and Bose-Einstein condensate.

In our discussion of matter, only three of the five states apply to air conditioning and refrigeration (Figure 7-1):

- Matter in the form of a solid will retain its shape and volume without a container. The molecules within a solid are compressed and bound together, and under normal conditions will not move at all.

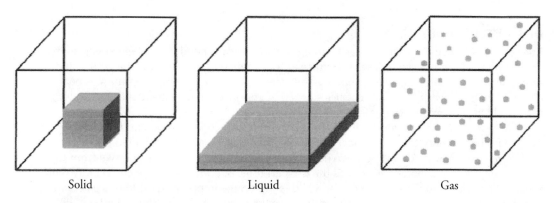

Solid   Liquid   Gas

FIGURE 7-1   The three states of matter that pertain to air conditioning and refrigeration.

- Matter in the form of a liquid will take the shape of its container, without losing any volume if not under pressure. Light-density fluids, such as water, will eventually lose some of their volume due to evaporation. The molecules within a liquid are spaced apart from each other and are constantly on the move.
- Matter in the form of a gas will take the shape and volume of its container and will expand to fill the container. The molecules within a gas are spaced far apart from each other.

Change of State

There are five principles regarding how matter can change from one form to another. When matter is heated, cooled, or an increase or decrease of pressure occurs, matter will be transformed to another state:

- Liquefaction occurs when matter changes from a solid to a liquid. For example, an ice cube will begin to change state when heat is applied.
- Solidification occurs when matter changes from a liquid to a solid. For example, when the temperature of water reaches its freezing point, it will freeze and become a solid.
- Vaporization occurs when a liquid matter like water is heated to its boiling point and transforms into a vapor.
- Condensation occurs when a vapor transforms into a liquid. For example, when water vapor is boiled off into steam in a closed vessel and then allowed to cool down, it will begin to turn back into a liquid.
- Sublimation can occur when a solid matter transforms into a vapor without passing through the liquid state. For example, carbon dioxide in a solid form (dry ice) at atmospheric pressure will transform into a vapor state at normal temperatures, thus bypassing the liquid state.

Law of Thermodynamics

Energy exists in many forms, such as heat, light, mechanical, chemical, and electrical energy. The first law of thermodynamics is the study of that energy. Energy itself is the ability to do work, and heat is only one form of energy—ultimately, all forms of energy end up as heat energy. The first law of thermodynamics states that energy cannot be created or destroyed; it can only be converted from one form to another.

The second law of thermodynamics as it pertains to air conditioning and refrigeration deals with heat energy travel. The law states that heat energy can only travel in one direction—from hot to cold. When two objects are placed together, heat will travel from the warmer object to the cooler object until the temperatures of both objects are equal; only then will heat transfer stop. The rate of travel will depend on the temperature difference between the objects. The greater the temperature difference between the objects, the faster heat will travel to the cooler object.

It's All about Heat

Heat is a form of energy that:

- Causes temperatures to rise in an object
- Has the capacity to do work
- Has the ability to flow from a warmer substance to a cooler one
- Can be converted to other forms of energy

Heat energy can be measured with a thermometer. The scales used are Fahrenheit (F) and Celsius (C). Heat intensity has three reference points: freezing point (32 degrees F or 0 degrees C), boiling point (212 degrees F and 100 degrees C), and absolute zero (-470 degrees F or -273 degrees C). This third reference point is believed to be where all molecular action ceases.

Another way to measure heat energy is in units called British Thermal Units (BTUs). A BTU represents the amount of heat required to raise the temperature of 1 pound of water 1 degree Fahrenheit at sea level. All refrigeration equipment and air conditioning equipment are rated in terms of how many BTUs of cooling or heating they can remove or put into a given area.

On larger room air conditioners and central air conditioning, it is necessary to calculate the heat load and heat gain of a given area to properly size the unit. The same is required for large refrigeration equipment and commercial refrigeration.

Methods of Heat Transfer

Heat energy is an important concern in refrigeration and air conditioning. The room air conditioner and domestic refrigerator must be able to remove the heat from a given area and transport it to an area outside of the cabinet or room. Heat transfer will take place using one or more of the following basic transfer methods:

FIGURE 7-2
The transfer of heat from a warmer object to a cooler object is known as conduction.

- Conduction occurs when two objects are in contact with each other and heat transfers from one object to another (Figure 7-2). The rate of heat transfer will depend on the makeup of the materials used. Conduction will also occur in a single object. For example, if one end of a solid metal rod is heated, the heat will travel toward the cooler end of the rod. Copper and aluminum are good conductors of heat and are used mostly in air conditioners and refrigerators.

- Convection occurs when heat transfers are caused by air movement or fluid movement, whether naturally or by the means of forced air movement (Figures 7-3 and 7-7). Using the refrigerator as an example, the forced air moving over the cold evaporator coil will cause the air to become colder and denser, and it begins to fall to the bottom of the cabinet. In doing so, the air will absorb the heat from the food and air within the refrigerator cabinet. With the heat absorbed in the colder air, it will begin to expand and become lighter; it will rise and, once again, it will be exposed to the colder evaporator coil temperatures, where heat will transfer to the cold coils. This cycle will continue until the temperature requirements are met. Convection also occurs in a liquid when heat transfers in and out of the liquid, as in a refrigerant used in the sealed system of a refrigerator or air conditioner.

- Radiation is the travel of heat energy through the atmosphere by means of radiant waves (light or radio frequency). It travels in a straight path, does not heat the air, and heats only a solid object. For example, if you are standing outside directly in the sunlight, it feels much hotter than if you were to stand in the shade outside. This type of heat transfer comes into play when the refrigerator door is opened or a window covering is opened in an air-conditioned room, allowing heat to enter by radiation. The amount of heat transferred into a refrigerator cabinet or air-conditioned room depends on the temperature difference between the atmosphere and the object.

FIGURE 7-3 The refrigeration cycle.

Refrigerants

Refrigeration involves the process of lowering the temperature of a substance or cooling a designated area. Manufacturers must consider safety, reliability, environmental acceptability, performance, and economics of each refrigerant that is used in refrigeration and air conditioning. The product that absorbs the heat from a substance to be cooled is known as a refrigerant. Refrigerants are classified into two main groups. The first group will absorb the heat by having the refrigerant go through a change of state; the second group involves the absorption of heat without changing the state of the refrigerant. A secondary refrigerant is a refrigerant that will transport the "cold" or "hot" from one location to another (water, for example). The most widely used method of refrigeration is known as the vapor compression cycle (refrigeration cycle using a compressor), which relies on the evaporation of a liquid. This type of method is widely used in domestic refrigeration and air conditioning.

Refrigerants are made up of chemical compounds that are used in a sealed refrigeration system of an air conditioner or refrigerator. They absorb the heat by the process of evaporation or boiling the refrigerant, thereby changing it from a liquid state to a gaseous state. Refrigerants have a much lower boiling point than water when it begins to change state. To remove the heat from the refrigerants, they must be able to change back into a liquid state by the process known as condensation.

Refrigerants are classified into three groups according to their flammability:

- Class 1 - non-flammable refrigerant
- Class 2 - moderately flammable refrigerant
- Class 3 - highly flammable refrigerant

A refrigeration system's reliability will also depend on the chemical stability of the refrigerant, its compatibility with the components within the refrigeration circuit, and the type of lubricant used. At all times, the refrigerant should be miscible at various temperatures with the oil used in the compressor to ensure that the oil will circulate throughout the refrigeration circuit and return to the compressor, where it belongs.

Tables 7-1, 7-2, and 7-3 illustrate the different types of refrigerants used in modern refrigeration and air conditioning equipment.

Beginning technicians or do-it-yourselfers should check out this Web site before attempting any sealed system repair: http://www.epa.gov/Ozone/title6/608/general/index.html.

Room Air Conditioner vs. Refrigerator/Freezer

A room air conditioner (RAC) consists of a cabinet, a sealed refrigeration system (see Figure 7-3), and the electrical circuitry, including the fan motor and other components. The RAC will circulate air (Figure 7-4), remove the humidity in a room, filter out the dust particles, and maintain a desired temperature within the room. Some RAC models are equipped with electric heating elements or reverse-cycle heating. Room air conditioners are less expensive than central air conditioning, and they are easier to install. The RAC refrigeration cycle is similar to the refrigeration cycle of a refrigerator/freezer. The only physical difference is in the packaging of the appliance (Figure 7-5).

A refrigerator/freezer consists of a cabinet, a sealed refrigeration system (Figure 7-6), and the electrical circuitry. It circulates the air (Figure 7-7), removes humidity, and controls the temperature within a closed cabinet that prevents food from spoiling. The refrigerator/freezer refrigeration cycle is similar to the refrigeration cycle of a room air conditioner.

The temperature range for a refrigerator or freezer will range from -20 degrees Fahrenheit to 45 degrees Fahrenheit. A room air conditioner's temperature range varies from 60 degrees to 86 degrees Fahrenheit.

The Basic Refrigeration Cycle

Mechanical refrigeration is accomplished by continuously circulating, evaporating, and condensing a fixed supply of refrigerant in a closed system. Evaporation occurs at a low temperature and low pressure, while condensation occurs at a high temperature and high pressure. Thus, it is possible to transfer heat from an area of low temperature (that is, the refrigerator cabinet) to an area of high temperature (that is, the kitchen).

Beginning the cycle (see Figure 7-3) at the evaporator inlet (1), the low-pressure liquid expands, absorbs heat, and evaporates, changing to a low-pressure gas at the evaporator outlet (2). The compressor (4) pumps this gas from the evaporator through

REFRIGERANTS		
Refrigerant ASHRAE #	**Chemical Name of Refrigerant**	**Color of Container**
R11	Trichlorofluoromethane	Orange
R12	Dichlorodifluromethane	White
R13	Chlorotrifluoromethane	Light Blue
R113	Trichlorotrifluoroethane	Dark Purple
R114	Dichlorotetrafluoroethane	Navy Blue
R12/114	Dichlorodifluromethane, Dichlorotetrafluoroethane	Light Gray
R13B1	Bromotrifluoromethane	Pinkish-Red
R22	Chlorodifluoromethane	Light Green
R23	Trifluoromethane	Light Blue Gray
R123	Dichlorotrifluoroethane	Light Blue Gray
R124	Chlorotetrafluoroethane	DOT Green
R134A	Tetrafluoroethane	Light Blue
R401A	Chlorodifluoromethane, Difluoroethane, Chlorotetrafluoroethane	Pinkish-Red
R401B	Chlorodifluoromethane, Difluoroethane, Chlorotetrafluoroethane	Yellow-Brown
R402A	Chlorodifluoromethane, Pentafluoroethane, Propane	Light Brown
R402B	Chlorodifluoromethane, Pentafluoroethane, Propane	Green Brown
R403B	Chlorodifluoromethane, Octafluoropropane, Propane	Light-Gray
R404A	Pentafluoroethane, Trifluoroethane, Tetrafluoroethane	Orange
R407C	Difluoromethane, Pentafluoroethane, Tetrafluoroethane	Brown
R408A	Chlorodifluoromethane, Trifluoroethane, Pentafluoroethane	Medium Purple
R409A	Chlorodifluoromethane, Chlorotetrafluoroethane, Chlorodifluoroethane	Medium Brown
R410A	Difloromethane, Pentafluoroethane	Rose
R414B	Chlorodifluoromethane, Chlorotetrafluoroethane, Chlorodifluoroethane, Isobutane	Medium Blue
R416A	Tetrafluoroethane, Chlorotetrafluoroethane, Butane	Yellow-Green
R417A	Pentafluoroethane, Tetrafluoroethane, Isobutane	Green
R500	Dichlorodifluoromethane, Difluoroethane	Yellow
R502	Chlorodifluoromethane, Chloropentafluoroethane	Light Purple
R503	Chlorotrifluoromethane, Trifluoromethane	Blue-Green
R507	Pentafluoroethane, Trifluoroethane	Aqua Blue
R508B	Trifluoromethane, Hexafluoroethane	Dark Blue

TABLE 7-1 Refrigerants.

OTHER REFRIGERANTS	
Refrigerant	**Chemical Name of Refrigerant**
R14	Tetrafluoromethane
R32	Difluoromethane
R41	Fluoromethane
R50	Methane
R125	Pentafluoroethane
R141B	Dichlorofluoroethane
R152A	Difluoroethane
R290	Propane
R600	Butane
R600A	Isobutane
R610	Ethyl Ether
R611	Methyl Formate
R630	Methyl Amine
R631	Ethyl Amine
R702	Hydrogen
R704	Helium
R717	Ammonia
R718	Water
R720	Neon
R728	Nitrogen
R732	Oxygen
R740	Argon
R744	Carbon Dioxide
R744A	Nitrous Oxide
R764	Sulfur Dioxide
Refrigerant Categories	
CFC	Chlorofluorocarbons - are being phased out due to Montreal Protocol agreement
HCH	Hydrocarbons
HCFC	Hydrochlorofluorocarbons - soon to be phased out - production has ceased
HFC	Hydrofluorocarbons - are relatively new for vapor compression systems
Azeotropes	A mixture of two or more refrigerants that retains the same composition in vapor form as in liquid form with similar boiling points for a given temperature and pressure and act as one refrigerant.
Zeotropes	A mixture of two or more refrigerants that shift in composition during the boiling or condensing process.

TABLE 7-2   Other Types of Refrigerants.

PART III

Refrigerant ASHRAE #	Trade Name	Refrigerant Components	Components Weight %	Type of Refrigerant	Type of Lubricant
R123		PURE	100%	HCFC	ALKYL BENZENE or MINERAL OIL
R124		PURE	100%	HCFC	ALKYL BENZENE or MINERAL OIL
R134A		PURE	100%	HFC	POLYOLESTER OIL
R22		PURE	100%	HCFC	ALKYL BENZENE or MINERAL OIL
R23		PURE	100%	HFC	POLYOLESTER OIL
R401A	MP39	R22 / R152A / R124	53% / 13% / 34%	HCFC BLENDED	ALKYL BENZENE or MO/AB MIXTURE
R401B	MP66	R22 / R152A / R124	61% / 11% / 28%	HCFC BLENDED	ALKYL BENZENE or MO/AB MIXTURE
R402A	HP80	R125 / R290 / R22	60% / 2% / 38%	HCFC BLENDED	ALKYL BENZENE or MO/AB MIXTURE
R402B	HP81	R125 / R290 / R22	38% / 2% / 60%	HCFC BLENDED	ALKYL BENZENE or MO/AB MIXTURE
R403B	ISCEON 69L	R290 / R22 / R218	5% / 56% / 39%	HCFC BLENDED	ALKYL BENZENE or MINERAL OIL
R404A	HP62, FX70	R125 / R143A / R134A	44% / 52% / 4%	HFC BLENDED	POLYOLESTER OIL
R407C	SUVA 9000	R32 / R125 / R134A	23% / 25% / 52%	HFC BLENDED	POLYOLESTER OIL
R408A	FX10	R125 / R143A / R22	7% / 46% / 47%	HCFC BLENDED	ALKYL BENZENE or MINERAL OIL
R409A	FX56	R22 / R124 / R142B	60% / 25% / 15%	HCFC BLENDED	ALKYL BENZENE or MINERAL OIL
R410A	AZ20, PURON	R32 / R125	50% / 50%	HFC BLENDED	POLYOLESTER OIL
R414B	HOT SHOT	R22 / R600A / R124 / R142B	50% / 1.5% / 39% / 9.5%	HCFC BLENDED	ALKYL BENZENE or MINERAL OIL
R416A	FRIGC FR12	R134A / R124 / R600	59% / 39% / 2%	HCFC BLENDED	POLYOLESTER OIL
R417A	NU-22	R125 / R134A / R600A	46.6% / 50% / 3.4%	HFC BLENDED	ALKYL BENZENE or MINERAL OIL or POLYOLESTER OIL
R507	AZ50	R125 / R143A	50% / 50%	HFC BLENDED	POLYOLESTER OIL
R508B	SUVA 95	R23 / R116	46% / 54%	HFC BLENDED	POLYOLESTER OIL

TABLE 7-3   Refrigerant Components, Types, and Lubricants.

FIGURE 7-4
The airflow patterns of
a room air conditioner.

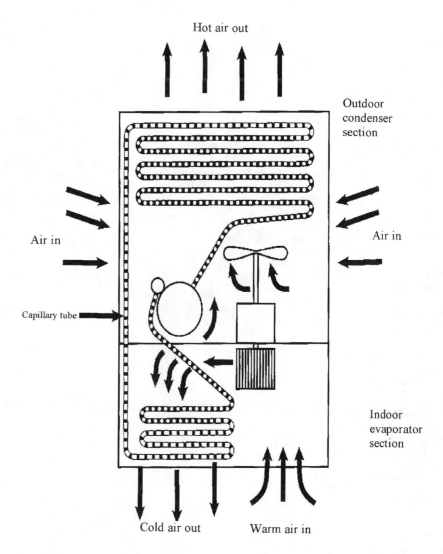

FIGURE 7-4
The airflow patterns of a room air conditioner.

the accumulator (3), increases its pressure, and discharges the high-pressure gas to the condenser (5). The accumulator is designed to protect the compressor by preventing slugs of liquid refrigerant from passing directly into the compressor. An accumulator should be included on all systems that are subjected to varying load conditions or to frequent compressor cycling. In the condenser, heat is removed from the gas, which then condenses and becomes a high-pressure liquid. In some systems, this high-pressure liquid drains from the condenser into a liquid storage or receiver tank (6). On other systems, both the receiver and the liquid line valve (7) are omitted.

A heat exchanger (8) between the liquid line and the suction line is also an optional item, which may or may not be included in a given system design. Between the condenser and the evaporator, an expansion device (10) is located. Immediately preceding this device

Compressor

Main PCB

Capacitor

Control panel
& display

Overload protector

Condenser

Fan
motor

Thermistor

Evaporator

FIGURE 7-5 Typical component locations in a room air conditioner (RAC).

FIGURE 7-6
Typical components of
a sealed refrigeration
system.

Capillary tube

Evaporator

Suction line

Compressor

Condenser

Discharge line

Heat exchanger

Around meat
keeper

FIGURE 7-7   The cold airflow pattern in side-by-side and two-door refrigerators.

is a liquid line strainer/drier (9), which prevents plugging of the valve or tube by retaining scale, dirt, and moisture. The flow of refrigerant into the evaporator is controlled by the pressure differential across the expansion device or, in the case of a thermal expansion valve, by the degree of superheat of the suction gas. Thus, the thermal expansion valve shown requires that a sensor bulb be located at the evaporator outlet. In any case, the flow of refrigerant into the evaporator normally increases as the evaporator load increases.

As the high-pressure liquid refrigerant enters the evaporator, it is subjected to a much lower pressure due to the suction of the compressor and the pressure drop across the expansion device. Thus, the refrigerant tends to expand and evaporate. In order to evaporate, the liquid must absorb heat from the air passing over the evaporator. Eventually, the desired air temperature is reached and the thermostat or cold control (11) will break the electrical circuit to the compressor motor and stop the compressor. As the temperature of the air running through the evaporator rises, the thermostat or cold control reestablishes the electrical circuit. The compressor starts, and the cycle continues.

In addition to the accumulator, a compressor crankcase heater (12) is included on many systems. This heater prevents accumulation of refrigerant in the compressor crankcase during the non-operating periods and prevents liquid slugging or oil pumpout on startup.

Additional protection to the compressor and system is afforded by a high- and low-pressure cutout (13). This control is set to stop the compressor in the event that the system pressures rise above or fall below the design operation range.

PART III

Other controls not indicated on the basic cycle that may be part of a system include: evaporator pressure regulators, hot gas bypass regulators, electric solenoid valves, suction pressure regulators, condenser pressure regulators, low-side or high-side float refrigerant controllers, oil separators, etc.

It is extremely important to analyze every system completely and understand the intended function of each component before attempting to determine the cause of a malfunction or failure.

Refrigeration Components

As a new technician, you may understand the process of a refrigeration cycle, but you must also understand how each component functions in the refrigeration cycle. Each component has a specific job to assist in maintaining the objective of removing heat from a given area.

Compressor

The compressor (see Figures 7-8a, b, and c) is the heart of the vapor compression system, also known as the pump. All refrigerant circulating within the sealed system must pass through the compressor. This component has two functions within the refrigeration cycle:

- Retrieve the refrigerant vapor from the evaporator, thereby maintaining a constant pressure and temperature

- Increase the refrigerant vapor pressure and temperature to allow the refrigerant to give up its heat in the condenser coil

(a)

FIGURE 7-8a The internal view and identification of the internal components in a hermetically sealed air conditioner compressor.

(b)

(c)

FIGURE 7-8b   An internal view of a rotary compressor.

FIGURE 7-8c   An internal view of a compressor.

Compressors used for domestic refrigeration and room air conditioners are either reciprocating or rotary in design and are hermetically sealed. A hermetically sealed compressor cannot be repaired in the field; it can only be replaced as a complete unit by a certified technician. The electric motor within the compressor can be tested using a multimeter to determine if the windings of the motor are good or bad.

Compressor Safety

In the interest of promoting safety in the refrigeration and air conditioning industry, Tecumseh Products Company has prepared the following information to assist service personnel in safely installing and servicing equipment. This section covers a number of topics related to safety. However, it is not designed to be comprehensive or to replace the training required for professional service personnel.

Trained Personnel Only

Refrigeration and air conditioning devices are extremely complicated by nature. Servicing, repairing, and troubleshooting these products should be done only by those with the necessary knowledge, training, and equipment.

⚠ **WARNING**

Never service, repair, or troubleshoot unless you are a professional air conditioning/ refrigeration service person. Improper servicing can lead to serious injury or death from fire, electric shock, or explosion.

Terminal Venting and Electrocution

Improperly servicing, repairing, or troubleshooting a compressor can lead to electrocution or fire due to terminal venting with ignition. The following precautions can help avoid serious injury or death from electrocution or terminal venting with ignition.

A. Fire Hazard from Terminal Venting with Ignition

Oil and refrigerant can spray out of the compressor if one of the terminal pins is ejected from the hermetic terminal. This "terminal venting" can occur as a result of a grounded fault (also known as a short circuit to ground) in the compressor. The oil and refrigerant spray from terminal venting can be ignited by electricity and produce flames that can lead to serious burns or even death. When spray from terminal venting is ignited, this is called "terminal venting with ignition." (See Figures 7-9a, b, and c for details.)

B. Terminal Venting and Electrocution Precautions

To reduce the risk of electrocution or serious burns or death from terminal venting with ignition:

- Be alert for sounds of arcing (sizzling, sputtering, or popping) inside the compressor. IMMEDIATELY GET AWAY if you hear these sounds.

- Disconnect ALL electrical power before removing the protective terminal cover. Make sure that all power legs are open. (*Note*: The system may have more than one power supply.)

- Never energize the system unless: 1) the protective terminal cover is securely fastened and 2) the compressor is properly connected to ground. Figures 7-9d, 7-9e, and 7-9f illustrate the different means of fastening protective terminal covers.

- Never reset a breaker or replace a fuse without first checking for a ground fault (a short circuit to ground). An open fuse or tripped circuit breaker is a strong indication of a ground fault. To check for a ground fault, use the procedure outlined in "Identifying Compressor Electrical Problems" on page 47 in the Tecumseh Hermetic Compressor Service Handbook (located at www.tecumseh.com).

- Always disconnect power before servicing, unless it is required for a specific troubleshooting technique. In these situations, use extreme caution to avoid electric shock.

Compressor Motor Starting Relays

A hermetic motor starting relay is an automatic switching device to disconnect the motor start capacitor and/or start winding after the motor has reached running speed. There are two types of motor starting relays used in refrigeration and air conditioning applications: the current responsive type and the potential (voltage) responsive type.

Never select a replacement relay solely by horsepower or other generalized rating. Select the correct relay as specified in the Tecumseh Electrical Service Parts Guide Book.

Current Type Relay

When power is applied to a compressor motor, the relay solenoid coil attracts the relay armature upward, causing the bridging contact and stationary contact to engage. This energizes the motor start winding.

Figure 7-9 (a) Compressor with (1) protective terminal cover and (2) bale strap removed to show (3) hermetic terminals. (b) Close-up view of hermetic terminal showing individual terminal pins with power leads removed. (c) Close-up view of hermetic terminal after it has vented. (d) Compressor with (1) protective terminal cover held in place by (2) metal bale strap. (e) Compressor with (1) protective terminal cover held in place by (2) nut. (f) Compressor with (1) snap-in protective terminal cover.

When the compressor motor attains running speed, the motor main winding current is such that the relay solenoid coil de-energizes, allowing the relay contacts to drop open and disconnecting the motor start winding.

The relay (Figure 7-10) must be mounted in true vertical position so that armature and bridging contacts will drop free when the relay solenoid is de-energized.

PTC Type Relay

Solid-state technology has made available another type of current sensitive relay: a PTC starting switch (see Figure 7-10). Certain ceramic materials have the unique property of greatly increasing their resistance as they heat up from current passing through them. A PTC solid-state starting device is placed in series with the start winding and normally has a very low resistance. Upon startup, as current starts to flow to the start winding, the resistance rapidly rises to a very high value, thus reducing the start winding current to a trickle and effectively taking that winding out of operation.

FIGURE 7-10
(a) Current type relay.
(b) PTC type relay.

(a)   (b)

FIGURE 7-10
(a) Current type relay.
(b) PTC type relay.

Usage is generally limited to domestic refrigeration and freezers. Because it takes 3 to 10 minutes to cool down between operating cycles, it is not feasible for short-cycling commercial applications.

Compressor Thermal Protectors (Overloads)

All hermetic compressors have either an internal or an external protector (overload) to protect the motor from overheating. The compressor overload in Figure 7-11a is attached to the compressor in Figures 7-11c and 7-11d. The overload is designed to react quickly by means of a bimetal disk to an increase in temperature or an increase in current.

The bimetal disk (Figure 7-11b) consists of two dissimilar metals combined together. Any change in temperature or current draw will cause it to deflect, actuating the switch contacts. When the bimetal cools, the reverse action takes place. The external protector is a non-serviceable part, and it should be replaced with a duplicate of the original. If the compressor has an internal overload, and it is diagnosed as being defective, the compressor will have to be replaced by a certified technician.

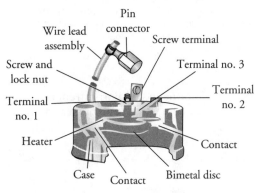

External thermal protector. (Models AE, TP, TH, AK, AJ, CAJ, AZ, RK, RG, TW, and some CL.)

(a)

FIGURE 7-11a The compressor overload protector.

Open   Closed

Bimetal disc.

(b)

FIGURE 7-11b An internal view of the compressor overload protector.

AE refrigeration compressor with the thermal protector and relay assembled.

(d)

FIGURE 7-11d An illustration of the components after they are attached to the exterior of the compressor.

AE refrigeration compressor showing (1) hermetic terminal, (2) thermal protector, (3) thermal protector clip, (4) push-on relay, (5) protective terminal cover, and (6) bale strap.

(c)

FIGURE 7-11c An illustration of the components before they are attached to the exterior of the compressor.

Condenser Coil

A condenser coil is manufactured out of copper, aluminum, or steel tubing, with metal fins attached to the tubing to assist in dissipating the heat (Figures 7-12a and b). The main purpose of the condenser coil is to dissipate the heat from the refrigerant along with the heat from the compression stage of the compressor. Condenser coils in refrigerators, freezers, and room air conditioners usually have a fan motor and fan blade circulating the air across the condenser coil to assist in removing the heat from the refrigerant. On some models of refrigerators and freezers, the condenser coils are static-cooled whereby the heat from the condenser coil will rise and dissipate into the room. In some models of freezers, the condenser coil is attached to the inside of the outer cabinet so that the cabinet is the heat transfer medium.

Evaporator Coil

The evaporator coil, manufactured in the same manner as the condenser coil, absorbs the heat from the food product or from the air in a room to be cooled (Figures 7-5, 7-12a, and 7-12b). Manufacturers design

(a)

FIGURE 7-12a The condenser coil in a refrigerator's sealed refrigeration system.

FIGURE 7-12b
The condenser coil in a freezer's sealed refrigeration system.

Gasket

Hinge

Compressor

Breaker strip

Evaporator coils

Condenser coils

Temperature control

Power indicator light

Compressor compartment cover

Terminal cover

Compressor mountings

(b)

evaporator coils in different shapes and designs to meet the needs of specific products. Most refrigerator models and room air conditioners use a fan motor and fan blade to circulate the air across the evaporator coil, which absorbs the heat. Other refrigerator/freezer models have rows of tubing placed between two plates, which are welded together to form an interior cabinet in the freezer compartment or shelving.

Metering Device

The most commonly used metering device in a domestic refrigerator and room air conditioner is a capillary tube, also called a cap tube (Figures 7-3, 7-4, and 7-6). This metering device is manufactured out of copper tubing, with a very small-diameter opening at both ends, and is premeasured at a specific length for maximum heat load efficiency for the product being cooled or refrigerated. The small-diameter opening in the cap tube will restrict the flow of refrigerant, causing a pressure drop from the high-pressure side (condenser) of the sealed system to the low-pressure side (evaporator). This will allow the refrigerant to enter the evaporator at the rate needed to remove the unwanted heat from the product being cooled or refrigerated. There are no moving parts in a cap tube, and it does not require any service. The capillary tube is located between the condenser coil and the evaporator coil at the end of the liquid line.

Diagnosing Sealed Systems

When servicing the refrigerator/freezer or room air conditioner, do not overlook the simple things that might be causing the problem. Before you begin testing and servicing an appliance, check for the most common faults. Inspect all wiring connections for broken or loose wires. Check and see if all components are running at the proper times during the cycle. Test the temperatures in the refrigerator/freezer cabinet or air conditioner inlet and outlet grills.

The following chart (Table 7-4) is intended to serve as an aid to assist the technician in determining if there is a sealed system malfunction, providing that all other components are functioning correctly within the appliance.

Sealed System Condition	Suction Line Pressure (TECHNICIANS ONLY)	Suction Line to Compressor	Compressor Discharge Line	Condenser Coil	Capillary Tube	Evaporator Coil	Amperage or Wattage
NORMAL OPERATION	Normal pressure readings.	Slightly below room temperature.	**WARNING** very hot to the touch.	**WARNING** very hot to the touch.	Warm.	Cold coil and maintaining proper temperatures.	Normal meter readings.
SEALED SYSTEM OVERCHARGED	Higher than normal pressure readings.	Heavily frosted, may be very cold to the touch.	Between slightly warm to hot.	Between hot to warm to the touch.	Cool below room temperature.	Cold coil and possibly not maintaining temperatures.	Higher than normal meter readings.
SEALED SYSTEM UNDERCHARGED	Pressure readings are lower than normal.	Warm to the touch, possibly near room temperature.	**WARNING** hot to the touch.	The entire coil feels warm.	Warm.	Inlet to coil feels extremely cold while the outlet from the coil will be below room temperature.	Lower than normal meter readings.
PARTIAL RESTRICTION WITHIN SEALED SYSTEM	Lower than normal pressure readings, possibly in a vacuum.	Warm to the touch, possibly near room temperature.	**WARNING** very hot to the touch.	The top passes in the coil are warm and the lower passes are cool, near to room temperature.	Feels like room temperature between cool to colder.	Inlet to coil feels extremely cold while the outlet from the coil will be below room temperature. Frost line will begin to grow in length.	Lower than normal meter readings.
COMPLETE RESTRICTION WITHIN SEALED SYSTEM	Pressure reading is in a deep vacuum.	Feels the same as room temperature.	Feels the same as room temperature.	Feels the same as room temperature.	Feels the same as room temperature.	No refrigeration or air conditioning.	Lower than normal meter readings.
OUT OF REFRIGERANT POSSIBLE LEAK IN SYSTEM	The pressure reading will be from 0 PSIG to 30" vacuum.	Feels the same as room temperature.	It can feel like cool to hot.	Feels the same as room temperature.	Feels the same as room temperature.	No refrigeration or air conditioning.	Lower than normal meter readings.

TABLE 7-4   Refrigeration Sealed-System Diagnosis Chart.

PART III

Parts

IV

PART

Electric Appliance and Air Conditioner Parts

Servicing a highly complex, electromechanical appliance or room air conditioner is not as hard as might be expected. Just keep in mind that an appliance or air conditioner is simply a collection of parts, located inside a cabinet, coordinated to perform a specific function (Figures 8-1a, 8-1b, 8-1c, 8-1d, and Figure 7-4). Before servicing an appliance or air conditioner, you must know what these parts are and how they function.

The Switch

The switch is a mechanical device used for directing and controlling the flow of current in a circuit. Simply put, the switch can be used for turning a component on or off (Figure 8-2). Internally, the switch has a set of contacts that close, allowing the current to pass; when opened, current is unable to flow through it. Built into the switch, a linkage mechanism actuates these contacts inside of the closed housing (Figure 8-3).

Switches come in a wide range of sizes and shapes, and can be used in many different types of applications (Figure 8-4). The voltage and amperage rating is marked on the switch or on the mounting bracket for the type of service the individual switch was designed to do. The switch housing is usually marked with the terminal identification numbers that correspond to the wiring diagram. These identify the contacts by number: normally open (NO) contacts, normally closed (NC) contacts, or common (COM) contacts. Internally, the switch can house many contact points for controlling more than one circuit.

When a switch failure is suspected, remember that there are only three problems that can happen to a switch:

- The contacts of the switch might not make contact. This is known as an *open switch*.
- The switch's contacts might not open, causing a *shorted switch*.
- The mechanism that actuates the contacts might fail. This is a *defective switch*.

When these problems arise, the switches are not repairable, and they should be replaced with a duplicate of the original.

(a)

FIGURE 8-1a   Typical locations of electric range parts.

(b)

FIGURE 8-1b   Typical locations of dishwasher parts.

(c)

FIGURE 8-1c   Typical locations of automatic washer parts.

(d)

FIGURE 8-1d   Typical locations of automatic dryer parts.

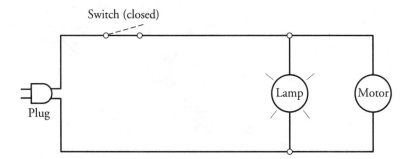

FIGURE 8-2
The wiring diagram illustrates a switch in the closed position. If the switch is closed, the light and motor are on. If the switch is open, the light and motor are off.

Pressure Switch

The *pressure switch* is a specialty switch, with a similar operation to those mentioned previously, but with one important exception: The pressure switch is actuated by a diaphragm that is responsive to pressure changes (Figure 8-5). This switch can be found in washing machines and in dishwashers, and it operates as a water level control. Other uses include: furnaces, gas heaters, computers, vending machines, sump pumps, and other low-pressure applications. The pressure switch is not serviceable, and should be replaced with a duplicate of the original.

Thermostat

The *thermostat* operates a switch. It is actuated by a change in temperature. The two most common heat-sensing methods used in appliances and air conditioners are the bimetal and the expansion thermostats (Figure 8-6).

FIGURE 8-3   The exploded view of a switch.

FIGURE 8-4   This is a sample of the many types of switches used in major appliances.

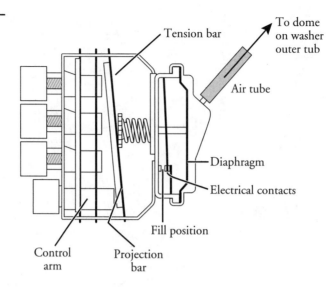

FIGURE 8-5
Construction of a
pressure switch.

Tension bar

To dome
on washer
outer tub

Air tube

Diaphragm

Electrical contacts

Fill position

Control
arm

Projection
bar

The *bimetal thermostat* (Figure 8-7) consists of two dissimilar metals combined together. Any change in temperature will cause it to deflect, actuating the switch contacts. When the bimetal cools, the reverse action takes place.

The *expansion (temperature control) thermostat* (Figure 8-8) uses a liquid in a tube that is attached to bellows. The liquid converts to a gas when heated and travels up the tube to the bellows. This causes the bellows to expand, thus actuating the switch contacts. When the gas cools, the reverse action occurs.

Thermostats are used in applications as diverse as gas and electric ranges, automatic dryers, room air conditioners, irons, waterbeds, spas, and in heating and refrigeration units.

Expansion type

Bimetal type

FIGURE 8-6   Bimetal and expansion thermostats.

FIGURE 8-7
Construction of a
bimetal thermostat.

Bimetal element

Switch
contacts

Auxiliary
heater

Terminals

Overload

Electromechanical Timer

Although the timer is the most complex component in the appliance, don't assume that it is
the malfunctioning part. Check all of the other components associated with the symptoms
as described by the customer.

Electromechanical timers are utilized for controlling performance in automatic washers,
automatic dryers, and dishwashers. Most of these timers are not serviceable and should be
replaced with a duplicate of the original.

The timer assembly is driven by a synchronous motor in incremental advances. It
controls and sequences the numerous steps and functions involved in each cycle of an
appliance (Figure 8-9). The timer directs the on and off times of the components in an
electrical circuit. It consists of three components assembled into one unit: the motor, the
escapement, and the cam switches (Figure 8-10).

FIGURE 8-8
Construction of
an expansion
(temperature-control)
thermostat.

Shaft

Cam

Switch
contacts

Bellows

Capillary tube

Temperature control

PART IV

Timer

Dust cover

Switch locating boss

#1 switch asm.

Tab "C"

Push-pull shaft

Motor

Front view

Mounting screws holes

#1 switch asm.

Tab "A"

Motor mounting screws

Dust cover

Tab "B"

#2 switch asm.

Rear view

Tabs

Dust cover

Tabs

Dust cover removed

FIGURE 8-9 An electromechanical timer.

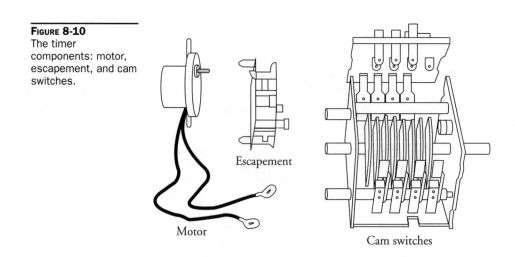

FIGURE 8-10
The timer components: motor, escapement, and cam switches.

Escapement

Motor

Cam switches

FIGURE 8-11
The construction of
the timer cam and
switch contact.

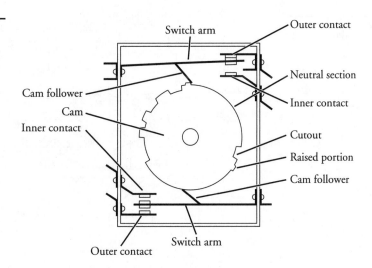

The Motor

The motor is a *synchronous motor*, geared to drive the escapement. In Figures 8-9 and 8-10, the motor is mounted on the timer assembly. These are specially designed motors, whose speed is controlled by the 60-cycle period of the current, rather than by the fluctuating supply voltage.

The Cam Switches

The number of switch contacts and cams varies with the number of functions an appliance performs. Some of the switches perform two or more functions (Figure 8-11). The particular shape of the cam varies with the number of switch contacts that it controls and with the length of time each switch contact opens and closes (Figure 8-12). A single metal strip, called a cam follower (see Figure 8-11), is anchored to each cam switch arm. As the cam turns, this metal strip follows the contour of the circumference of the cam, causing the cam switch to open or close at the proper time.

The Escapement

The escapement is a spring-controlled mechanism that limits the cam shaft rotation to a set number of degrees in each increment (see Figure 8-10). The cam follower moves rapidly to ensure a snap action of the switch contacts and also to prevent any arcing of the points. Not all timers have an escapement. For example, some dryer timers, and all defrost and electronic timers, have no escapement.

FIGURE 8-12
How a switch contact
opens and closes on
the cam.

FIGURE 8-13
Construction of a
relay.

Solenoid
coil
Armature

Switch contacts

Relays

A relay is an electrically operated switch. It contains an electromagnet with a fixed coil and a movable armature, which actuates a set of contacts to open and close electric circuits (Figure 8-13). Relays have heavy-duty switching contacts, and they are used to operate larger components, such as motors and compressors. The most common failures of relays are:

- Open relay coil
- Burned switch contacts
- Armature not actuating the contacts

A broken relay is not serviceable and should be replaced with a duplicate of the original.

Solenoids

A solenoid is a device used to convert electrical energy into mechanical energy (Figure 8-14). When a solenoid is energized, it acts like an electromagnet and is positioned to move a pre-designated metal object. The work performed by the moving plunger makes the solenoid coil useful in appliances. Some solenoids are equipped with a free-moving armature or plunger. Common failures of a solenoid are:

- Open coil
- Shorted coil
- Jammed armature

FIGURE 8-14
The solenoid coil and plunger. When the coil is activated, the plunger will be drawn to the center of the magnetic field.

FIGURE 8-15
Construction of a
water valve.

Solenoid

Armature

Diaphragm

Inlet

Screen

Outlet

Flow washer

A solenoid is not a serviceable part and should be replaced with a duplicate of the original. These devices are manufactured in a variety of designs for various load force and operational requirements. Solenoids are found in automatic washers and dryers, gas and electric ranges, automatic dishwashers, refrigerators, freezers, automatic ice machines, and in heating and air conditioning units.

Water Valves

The water inlet valve controls the flow of water into an appliance, and is solenoid-operated (Figure 8-15). When it is energized, water in the supply line will pass through the valve body and into the appliance. Some of the different types of water inlet valves that are used on appliances include:

- Single water inlet valve (Figure 8-16): Used on dishwashers, ice makers, refrigerators, undercounter appliances and ice machines.

- Dual water inlet valve (Figure 8-17): Used on washing machines, refrigerators, and ice makers. Some dishwasher models also use dual water inlet valves. The inlet side of the valve has a fine mesh screen to prevent foreign matter from entering the valve. Some water valves also have a "water hammer" suppression feature built into them.

Drain valves (Figure 8-18) are used on some dishwasher and washing machine models to control the drainage of the water in the tub and its expulsion into the sewage system of the residence.

The water valve should not be serviced. Replace it with a duplicate of the original.

Motors

The two major assemblies that form an electric motor are the rotor and the stator (Figure 8-19). The rotor is made up of the shaft, rotor core, and (usually) a fan. The stator is formed from steel laminations, stacked and fastened together so that the notches form a continuous

FIGURE 8-16
The single water valves are just some of the different types available that are used in major appliances.

FIGURE 8-17
These dual water valves are designed for dual water inlet connections.

FIGURE 8-18
The drain water valves.

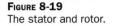

Figure 8-19
Figure 8-19
The stator and rotor.

lengthwise slot on the inside diameter. Insulation is placed so as to line the slots; and then coils, wound with many turns of wire, are inserted into the slots to form a circuit. The wound stator laminations are pressed into, or otherwise assembled within, a cylindrical steel frame to form the stator (Figure 8-20). The end bells, or covers, are then placed on each end of the motor. One important function of the end bells is to center the rotor or armature accurately within the stator to maintain a constant air gap between the stationary and moving cores (Figures 8-19 and 8-21).

These coils of wire are wound in a variety of designs, depending upon the electrical makeup of the motor. They provide two or more paths for current to flow through the stator windings. When the coils have two centers, they form a two-pole motor; when they have four centers, they form a four-pole motor. In short, the number of coil centers determines the number of poles that a motor has (Figure 8-22).

Thermal protection in a motor is provided by a temperature-sensitive element, which activates a switch. This switch will stop the motor if it reaches the pre-set temperature limit. The thermal protector in a motor is a non-replaceable part, and the motor will have to be replaced as a complete component. There are two types of thermal protection switches:

- Automatic reset: It automatically resets the switch when the temperature has been reduced.

- Manual reset: It has a small reset button on the motor on the opposite end from the shaft.

Figure 8-20   Stator and rotor construction.

Figure 8-21   End bells position the motor shaft in the center of the stator.

FIGURE 8-22
Two-pole motor and
four-pole motor.

Two-pole stator motor.

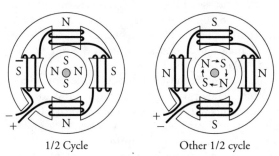

1/2 Cycle   Other 1/2 cycle

Four-pole stator motor.

Several types of motors that are used for different types of applications include the following:

- Synchronous motors are permanent magnet-timing motors, often used in automatic ice cube makers, water softeners, and humidifiers. In addition, they are integral to timers for automatic washers, automatic dryers, and dishwashers.

- Shaded pole motors are used as continuous duty motors, with limited or adjustable speeds. They are used for small fans and clocks.

- Split phase motors are used as continuous duty motors, with fixed speeds. They are often used in automatic washer and dryer drive motors.

- Capacitor start motors are similar to the split-phase motors, and they are used in hard-to-start applications, such as compressors and pumps.

- Permanent split capacitor motors are used in a variety of direct-drive air-moving applications—for example, air conditioner fans.

- Three-phase motors are used in industrial or large commercial applications where three-phase power is available.

- Multispeed, split-phase motors are used in fans, automatic dryers, automatic washers, and many other appliances.

Figure 8-23 illustrates some of these motors. Appliance motors are not repairable, and they should be replaced with a duplicate of the original.

FIGURE 8-23
Motors are available in different sizes and shapes.

Compressors

The compressor is an electric motor that drives a mechanical compression pump designed to compress the refrigerant vapors and to circulate the refrigerant within a sealed system. Domestic refrigeration and room air conditioners use a hermetic compressor. The electric motor and mechanical compression pump are sealed within the same housing (Figure 7-7), and it is a non-serviceable part. If the compressor fails, it must be replaced with a duplicate of the original by a certified technician. Two types of compressors are used in domestic refrigeration and room air conditioners: reciprocating and rotary. For more information on compressor construction, visit the following compressor manufacturer's Web sites:

- www.copeland-corp.com
- www.tpc-nacg.com
- www.compressors.danfoss.com
- www.embraco.com

FIGURE 8-24
(a) The capacitor is rated by voltage and by capacitance (in microfarads). (b) This built-in disconnect device is also known as a fail-safe.

Normal   Fail-safe mode

(a)   (b)

Capacitors

A capacitor is a device that stores electricity to provide an electrical boost for motor starting (Figure 8-24). Most high-torque motors need a capacitor connected in series with the start winding circuit to produce the desired rotation under a heavy starting load.

There are two types of capacitors:

- Start capacitor: This type of capacitor is usually connected to the circuit between the start relay and the start winding terminal of the motor. Start capacitors are used for intermittent (on and off) operation.

- Run capacitor: The run capacitor is also in the start winding circuit, but it stays in operation while the motor is running (continuous operation). The purpose of the run capacitor is to improve motor efficiency during operation.

Capacitors are rated by voltage and by their capacitance value in microfarads. This rating is stamped on the side of the capacitor. A capacitor must be accurately sized to the motor and the motor load. Both run and start capacitors can be tested by means of an ohmmeter.

Heating Elements

Most heating elements are made with a nickel-chromium wire, having both tensile strength and high resistance to current flow. The resistance and voltage can be measured with a multimeter to verify if the element is functioning properly. Heating elements are available in many sizes and shapes (Figure 8-25). They are used for:

- Cooking food
- Heating air for drying clothes
- Heating water to wash clothes, dishes, etc.
- Environmental heating

Heating elements are not repairable, and they should be replaced with a duplicate of the original.

FIGURE **8-25**
Heating elements.

Mechanical Linkages

The mechanical linkages are those devices (connecting rods, gears, cams, belts, levers, pulleys, etc.) that are used on appliances and air conditioners in order to transfer mechanical energy from one point to another. Figure 8-26, the automatic ice maker, is an excellent example of this. Some other examples are:

- In the automatic dryer, the motor is turning a pulley, which moves a belt, which turns the drum.
- In the automatic washing machine, the motor turns a pulley, which moves the belt, which turns the transmission gears, which performs the agitation or spin cycle.
- In the automatic ice maker, the timer gear turns the drive gear, which moves the cam, which actuates the switches and rotates the ice ejector.

Wires

The wiring, which connects the different components in an appliance or air conditioner, is the highway that allows current to flow from point A to point B. Copper and aluminum are the most common types of wires that are used in appliances. They are available as solid or stranded. Wires are enclosed in an insulating sleeve, which might be rubber, cotton, or one of the many plastics. Wires are joined together or to the components by:

- Solderless wire connectors
- Solderless wire terminal connectors
- Solderless multiple-pin plug connectors
- Soldering

PART IV

Motor, ice maker drive

Brass drive gear

Conductor

Shaft

Ice maker
weigh switch shaft

Contact
ice level control

Gear
ice maker drive

FIGURE 8-26 The ice maker is a perfect example of mechanical linkages in use.

Never join copper and aluminum wires together, because the two dissimilar metals will corrode and interrupt the flow of current. The standard wire-gauge sizes for copper wire are listed in Table 8-1. As the gauge size increases from 1 to 20, the diameter decreases and the amperage capacity (ampacity) will decrease also (see Table 8-2).

Gauge No.	Diameter, Mil	Circular-Mil Area	Ohms Per 1000 ft. of Copper Wire at 25°C*
1	289.3	83,690	0.1264
2	257.6	66,370	0.1593
3	229.4	52,640	0.2009
4	204.3	41,740	0.2533
5	181.9	33,100	0.3195
6	162.0	26,250	0.4028
7	144.3	20,820	0.5080
8	128.5	16,510	0.6405
9	114.4	13,090	0.8077
10	101.9	10,380	1.018
11	90.74	8234	1.284
12	80.81	6530	1.619
13	71.96	5178	2.042
14	64.08	4107	2.575
15	57.07	3257	3.247
16	50.82	2583	4.094
17	45.26	2048	5.163
18	40.30	1624	6.510
19	35.89	1288	8.210
20	31.96	1022	10.35

TABLE 8-1   Standard Wire-Gauge Sizes for Copper Wire

Size (AWG)	Ampacity
18	6
16	8
14	17
12	23
10	28

TABLE 8-2   Wire Size and Ampacity

CHAPTER

Electronic Parts and Fault Codes

Did you ever wonder what happens when you touch a function on the touch panel keypad on a microwave, range, washer, or room air conditioner? In this chapter, I will explain the sequence of events from behind the control panel (Figure 9-1) that occurs when you turn on the appliance or room air conditioner for the first time. In addition, I will discuss the service techniques and fault codes needed to service electronic components in appliances and air conditioners.

With the introduction of electronic components in appliances and room air conditioners, there are consumers, first-time repairers, and even some technicians who do not have a clue as to the operation of the sequence of events that takes place after the product is programmed.

On standard appliances and room air conditioners, consumers will turn knobs and press buttons to set the functions, and sometimes they will have to manually turn their appliances on or off. Appliances and room air conditioners with electronic touch panels (see Figure 9-1) can now be programmed to perform a single event or multiple events and to automatically turn on or off.

Electronic Components in General

Much of the information in this chapter covers electronic components in general, rather than specific models, in order to present a broad overview of operation and service techniques. The pictures and illustrations that are used in this chapter are for demonstration purposes to clarify the description of how to service these appliances and room air conditioners. They in no way reflect a particular brand's reliability.

How Electronic Appliances and Air Conditioners Operate

Beginning with the electronic touch panel on the product, the individual will place his or her finger on a function or number to begin the process of programming the product or telling it what action to perform. The electronic touch panel is made up of a thin membrane with a matrix configuration of pressure-sensitive resistive elements that are sealed.

When you touch any key pad, you are closing a circuit in the touch panel membrane to be transmitted to the printed circuit board (PCB), called a display board (Figure 9-2). The display board consists of LEDs (light-emitting diodes) or an LCD (liquid crystal display) that shows the consumer what functions have been stored. In Figure 9-3, the schematic illustrates the matrix configuration of the touch panel membrane. A technician can test the individual key pad functions with the electricity off to the product. For example, if you hold down the cook time key pad and place the ohmmeter probes on pins 11 and 13 on the ribbon connector, you should measure a resistance from 50 to 200 ohms. If you measure zero ohms, the touch panel membrane is faulty and must be replaced. Depending on what model you are servicing, the display board may be part of the main processor PCB or it may be a separate PCB entirely. This PCB is powered by a low-voltage transformer that is either

Touch pad membrane

Display board

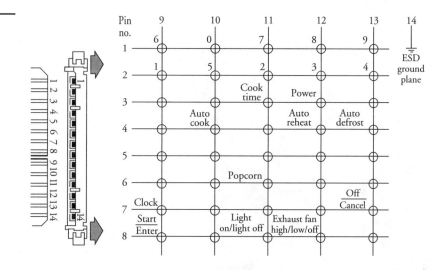

FIGURE 9-3
A technician can test the matrix configuration of the touch panel membrane at the ribbon connector.

mounted on the PCB or mounted somewhere within the appliance. The touch panel and display board are connected to the main processing board (Figure 9-4) by a ribbon cable that is plugged into the main processor board. On some models, all of the components discussed so far may be assembled on to one printed circuit board. After the commands have been entered into the touch panel, the signal is transferred to the main processing PCB to be stored in the main microcomputer chip (CPU) that is mounted on the board (see Figure 9-4). When the user presses the start button, a signal goes to the main CPU chip on the main PCB to initiate the start cycle. The main CPU chip will then select the correct relays to turn on or off the functions that were programmed into it. On some models, the main processor board will send a signal to other PCBs within the appliance to initiate the cycle of events.

Electrostatic Discharge (ESD)

To prevent ESD from damaging expensive electronic components, simply follow these steps:

- Turn off the electricity to the appliance or air conditioner before servicing any electronic component.

- Before servicing the electronics in an appliance or air conditioner, discharge the static electricity from your body by touching your finger repeatedly to an unpainted surface on the appliance or air conditioner. Another way to discharge the static electricity from your body is to touch your finger repeatedly to the green ground connection on that product.

- The safest way to prevent ESD will be to wear an anti-static wrist strap.

- When replacing a defective electronic part with a new one, touch the anti-static package that the part comes in to the unpainted surface of the appliance or air conditioner or to the green ground connection of the appliance or air conditioner.

- Always avoid touching the electronic parts or metal contacts on an integrated board.

- Always handle integrated circuit boards by the edges.

Figure 9-4   A pictorial diagram of a main processor printed circuit board.

Testing Printed Circuit Boards

Before disassembling or testing can begin, look for the technical data sheets. These are typically attached to the outer cabinet under the appliance, behind the control panel, or behind an access panel. These data sheets will provide you with a lot of helpful information when diagnosing and testing procedures for the appliance or room air conditioner. Most technical data sheets will provide you with a self-diagnostic test sequence that can be programmed through the touch panel. On some models, you can isolate and operate the components through the electronic control to see if they operate. When you initiate the diagnostic test, the main PCB will respond, in most cases, with a code that will indicate where the problem lies. On other models, the fault code appears when a malfunction occurs. In addition, the technical data sheet provides other important information, such as the position of the switch contacts, color-coding of wires, performance data tests, a wiring schematic, and other information that might be helpful to the technician.

The most common problem with electronic components in appliances is loose plug connections and corrosion. Before you begin to replace any component, it is recommended that you disconnect the plug connections from the circuit boards and reconnect them. This process will eliminate any corrosion buildup on the plug connectors or pin connections on the circuit board. In addition, if any plug connections were loose, they will be reattached when you plug them back into the circuit board. Most printed circuit boards have fuses soldered to the circuit board. These fuses must be tested first before condemning the component.

Touch Panel Membrane

Before condemning the touch panel, you need to perform certain inspections first. The following is a list of procedures to follow:

- Examine the touch panel membrane (see Figure 9-1) for dents or scratches in the panel. This might cause a short in one or more of the touch pads.
- Inspect the ribbon cable from the touch panel to the display board. Look for evidence of corrosion, tarnishing, or wear on the cable.
- Test all of the keypads and check to see if all functions are working properly.
- If you have to press hard on the touch panel to activate a function, the touch panel will have to be replaced.
- If you press a number and the display shows a different number, the touch panel may have to be replaced.

Transformer

A transformer is an electrical device that can increase (step up) or decrease (step down) the voltage and current. It works on the principle of transferring electrical energy from one circuit to another by electromagnetic induction (Figure 9-5). The primary side of the transformer is the high-voltage side, with the voltage ranging from 120 volts AC to 240 volts AC. On the secondary, or low-voltage, side, the voltage will range from 5 volts AC to 24 volts AC, depending on the amount of voltage and current needed to operate the circuit boards. Some circuit boards require DC voltage to operate, depending on the manufacturer's requirements for the product.

FIGURE 9-5
A transformer. Most
appliances and room
air conditioners use a
step-down transformer
to supply a low voltage
to electronic PCBs.

Circuit Board

When diagnosing the circuit board, the wiring schematic for the appliance will be helpful in
diagnosing, understanding wire color codes, and reading the correct voltages. For example,
in Figure 9-6, the main PCB controls the on/off functions and the temperature for the

FIGURE 9-6
A sample RAC wiring
schematic.

air conditioner. To determine if the main PCB is defective, you would check for the correct supply voltage coming into the PCB. In this case, the voltage should be 120 V AC. In addition, you can also check the voltage at the primary winding of the transformer mounted on the PCB for 120 V AC. Next, test the secondary side of the transformer for output voltage. There is a line fuse on this printed circuit board. Turn off the electricity and check the fuse for continuity. If all checks out, test the relays on the PCB for voltage to the relay coils, or test if the switch contacts on the relays are opening and closing.

A good rule to remember when testing any printed circuit board is that there must be voltage supplied to the board and there must be voltage leaving the board to turn a function on.

Error/Fault or Function Codes

Over the years, appliance manufacturers have incorporated fault codes, diagnostic and test information, and appliance usage information into their appliances. All appliances with electronic displays and electronic components have fault codes programmed into the microprocessor on the main printed circuit board. These codes act as a service aid and starting point for the technician when diagnosing a fault with the appliance. An error/fault code will appear on the display when the appliance encounters some type of error detection. This can be in the form of a series of blinking lights or an actual alphanumeric code.

Before you begin diagnostics, locate the technical data sheet and the manufacturer's use and care guide for the product you are about to service. The technical data sheet will have the fault codes and the information on how to get into the appliance test mode. The test mode is another diagnostic tool that allows the technician to advance to a specific function, or the technician can turn on or off a single component. The use and care guide will provide you with the usage information and some fault codes as well. There are times when an error/fault code will appear that is not on the technical data sheet or in the use and care guide. If this occurs, go online to the manufacturer's Web site or contact the appliance manufacturer's customer service or technical department for the code description.

Do not assume that a service problem is directly caused by the main printed circuit board. The problem might exist with the voltage supply to the product or a line voltage component (including the power cord and the appliance wiring) that has shorted, grounded, opened, or otherwise malfunctioned.

What follows is a sampling of some of the appliances that have error/fault and function codes. Not all of the codes listed in this chapter will cover every appliance manufacturer or model. Rely on the technical data sheet for most of the information for the product you are servicing.

ASKO Dishwasher Error/Fault or Function Codes

The error/fault or function codes listed here will vary from model to model. The codes on the model you are servicing might be different from what is listed. If this is the case, check the technical data sheet, the use and care guide, the manufacturer's Web site, or call the manufacturer for the description of the error/fault code you are experiencing. Sometimes, by disconnecting the electricity to the product and waiting for two minutes and then turning on the electricity, this may reset the error/fault or function code, allowing you to reprogram and test the operation of the appliance.

PART IV

For your safety and the safety of others, all testing and physical inspections of the product must be performed with the electricity turned off. On occasion, you will have to test a component with the electricity on. Extreme care must be taken to prevent electrical mishaps.

F1   *Description*   A problem has occurred in the water heating cycle. The water temperature in the dishwasher did not reach the predetermined temperature within 60 minutes.

Solution   Check the water temperature at the sink next to the dishwasher. It should be around 120–140 degrees Fahrenheit. Check the water heater temperature also. Check the dishwasher heater and circuit components. Do not take showers or baths or run the automatic washer before running the dishwasher—this will deplete the hot water supply in the water heater.

F2   *Description*   Dishwasher overfilled or float stuck and/or float switch failure or clogged sump or drain.

Solution   Check for debris in the sump area of the tub. Check the water float and switch assembly. Check the water inlet valve.

F3   *Description*   Thermistor malfunction.

Solution   Check the thermistor and connecting wiring circuit.

F4   *Description*   Water fill problem. Dishwasher models with level-controlled fill only. *Note:* If the control lights are blinking, the pump is running, and the controls are dead, the fault will indicate an overfill problem.

Solution   Check for water in the base pan. The dishwasher will not operate if water is in the pan. Remove the water, reset the dishwasher, and inspect for leaks. Check float assembly.

F5   *Description*   Valve leakage fault code.

Solution   Inspect valve for leaks.

F7   *Description*   Blocked drain fault code.

Solution   The drain pump can be cleaned out. Refer to the use and care guide for the cleanout procedure.

F8   *Description*   Clogged filter fault code. One or all three filters are clogged.

Solution   Before beginning to disassemble the filters, refer to the use and care guide for proper removal and cleaning instructions for the filters.

Some Asko dishwasher models use flashing indicator lights as error/fault or function codes. See Table 9-1.

ASKO Washer Error/Fault or Function Codes

The error/fault or function codes listed here will vary from model to model. The codes on the model you are servicing might be different from what is listed. If this is the case, check the technical data sheet, the use and care guide, the manufacturer's Web site, or call the manufacturer for the description of the error/fault code you are experiencing. Sometimes, by

Flashing Indicator Light(s)	Asko Dishwasher Error/Fault or Function Code
Heavy Wash	Dishwasher overfilled. Check float assembly and float switch. Check for clogged filters or clogged drain.
Delicate Wash	No water to dishwasher. Check water supply and water inlet valve.
Quick Wash	Valve leakage fault. Check valves for leaks.
Pots and Pans and Heavy Wash	The drain is blocked. Refer to the use and care guide for cleanout procedure.
Pots and Pans and Normal Wash	The filter is clogged. Refer to the use and care guide for proper removal and cleaning instructions for the filters.

TABLE 9-1   ASKO Dishwasher Error/Fault or Function Codes

disconnecting the electricity to the product and waiting for two minutes and then turning on the electricity, this may reset the error/fault or function code, allowing you to reprogram and test the operation of the appliance.

For your safety and the safety of others, all testing and physical inspections of the product must be performed with the electricity turned off. On occasion, you will have to test a component with the electricity on. Extreme care must be taken to prevent electrical mishaps.

F1   *Description*   An overfill condition was detected.

    Solution   Check the water inlet valve. Replace if necessary.

F3   *Description*   The washer has a draining problem.

    Solution   Check drain filter, pump, and drain hose. Replace defective part.

F4   *Description*   A water inlet error was detected.

    Solution   Make sure that the water is turned on to the washer. Check water inlet valve. Replace defective part.

F6   *Description*   The control detected a door-open condition.

    Solution   Check door latch assembly. Check wiring connectors and wiring.

ASKO Dryer Error/Fault or Function Codes

The error/fault or function codes listed here will vary from model to model. The codes on the model you are servicing might be different from what is listed. If this is the case, check the technical data sheet, the use and care guide, the manufacturer's Web site, or call the manufacturer for the description of the error/fault code you are experiencing. Sometimes, by disconnecting the electricity to the product and waiting for two minutes and then turning on the electricity, this may reset the error/fault or function code, allowing you to reprogram and test the operation of the appliance.

For your safety and the safety of others, all testing and physical inspections of the product must be performed with the electricity turned off. On occasion, you will have to test a component with the electricity on. Extreme care must be taken to prevent electrical mishaps.

F1   *Description*   The condenser drain hose has detected a blockage.

    Solution   Clear the blockage and restart.

F2   *Description*   Dryer takes too long to dry.

    Solution   Check dryer installation, dryer venting system, and heater. Replace defective part if necessary.

Friedrich Room Air Conditioner Error/Fault or Function Codes

The error/fault or function codes listed here will vary from model to model. The codes on the model you are servicing might be different from what is listed. If this is the case, check the technical data sheet, the use and care guide, the manufacturer's Web site, or call the manufacturer for the description of the error/fault code that you are experiencing. Sometimes, by disconnecting the electricity to the product and waiting for 10 minutes and then turning on the electricity, this may reset the error/fault or function code, allowing you to reprogram and test the operation of the appliance.

For your safety and the safety of others, all testing and physical inspections of the product must be performed with the electricity turned off. On occasion, you will have to test a component with the electricity on. Extreme care must be taken to prevent electrical mishaps.

E1   *Description*   The electronic control board detected a short-cycle malfunction.

    Solution   Check for conditions that caused the short-cycle situation. Check thermostat setting, air flow conditions, refrigeration cycle, and fan motor.

E2   *Description*   The electronic control board detected a touch keypad error.

    Solution   Reset the air conditioner controls and test. Check for stuck keys. If error code returns, replace the touch keypad assembly.

E3   *Description*   The electronic control detected the frost probe is open.

    Solution   Test the valve of the frost probe. If the probe tested bad, replace the frost probe. If good, replace the electronic control board.

E4   *Description*   The electronic control board detected a short in the frost probe.

    Solution   Replace the electronic control board.

E5 or E6   *Description*   The electronic control board detected an open or shorted indoor temperature sensor.

    Solution   Replace the electronic control board.

CHECK FILTER INDICATOR LAMP   *Description*   The check filter indicator lamp will come on after 250 hours of operation. It is a reminder for the consumer to clean the air conditioner filter. This code can be reset to start the countdown to the next filter cleaning.

Frigidaire/Electrolux Dishwasher Error/Fault or Function Codes

The error/fault or function codes listed here will vary from model to model. The codes on the model you are servicing might be different from what is listed. If this is the case, check the technical data sheet, the use and care guide, the manufacturer's Web site, or call the manufacturer for the description of the error/fault code you are experiencing. Sometimes, by disconnecting the electricity to the product and waiting for two minutes and then turning on the electricity, this may reset the error/fault or function code, allowing you to reprogram and test the operation of the appliance.

For your safety and the safety of others, all testing and physical inspections of the product must be performed with the electricity turned off. On occasion, you will have to test a component with the electricity on. Extreme care must be taken to prevent electrical mishaps.

* or Lo or status LEDs   *Description*   Wetting agent level is low.

> *Solution*   Check the wetting agent level in the dispenser. After five wash cycles, the indicator light will turn off when the dispenser is not filled up.

PF   *Description*   A power failure occurred. The high-temp wash and the no-heat dry/power-dry off lights will begin to flash when the power resumes.

> *Solution*   Press the start/cancel pad and reset the controls. If the code persists, turn the electricity off for five minutes. Check all electrical connections to the electronic control panel.

HO   *Description*   Delay in the water heating cycle.

> *Solution*   This is a normal condition until the water reaches the predetermined temperature. Check the water temperature at the sink next to the dishwasher—it should be around 120 to 140 degrees Fahrenheit. Check the water heater temperature also. Check dishwasher heater and circuit components. Do not take showers or baths or run the automatic washer before running the dishwasher—this will deplete the hot water supply in the water heater.

CL   *Description*   The door is open.

> *Solution*   Close and latch the dishwasher door.

01-24   *Description*   On some models, this is the amount of time in hours left before the dishwasher begins the cycle.

01-25 or 01-10   *Description*   On some models, this is the amount of time in hours left before the dishwasher begins the cycle.

SENSING   *Description*   On some models, a thermistor called the turbidity sensor is checking the condition of the wash/rinse water.

> *Solution*   Check thermistor and electronic control board.

PART IV

Frigidaire/Electrolux Washer Error/Fault or Function Codes

The error/fault or function codes listed here will vary from model to model. The codes on the model you are servicing might be different from what is listed. If this is the case, check the technical data sheet, the use and care guide, the manufacturer's Web site, or call the manufacturer for the description of the error/fault code you are experiencing. Sometimes, by disconnecting the electricity to the product and waiting for two minutes and then turning on the electricity, this may reset the error/fault or function code, allowing you to reprogram and test the operation of the appliance.

For your safety and the safety of others, all testing and physical inspections of the product must be performed with the electricity turned off. On occasion, you will have to test a component with the electricity on. Extreme care must be taken to prevent electrical mishaps.

E11	*Description* Washer taking too long to fill up.
	Solution Check water inlet supply. Check the water pressure—it should be at or above 30 PSI. Check the washer fill valve, pressure switch or pressure sensor, and the control board. If the washer fills in the off mode and the electricity is turned off, replace the water inlet valve.
E13	*Description* The electronic control detected a water leak in the tub or an air leak in the air bell.
	Solution Check for water leaks. Check the air bell system, pressure switch, or pressure sensor. Check the control board. Repair any water leaks. Replace any defective part.
E21	*Description* The water in the washer is not pumping out fast enough.
	Solution Check washer installation. Make sure that the drain hose is not restricted. Check the drain pump for 120 volts in the drain cycle. If the voltage reading is zero, replace the control board. If the voltage is correct, check the drain pump for a restriction or blockage. If none is found, replace the pump.
E23 or E24	*Description* The electronic control detected that the drain pump relay malfunctioned.
	Solution Test the drain pump relay on the electronic control board. If the test failed, replace the electronic control board. Check all wiring connections and wiring to the drain pump.
E31	*Description* A communication error was detected between the pressure sensor and the electronic control board.
	Solution Check the wiring connections and wiring for the pressure sensor to the electronic control board. Test the pressure sensor valve and replace if necessary.

E35   *Description*   The pressure sensor has detected a water overfill problem.

Solution   Check to see if the water level is above the manufacturer's recommendations. Check the water inlet valve for proper operation. Check the wiring connections and the wiring from the electronic control to the pressure sensor and to the water inlet valve. Check the electronic control board for proper operation. Replace any defective part.

E36, E43, E44, E45, or E46   *Description*   The electronic control board has detected a malfunction.

Solution   Check all the wiring connections to the electronic control board. If okay, replace the electronic control board.

E41   *Description*   The electronic control board has detected a door-open code.

Solution   Check that the door closes properly. Check the door strike and door switch assembly. Check the electronic control. Replace any defective part.

E47 or E48   *Description*   The electronic control board detected a problem in the door thermistor circuit.

Solution   Check the door lock assembly and measure the resistance of the thermistor. If you detect a short or an open, replace the door lock assembly. If your measurement of the thermistor is at least 1500 ohms, replace the electronic control board.

E52   *Description*   The electronic control board received a bad signal from the tacho generator.

Solution   Test the drive motor for 105 to 130 ohms at pins 4 and 5 on the motor only. If the readings are other than 105 to 130 ohms, replace the speed control board.

E56, E57, or E58   *Description*   The electronic control board detected a motor problem.

Solution   Check the following components: motor, tub bearings, speed control. After removing the belt from the motor, spin the motor—it should turn freely. If it does not, replace the motor. With the belt off, spin the tub—it should spin freely. If it does not, replace the tub bearings. Test the motor windings and test pins 1 and 2, pins 1 and 3, and pins 2 and 3 on the motor. It should read about 6 ohms. If this is not the case, replace the motor.

E59   *Description*   The tacho generator lost a signal.

Solution   Check the following components: motor, tub bearings, speed control. After removing the belt from the motor, spin the motor—it should turn freely. If it does not, replace the motor. With the belt off, spin the tub—it should spin freely. If it does not, replace the tub bearings. Test the motor windings and test pins 1 and 2, pins 1 and 3, and pins 2 and 3 on the motor. It should read about 6 ohms. Test the motor for 105 to 130 ohms at pins 4 and 5 on the motor only. If the readings are other than 105 to 130 ohms, replace the speed control board.

E5A, E5B, or E5C   *Description*   The heat sink overheated on high temperature.

Solution   Test the heat sink. If it is bad, replace the speed control board.

E5D, E5E, or E5F

Description A communication error has occurred between the electronic control board and the speed control board.

Solution Check the wiring connections and the wiring between the electronic control board and the speed control board. Check the electronic control board and the speed control board. Replace the defective part.

E66 or E68

Description The electronic control detected a heater problem.

Solution Check the wiring connections and the wiring for the heater circuit. Check the heater element for 14 ohms. Check for a grounded heater. Replace the heater if the readings are incorrect.

E67

Description The electronic control detected the wrong input voltage to the microprocessor.

Solution Check the wiring connections and wiring. Check the tub thermistor for 4.8 ohms. If incorrect, replace the heating element.

E71

Description Wash thermistor (tub heater) malfunction.

Solution Check the wiring. If okay, replace the electronic control board.

E74

Description The wash water will not heat up.

Solution Place the thermistor in the correct position.

E75

Description Water temperature sensor detection.

Solution Check the thermistor for resistance—it should be 50 ohms. If it is not, replace the water inlet valve assembly. If the resistor measured 50 ohms, replace the electronic control board.

E76

Description The thermistor temperature is off for the cold water inlet valve.

Solution The water fill hoses need to be installed correctly.

E82, E83, E93, E94, E95, E97, E98, EBE, or EBF

Description The electronic control board has a communications problem.

Solution Check all wiring connections and wiring. If okay, replace the electronic control board.

E91

Description There is a communication error between the electronic control board and the UI board.

Solution Check all wiring connections and wiring.

EB1

Description Incoming voltage frequency problem.

Solution Have your local electric company come out and check the input voltage frequency to the house. If okay, replace the electronic control board.

EB2

Description Incoming voltage too high.

Solution If the voltage at the washer receptacle is below 130 volts, replace the electronic control board.

EB3

Description Incoming voltage too low.

Solution If the voltage at the washer receptacle is above 90 volts, replace the electronic control board.

EF1   *Description*   The drain pump is clogged.

                Solution   Clear the blockage from the drain pump.

EF2   *Description*   The electronic control board detected too much soap in the washer.

                Solution   Advise the consumer on proper soap usage.

EF5   *Description*   The thermistor temperature is off for the hot water inlet valve.

                Solution   The water fill hoses need to be installed correctly.

1 beep/E10   *Description*   The washer is not filling up with water or water is possibly leaking onto the floor.

                Solution   Make sure that the water is turned on to the washer. Check the water pressure, fill hoses for kinks, and the drain hose installation. If you found water leaking, check the household drain for a blockage, and check for too much soap usage.

2 beeps/E20   *Description*   The water will not drain out.

                Solution   Check the drain hose installation and check for a drain blockage in the household drain.

3 beeps/E30   *Description*   The electronic control has detected that the drum is overfilled with water.

                Solution   Check the water inlet valve and pressure switch for proper operation. Check the electronic control board. Replace the defective part. Drain the water from the washer and reset the controls.

4 beeps/E40   *Description*   The electronic control detected that the door is open.

                Solution   Close the door and reset the controls. The washer will not operate when the door is open.

5 beeps/E50   *Description*   The electronic control detected a motor overheat problem.

                Solution   Let the motor cool down. Check the wiring connections and wiring. Reset the controls and restart the washer.

7 beeps/E70   *Description*   The cold water is not entering the washer.

                Solution   Make sure that the cold water is turned on all the way. Check the water fill hose installation. Check the water inlet valve.

15 beeps/EF0   *Description*   The electronic control detected an oversudsing condition or no hot water entering the washer.

                Solution   Check for too much detergent added to the load. Advise the consumer on proper detergent usage. Read the use and care guide. Check for hard water conditions, water temperature, and the size of the load. Make corrections if necessary. For no hot water, check and see if the hot water is turned on all the way. Also check the water fill hose installation.

Frigidaire/Electrolux Dryer Error/Fault or Function Codes

The error/fault or function codes listed here will vary from model to model. The codes on the model you are servicing might be different from what is listed. If this is the case, check the technical data sheet, the use and care guide, the manufacturer's Web site, or call the manufacturer for the description of the error/fault code you are experiencing. Sometimes, by disconnecting the electricity to the product and waiting for two minutes and then turning on the electricity, this may reset the error/fault or function code, allowing you to reprogram and test the operation of the appliance.

For your safety and the safety of others, all testing and physical inspections of the product must be performed with the electricity turned off. On occasion, you will have to test a component with the electricity on. Extreme care must be taken to prevent electrical mishaps.

E10, E11, *Description*   There is a communication error in the electronic control
or E12   microprocessor.

> *Solution*   Cancel the cycle by pressing cancel. Reset the controls and test the cycle. If the problem persists, replace the electronic control.

E24   *Description*   Thermistor sensor malfunction.

> *Solution*   Check the thermistor sensor for a short. Check all wiring between the thermistor and the electronic control. Test the electronic control for proper operation.

E25   *Description*   Thermistor sensor problem.

> *Solution*   Test the thermistor sensor. Check the wiring from the thermistor to the electronic control. Test the electronic control for proper operation.

E4A   *Description*   The drying cycle was interrupted because the drying time exceeded the time that was programmed.

> *Solution*   Inspect the clothes. If they are still wet, check for proper installation of the dryer. Check vent system, blower assembly, and motor. Check the moisture sensor bars for dirt. Check and test the heater. Check all the wiring and wiring connections. Check the electronic control for proper operation.

E5B   *Description*   The thermistor is sensing no heat and the cycle is stalled.

> *Solution*   Check and test the entire heating circuitry, including the heater. Check the electronic control for proper operation. See the technical data sheet for the specific procedure for the model you are working on.

E68   *Description*   A stuck button fault code.

> *Solution*   Inspect the buttons and run the test mode to determine which button is stuck in the active mode. Check the electronic control for proper operation.

E8C   *Description*   The high limit safety thermostat has tripped.

> *Solution*   Check for the proper installation of the dryer. Check the entire exhaust system for proper installation or blockage. Check the blower assembly and motor for proper operation. Check the drum seals. Check the lint filter. Check the door seal.

EAF   *Description*   The microprocessor in the electronic control has been reset by the watchdog timer.

        Solution   Clear the code and reset the controls. Test the dryer. If the problem persists, replace the electronic control.

PF   *Description*   A power failure occurred.

        Solution   Correct the problem by pressing the cancel button and reset the controls. If the code persists, turn the electricity off for five minutes. Check all electrical connections to the electronic control.

ERR   *Description*   When programming the control, a mistake was made.

        Solution   Press cancel and reprogram the controls.

LOC   *Description*   The control lock feature is activated. This is a function display code during normal operation.

PAU   *Description*   The drying cycle has been interrupted.

        Solution   The customer might have opened the door in the middle of the cycle. If not, make repairs to the door and latch.

"ad"   *Description*   This is a function display code indicating that the dryer is in the auto-dry cycle.

"dn"   *Description*   This is a function display code indicating that the dryer has completed the auto-dry cycle.

Frigidaire/Electrolux Refrigerator Error/Fault or Function Codes

ALARM INDICATOR LIGHT   *Description*   When the light is on, the alarm system is activated.

        Solution   This is a normal condition. If the refrigerator does not reach the set temperature setting, an alarm will sound. Check refrigeration cycle.

DOOR AJAR INDICATOR LIGHT   *Description*   The refrigerator door was left open for over five minutes. The indicator light will be flashing and the audible alarm will sound.

        Solution   Close the doors. Rest the alarm. If problem persists, check wiring connectors and wiring.

HIGH TEMP INDICATOR LIGHT   *Description*   The refrigerator temperature has been above the temperature setting for over an hour. The indicator light will begin to flash amber and an audible alarm will sound.

        Solution   Reset the alarm, and take a temperature reading to confirm. Check the refrigeration cycle, the defrost cycle, and components for proper operation.

POWER FAILURE INDICATOR LIGHT   *Description*   A power failure was detected, and the amber indicator light will begin to flash.

        Solution   Press the reset button and check the food storage temperature.

General Electric Room Air Conditioner Error/Fault or Function Codes

The error/fault or function codes listed here will vary from model to model. The codes on the model you are servicing might be different from what is listed. If this is the case, check the technical data sheet, the use and care guide, the manufacturer's Web site, or call the manufacturer for the description of the error/fault code you are experiencing. Sometimes, by disconnecting the electricity to the product and waiting for two minutes and then turning on the electricity, this may reset the error/fault or function code, allowing you to reprogram and test the operation of the appliance.

For your safety and the safety of others, all testing and physical inspections of the product must be performed with the electricity turned off. On occasion, you will have to test a component with the electricity on. Extreme care must be taken to prevent electrical mishaps.

E1   *Description*   The electronic control detected that the filter needs cleaning.

       Solution   Clean the filter.

ER   *Description*   The electronic control detected a temperature sensor malfunction.

       Solution   Check the value of the temperature sensor. Check for debris buildup on the probe. Check the sensor for proper positioning. Replace if necessary.

General Electric Dishwasher Error/Fault or Function Codes

The error/fault or function codes listed here will vary from model to model. The codes on the model you are servicing might be different from what is listed. If this is the case, check the technical data sheet, the use and care guide, the manufacturer's Web site, or call the manufacturer for the description of the error/fault code you are experiencing. Sometimes, by disconnecting the electricity to the product and waiting for two minutes and then turning on the electricity, this may reset the error/fault or function code, allowing you to reprogram and test the operation of the appliance.

For your safety and the safety of others, all testing and physical inspections of the product must be performed with the electricity turned off. On occasion, you will have to test a component with the electricity on. Extreme care must be taken to prevent electrical mishaps.

PF   *Description*   Power failure.

       Solution   Reset the electronic control by pressing the cleaning options and start key pad. If this does not correct the problem, press the cancel button and reset the control panel. If the code persists, turn the electricity off for five minutes. Check all electrical connections to the electronic control.

C1   *Description*   The drain cycle exceeded two minutes.

       Solution   Check for obstruction in the drain line to the sink. If there is a garbage disposer, run it and clean it out. If clear, check the drain line valve.

C2   *Description*   The motor/pump did not completely drain the water from the tub within seven minutes.

Solution Press the reset button to halt the alarm. Check the drain valve, drain hose, and garbage disposer for obstruction.

C3 *Description* The motor/pump will not drain the water from the tub.

Solution Disconnect the electricity to the dishwasher for two minutes and then reset the electronic control. Go into the test mode to run the drain cycle over and check the motor/pump assembly.

C4 *Description* Dishwasher overfilled or float stuck; float switch failure or clogged sump or drain.

Solution Reset the electronic control by touching the start key pad and then the cancel/reset key pad.

Sometimes, after a power failure, the dishwasher will fill the tub again. Check for debris in the sump area of the tub. Check the water float and switch assembly.

C5 *Description* Dishwasher drain cycle is too short.

Solution Reset the electronic control and restart the cycle. Check water supply and float/switch assembly.

C6 *Description* Low water temperature in the tub.

Solution Check the water temperature at the sink next to the dishwasher—it should be around 120 to 140 degrees Fahrenheit. Check the water heater temperature also. Check dishwasher heater and circuit components. Do not take showers or baths or run the automatic washer before running the dishwasher. This will deplete the hot water supply in the water heater.

C7 *Description* A fault has occurred in the water temperature sensor.

Solution Turn the electricity off to the dishwasher for two minutes and then reset the electronic control. Check the water temperature sensor.

C8 *Description* Detergent cup malfunction or blocked.

Solution Remove the obstruction, close the dispenser, and reset the electronic control.

CUP OPEN *Description* The detergent dispenser cup was left open.

Solution Close the detergent dispenser cup and press start.

General Electric Microwave Error/Fault or Function Codes

The error/fault or function codes listed here will vary from model to model. The codes on the model you are servicing might be different from what is listed. If this is the case, check the technical data sheet, the use and care guide, the manufacturer's Web site, or call the manufacturer for the description of the error/fault code you are experiencing. Sometimes, by disconnecting the electricity to the product and waiting for two minutes and then turning on the electricity, this may reset the error/fault or function code, allowing you to reprogram and test the operation of the appliance.

For your safety and the safety of others, all testing and physical inspections of the product must be performed with the electricity turned off. On occasion, you will have to test a component with the electricity on. Extreme care must be taken to prevent electrical mishaps.

PF or RESET   *Description*   A power failure occurred.

Solution   Correct the problem by pressing the cancel button and reset the control. If the code persists, turn the electricity off for five minutes. Check all electrical connections to the electronic control.

F1   *Description*   Thermal sensor problem; excessive heat from within the microwave cavity.

Solution   Test the thermal sensor. Check the fan motor and air circulation pattern for a blockage (dirty filter).

F2   *Description*   Thermal sensor problem.

Solution   Test the thermal sensor for a short. Check all circuits and components for proper operation.

F3   *Description*   The touch panel may have a short.

Solution   Inspect and test the touch panel and connectors.

F4   *Description*   Humidity sensor problem.

Solution   Test the humidity sensor. Check the fan motor and air circulation pattern for a blockage (dirty filter).

F5   *Description*   Humidity sensor problem.

Solution   Test the humidity sensor for a short. Check all circuits and components for proper operation.

F6 or PROBE   *Description*   Temperature probe problem.

Solution   Test the temperature probe value. Check the probe receptacle wiring.

F10   *Description*   Touch screen problem.

Solution   Test touch screen for a short. Check connecting wiring harness.

General Electric Range Error/Fault or Function Codes

The error/fault or function codes listed here will vary from model to model. The codes on the model you are servicing might be different from what is listed. If this is the case, check the technical data sheet, the use and care guide, the manufacturer's Web site, or call the manufacturer for the description of the error/fault code you are experiencing. Sometimes, by disconnecting the electricity to the product and waiting for two minutes and then turning on the electricity, this may reset the error/fault or function code, allowing you to reprogram and test the operation of the appliance.

For your safety and the safety of others, all testing and physical inspections of the product must be performed with the electricity turned off. On occasion, you will have to test a component with the electricity on. Extreme care must be taken to prevent electrical mishaps.

F0 or F1	*Description* Key panel problem.
	Solution Check key panel and wiring. Replace key panel.
F2	*Description* Oven temperature malfunction.
	Solution Test the oven temperature during operation for preset temperature. Check temperature sensor.
F3 or F4	*Description* Oven sensor problem.
	Solution Test oven sensor value and replace.
F5	*Description* Electronic control problem.
	Solution Check the wiring connections to the electronic control. Replace the electronic control.
F6 or F7	*Description* Key panel problem.
	Solution Check key panel and wiring. Replace key panel.
F8	*Description* Electronic control problem.
	Solution Check the wiring connections to the electronic control. Replace the electronic control.
F9 or FC	*Description* Door lock component malfunction.
	Solution Check the door latch assembly and switches. Check wiring connections. Replace failed component.
FD	*Description* Probe and/or receptacle malfunction.
	Solution Test the probe value and replace if necessary. Check the receptacle and make repairs or replace the receptacle.
FF	*Description* Electronic control problem.
	Solution Check the wiring connections to the electronic control. Replace the electronic control.
BAD LINE	*Description* This indicates that the main electric supply to the oven is wired incorrectly.
	Solution Check the voltage supply to the product. Make sure that the product is wired correctly, and normal operation should resume.
UNLOCK DOOR	*Description* Door lock malfunction at the end of the cleaning cycle.
	Solution Check the door latch assembly and circuitry. Read the use and care guide and check the GE Web site for the steps necessary to open the door lock.
LOC	*Description* A safety feature added to the programming to prevent children from turning on the product.
	Solution Read the use and care guide to learn about the programming for each model.

PART IV

ERR   *Description*   When programming the control, a mistake was made.

            Solution   Press cancel and reprogram the controls.

            Description   A holiday program function.

            Solution   A backwards "C" means that your oven was programmed for the Sabbath Mode. Read the use and care guide for correction, or go to the GE Web site for instructions.

Maytag Laundry Center Error/Fault or Function Codes

The error/fault or function codes listed here will vary from model to model. The codes on the model you are servicing might be different from what is listed. If this is the case, check the technical data sheet, the use and care guide, the manufacturer's Web site, or call the manufacturer for the description of the error/fault code you are experiencing. Sometimes, by disconnecting the electricity to the product and waiting for two minutes and then turning on the electricity, this may reset the error/fault or function code, allowing you to reprogram and test the operation of the appliance.

For your safety and the safety of others, all testing and physical inspections of the product must be performed with the electricity turned off. On occasion, you will have to test a component with the electricity on. Extreme care must be taken to prevent electrical mishaps.

L or 1E   *Description*   A lid lock error was detected.

          Solution   Check the lock assembly for proper operation. Check all wiring connections and wiring. Check the electronic control board. Replace defective part if necessary.

U   *Description*   An unbalanced load was detected.

          Solution   Check the unbalance switch, wiring, and electronic control board. Advise the consumer on loading the washer. Replace defective part if necessary.

6E or 7E *Description*   The electronic control detected a temperature error in the dryer.

          Solution   Check the temperature sensor, thermostats, electronic control board, and the wiring connections and wiring. Check the installation of the appliance. Replace part if necessary.

8E   *Description*   An error was detected in the motor and circuitry in the dryer.

          Solution   Check the motor, relay, electronic control board, and wiring connections and wiring. Replace defective part if necessary.

9E   *Description*   The electronic control board detected a door switch failure in the dryer.

          Solution   Check the door switch, electronic control board, and the wiring connections and wiring. Replace defective part if necessary.

Maytag Washer Error/Fault or Function Codes

The error/fault or function codes listed here will vary from model to model. The codes on the model you are servicing might be different from what is listed. If this is the case, check the technical data sheet, the use and care guide, the manufacturer's Web site, or call the manufacturer for the description of the error/fault code you are experiencing. Sometimes, by disconnecting the electricity to the product and waiting for two minutes and then turning on the electricity, this may reset the error/fault or function code, allowing you to reprogram and test the operation of the appliance.

For your safety and the safety of others, all testing and physical inspections of the product must be performed with the electricity turned off. On occasion, you will have to test a component with the electricity on. Extreme care must be taken to prevent electrical mishaps.

do   *Description*   Door lock problem detected.

      Solution   Make sure that the door closes properly. Check door lock assembly.

FL   *Description*   The control detected that the washer did not lock the door.

      Solution   Make sure that the door closes properly. Check door lock assembly.

Sd   *Description*   The washer detected an oversudsing problem.

      Solution   The suds recovery cycle should correct the problem. Advise the consumer on proper detergent usage.

Pf   *Description*   A power failure has occurred.

      Solution   Restart the washer.

nd   *Description*   The washer did not drain the water out of the machine.

      Solution   Check the drain hose for proper installation. Check the household drain for blockage. Check the pump for blockage. Replace defective part if necessary.

LO   *Description*   The door lock will not open.

      Solution   Make sure the door closes properly. Push the off button and wait a few minutes for the door to unlock. Disconnect the electricity to the washer for a few minutes and try to unlock the door. Check door lock assembly. Replace defective part if necessary.

Lr   *Description*   The control detected a motor problem.

      Solution   Check the motor windings. Check the wiring connections and wiring. Replace defective part if necessary.

nf   *Description*   Water will not fill the washer.

      Solution   Make sure that the water is turned on all the way. Check for kinked fill hoses. Check water inlet valve. Replace defective part if necessary.

od   *Description*   This is a safety feature that will not allow the washer to start on another cycle unless the door is opened first.

PART IV

Maytag Dryer Error/Fault or Function Codes

The error/fault or function codes listed here will vary from model to model. The codes on the model you are servicing might be different from what is listed. If this is the case, check the technical data sheet, the use and care guide, the manufacturer's Web site, or call the manufacturer for the description of the error/fault code you are experiencing. Sometimes, by disconnecting the electricity to the product and waiting for two minutes and then turning on the electricity, this may reset the error/fault or function code, allowing you to reprogram and test the operation of the appliance.

For your safety and the safety of others, all testing and physical inspections of the product must be performed with the electricity turned off. On occasion, you will have to test a component with the electricity on. Extreme care must be taken to prevent electrical mishaps.

PF   *Description*   A power failure has occurred.

            Solution   Reset the control.

Hp or hl   *Description*   The temperature sensor has detected an error.

            Solution   Check all the components in the heat circuit. Replace defective part if necessary.

hll   *Description*   The hanging rod in the dryer cabinet has a problem.

            Solution   Check all the components in the dryer cabinet for proper operation.

Maytag/Jenn-Air Oven Error/Fault or Function Codes

The error/fault or function codes listed here will vary from model to model. The codes on the model you are servicing might be different from what is listed. If this is the case, check the technical data sheet, the use and care guide, the manufacturer's Web site, or call the manufacturer for the description of the error/fault code you are experiencing. Sometimes, by disconnecting the electricity to the product and waiting for two minutes and then turning on the electricity, this may reset the error/fault or function code, allowing you to reprogram and test the operation of the appliance.

For your safety and the safety of others, all testing and physical inspections of the product must be performed with the electricity turned off. On occasion, you will have to test a component with the electricity on. Extreme care must be taken to prevent electrical mishaps.

F0, F1,   *Description*   The electronic control detected a stuck or shorted key on the
or F7   touch panel.

            Solution   Check the touch key pad for proper operation. Check all wiring connections and wiring. Replace touch pad.

F2   *Description*   The electronic control has detected a temperature problem.

            Solution   Check the sensor, relay board, and wiring connections and wiring. Replace defective part if necessary.

F3 or F4   *Description*   The electronic control has detected a sensor problem.

            Solution   Check the value of the sensor. Check the wiring connections and wiring. Replace the sensor.

F5   *Description*   A communication error was detected.

          Solution   Replace the clock.

F6   *Description*   A voltage problem has been detected in the oven.

          Solution   Check the voltage to the oven.

F8   *Description*   An analog/digital error occurred.

          Solution   Replace the clock.

F9   *Description*   The electronic control detected a door lock problem.

          Solution   Check the door latch assembly. Check the wiring connections and wiring. Replace the clock if necessary.

KitchenAid/Roper/Whirlpool Gas and Electric Range Error/Fault or Function Codes

The error/fault or function codes listed here will vary from model to model. The codes on the model you are servicing might be different from what is listed. If this is the case, check the technical data sheet, the use and care guide, the manufacturer's Web site, or call the manufacturer for the description of the error/fault code you are experiencing. Sometimes, by disconnecting the electricity to the product and waiting for two minutes and then turning on the electricity, this may reset the error/fault or function code, allowing you to reprogram and test the operation of the appliance.

Listed are the code designations for a four-digit display (**/** or ****) and for a three-digit display (***).

For your safety and the safety of others, all testing and physical inspections of the product must be performed with the electricity turned off. On occasion, you will have to test a component with the electricity on. Extreme care must be taken to prevent electrical mishaps.

F0/E0, F1/E0, or E0 or F1   *Description*   The microprocessor EEPROM on the PCB received a communication error and the heating LED will begin flashing.

          Solution   Turn off the electricity to the appliance and wait for two minutes before turning it back on. If the code reappears, replace the electronic control.

F1/E1, F1/E2, F1/E4, or F1   *Description*   An electronic control error occurred and the heating LED will begin flashing.

          Solution   Turn off the electricity to the appliance and wait for two minutes before turning it back on. If the code reappears, replace the electronic control.

F1/E5   *Description*   The electronic control detected a calibration error.

          Solution   Check operating temperatures. Turn off the electricity to the appliance and wait for two minutes before turning it back on. If the code reappears, replace the electronic control.

PART IV

F1/E9

Description   A stack overflow error was detected by the electronic control.

Solution   Turn off the electricity to the appliance and wait for two minutes before turning it back on. If the code reappears, replace the electronic control.

F1/E6

Description   The electronic control received a latch error.

Solution   Turn off the electricity to the appliance and wait for two minutes before turning it back on. If the code reappears, replace the electronic control.

F2/E0, F2/E1, F2/E5, or F1

Description   The touch key pad has possibly shorted out.

Solution   Turn off the electricity to the appliance and wait for two minutes before turning it back on. If the code reappears, replace the touch key pad. Check the touch pad connector to the electronic control PCB.

F3/E0 or F3

Description   The oven sensor might be open in the top oven. The oven will be on and the heating LEDs will begin to flash continually.

Solution   Test the oven sensor valve. Turn off the electricity to the appliance and wait for two minutes before turning it back on. If the code reappears, replace the oven sensor.

F3/E1 or F3

Description   The oven sensor might be shorted out in the top oven. The oven will be on and the heating LEDs will begin to flash continually.

Solution   Test the oven sensor valve. Turn off the electricity to the appliance and wait for two minutes before turning it back on. If the code reappears, replace the oven sensor.

F3/E2 or F3

Description   The oven sensor will not maintain temperature in the bake cycle. The oven will be on and the heating LEDs will begin to flash continually.

Solution   Test the oven sensor valve. Turn off the electricity to the appliance and wait for two minutes before turning it back on. If the code reappears, replace the oven sensor.

F3/E3 or F3

Description   The oven sensor will not maintain temperature in the clean cycle. The oven will be on and the heating LEDs will begin to flash continually.

Solution   Test the oven sensor valve. Turn off the electricity to the appliance and wait for two minutes before turning it back on. If the code reappears, replace the oven sensor.

F3/E4

Description   The oven sensor might be defective in the bottom oven.

Solution   Test the oven sensor valve. Turn off the electricity to the appliance and wait for two minutes before turning it back on. If the code reappears, replace the oven sensor.

F3/E5

Description   The oven sensor might be shorted out in the bottom oven.

Solution   Test the oven sensor valve. Turn off the electricity to the appliance and wait for two minutes before turning it back on. If the code reappears, replace the oven sensor.

F3, E6, or E7	*Description* The warming drawer sensor might be defective or shorted out.
	Solution Check and test the warming drawer sensor and switches. Check the door alignment.
	Turn off the electricity to the appliance and wait for two minutes before turning it back on. If the code reappears, replace the control.
F5/E0, F5/E1, F5/E2, or F5	*Description* In the clean cycle, an error occurred in the door latch assembly. The door will be locked and cleaning LEDs will begin to flash.
	Solution Test the electronic control. Check the oven latch assembly, switch actuation, and the clearance between the latch and door. Check the wiring and connectors. Check the door alignment. Turn off the electricity to the appliance and wait for two minutes before turning it back on. If the code reappears, test and/or replace the control.
F5/E5	*Description* The electronic control detected the self-clean temperature was not reached in time.
	Solution Check oven temperatures, thermostat, latch assembly, and latch switch actuation. Turn off the electricity to the appliance and wait for two minutes before turning it back on. If the code reappears, test and/or replace the control.
F5/E7	*Description* The electronic control detected that the door latch will not unlock.
	Solution Test the electronic control. Check the oven latch assembly, switch actuation, and the clearance between the latch and door. Check the wiring and connectors. Check the door alignment. Turn off the electricity to the appliance and wait for two minutes before turning it back on. If the code reappears, test and/or replace the control.
F8/E0	*Description* A refrigeration problem was detected.
	Solution Check all wiring connectors and wiring. Check electronic control. Check the refrigeration unit.
PF	*Description* A power failure occurred.
	Solution Reset the controls and test. Check wiring connections.
OVEN LED ON AND FLASHING	*Description* The oven knob was not in the "off " position when the oven went into cycle.
	Solution Turn the knob to the "off" position and then reset the temperature.
DOOR LOCKED/ CLEANING AND HEATING LEDs FLASHING	*Description* There is a voltage failure.
	Solution Check voltages and all connectors and wiring. Turn off the electricity to the appliance and wait for two minutes before turning it back on. If the code reappears, replace the control.

Sears Appliances

The error/fault or function codes for Sears appliances will vary from model to model. Sears is a department store and does not manufacture appliances. See Table 9-2 for appliance manufacturers. Then check the technical data sheet, the use and care guide, the manufacturer's

Manufacturer	Model Number									
Absocold	126									
Admiral (Inglis)	C646	C880								
Amana	335	596								
Bosch	630									
Broan	233									
Brown Stoveworks	196									
Caloric	174	960								
Camco	C362	C363	C978							
Carrier	416									
D&M	587									
Defiance	789									
Electrolux	336	790								
Fedders	484									
Freidrich	840									
Frigidaire	119	790	970							
General Electric	362	363	464							
Gibson	253									
Goldstar (LG Electronics)	580	721	795							
Haier	183									
Hitachi	934									
Hoover	473									
Inglis	C106	C110	C646							
In-Sink_Erator	175									
Jenn-air	629									
Keeprite	867	C938								
Kelvinator	417	622	628	662	C933	C970				
Kitchenaid	666									
Lennox	292									
Litton	747									
Maytag	925									
Moffat (Camco)	C968									
Monarch/ Simpson	143									

TABLE 9-2   Manufacturers of Appliances for Sears. (*continued*)

Manufacturer	Model Number								
Panasonic	568	586							
Phico Italy	683								
Preway	155	850							
Ranney/Marvel, Imperial	757								
RCA	274								
Rheem/Rudd	879								
Roper	103	155	278	647	835	911	917		
Royal Chef	119								
Samsung	401								
Sanyo/Fisher	143	564	565	566					
Sharp	575								
Speed Queen	651								
Tappen	719	791							
Toshiba	562								
Trane	144								
Whirlpool	106	110	154	198	484	562	664	665	850
Woods	C675								
York	342								

Sears is not a manufacturer of appliances. It is a department store. Sears appliances are manufactured for Sears and are sold under the Sears/Kenmore label. To locate which manufacturer built the appliance for Sears, locate the **first three digits** of the model number and match them with the list above.

TABLE 9-2   Manufacturers of Appliances for Sears.

Web site, or call the manufacturer for the description of the error/fault code you are experiencing. Sometimes, by disconnecting the electricity to the product and waiting for two minutes and then turning on the electricity, this may reset the error/fault or function code, allowing you to reprogram and test the operation of the appliance.

For your safety and the safety of others, all testing and physical inspections of the product must be performed with the electricity turned off. On occasion, you will have to test a component with the electricity on. Extreme care must be taken to prevent electrical mishaps.

Whirlpool Room Air Conditioner Error/Fault or Function Codes

The error/fault or function codes listed here will vary from model to model. The codes on the model you are servicing might be different from what is listed. If this is the case, check the technical data sheet, the use and care guide, the manufacturer's Web site, or call the manufacturer for the description of the error/fault code you are experiencing. Sometimes, by disconnecting the electricity to the product and waiting for two minutes and then turning on the electricity, this may reset the error/fault or function code, allowing you to reprogram and test the operation of the air conditioner.

For your safety and the safety of others, all testing and physical inspections of the product must be performed with the electricity turned off. On occasion, you will have to test a component with the electricity on. Extreme care must be taken to prevent electrical mishaps.

FF   *Description*   The electronic control has detected an open thermistor or a temperature above 140 degrees Fahrenheit.

    Solution   Test the thermistor for an open circuit. Also check the wiring connections and wiring and the location of the thermistor-sensing bulb. Replace if necessary.

00   *Description*   The electronic control has detected a shorted thermistor or a temperature below 32 degrees Fahrenheit.

    Solution   Test the thermistor for a short. Also check the wiring connections and wiring and the location of the thermistor-sensing bulb. Replace if necessary.

Whirlpool and KitchenAid Microwave Error/Fault or Function Codes

The error/fault or function codes listed here will vary from model to model. The codes on the model you are servicing might be different from what is listed. If this is the case, check the technical data sheet, the use and care guide, the manufacturer's Web site, or call the manufacturer for the description of the error/fault code you are experiencing. Sometimes, by disconnecting the electricity to the product and waiting for two minutes and then turning on the electricity, this may reset the error/fault or function code, allowing you to reprogram and test the operation of the appliance.

Listed are the code designations for a four-digit display (**/** or ****) and for a three-digit display (***).

For your safety and the safety of others, all testing and physical inspections of the product must be performed with the electricity turned off. On occasion, you will have to test a component with the electricity on. Extreme care must be taken to prevent electrical mishaps.

F1/E1   *Description*   The relay control board is faulty.

    Solution   Replace the relay control board and test the microwave.

F2   *Description*   The electronic control detected a touch pad error.

    Solution   Test the touch pad assembly and the electronic control board.

F2/H1 or F2/H2   *Description*   A problem occurred in the hood light switch and keys.

    Solution   Test the hood light switch and replace if necessary.

F2/Q1   *Description*   A problem occurred in the touch control key pad.

    Solution   Check the user interface board and the door control assembly for a malfunction. Turn off the electricity to the microwave and wait for two minutes before turning it back on. If the code reappears, replace the defective part.

F3/H1 or F3H   *Description*   There is a problem with the humidity sensor.

Solution   Test the value of the humidity sensor. Check all connections from the humidity sensor to the relay control board. Test the relay control board. Turn off the electricity to the microwave and wait for two minutes before turning it back on. If the code reappears, replace the defective part.

F3T   *Description*   The electronic control detected a temperature sensor failure.

Solution   Test the microwave oven cavity temperature. The sensor will not operate below 41 degrees Fahrenheit or above 140 degrees Fahrenheit. If tests are okay, replace the electronic control board.

F3/T1   *Description*   A user interface board thermistor error occurred.

Solution   Check all wiring connections and wiring from the user interface board, relay control board, and relay board. Test the user interface board and thermistor. Turn off the electricity to the microwave and wait for two minutes before turning it back on. If the code reappears, replace the defective part.

F3/T2   *Description*   The FC thermistor detected a problem.

Solution   Check the wiring connections and wiring from the FC thermistor to the relay control board. Test the value of the FC thermistor. Test the relay control board. Turn off the electricity to the microwave and wait for two minutes before turning it back on. If the code reappears, replace the defective part.

F3/T4   *Description*   The magnetron thermistor detected a problem.

Solution   Check the wiring connections and wiring from the magnetron to the relay control board. Turn off the electricity to the microwave and wait for two minutes before turning it back on. If the code reappears, replace the magnetron thermistor.

F6   *Description*   The electronic control detected a microwave relay problem.

Solution   Check all wiring connections and wiring to the inverter and the relay control board. Test the relay board. Turn off the electricity to the microwave and wait for two minutes before turning it back on. If the code reappears, replace the relay control board.

F7   *Description*   The magnetron stopped operating due to voltage interruption.

Solution   Check the cooling fans, magnetron, inverter, relay control board, and interlock switches. Check all wiring connections and wiring between these components. Replace the defective component and retest the microwave.

F9   *Description*   The electronic control received a communication error from the relay control board.

Solution   Check all wiring connections and wiring between the relay control board and the relay board. Turn off the electricity to the microwave and wait for two minutes before turning it back on. If the code reappears, replace the relay board and the relay control board.

F9/Q   *Description*   A communication error was detected at the touch panel.

Solution   Check all wiring connections and wiring from the user interface board to the touch panel. Turn off the electricity to the microwave and wait for two minutes before turning it back on. If the code reappears, replace the door control assembly.

Whirlpool Front-Load Washer Error/Fault or Function Codes

The error/fault or function codes listed here will vary from model to model. The codes on the model you are servicing might be different from what is listed. If this is the case, check the technical data sheet, the use and care guide, the manufacturer's Web site, or call the manufacturer for the description of the error/fault code you are experiencing. Sometimes, by disconnecting the electricity to the product and waiting for two minutes and then turning on the electricity, this may reset the error/fault or function code, allowing you to reprogram and test the operation of the appliance.

For your safety and the safety of others, all testing and physical inspections of the product must be performed with the electricity turned off. On occasion, you will have to test a component with the electricity on. Extreme care must be taken to prevent electrical mishaps.

F/H   *Description*   The electronic control detected no water entering the washer, or the pressure switch did not detect the water.

   Solution   Verify if the washer has water in the tub. If it does, check the following components for proper operation: drain pump, pressure switch and hose connection, flow meter, and electronic control board. In addition, check the washer installation, drain hose installation, and all wiring connections and wiring. If the tub has no water in it, check the water inlet valve and screen, make sure that the water is turned on, and check the inlet hoses for kinks. To clear the error code, press the pause/cancel button two times.

F/02   *Description*   The drain time exceeded the eight-minute time limit.

   Solution   Check the drain hose and the drain pump filter for obstructions. Check the electrical connections and wiring from the drain pump to the electronic control board. If no problems were found, then replace the drain pump.

F/05   *Description*   The electronic control has detected a water temperature fault.

   Solution   Check the heating element, relay, and the water temperature sensor. Check the wiring connections and wiring.

F/06   *Description*   The drive motor tachometer, which is mounted on the motor, has detected a problem with the speed of the motor.

   Solution   Check all of the wiring connections and wiring between the motor, motor control board, and the electronic control board. Test the drive motor for proper operation. Check the electronic control board and the motor control board.

F/07   *Description*   The electronic control board has detected a possible short in the circuitry with the motor control board.

   Solution   Check the wiring connections and wiring between the motor control board, drive motor, and the electronic control board. Check the following components for proper operation: drive motor, drive motor control board, and electronic control board.

F/09   *Description*   A water overflow error has been detected.

Solution   Check the wiring connections and wiring between the electronic control board, drain pump, and the pressure switch. Check the drain hose installation for blockages or a kinked hose. Clean the drain pump filter and check the drain pump for proper operation. Check the water inlet valve for proper operation. Check the pressure switch for proper operation. Replace any defective part found.

F/10   *Description*   The electronic control board received a signal from the motor control board that the thermal heat sink has tripped.

Solution   Check washer installation for proper ventilation. Check the wiring connections and wiring between the electronic control board and the motor control board. Check for worn or failed components in the drive mechanism. Check the electronic control board and the drive motor for proper operation.

F/11   *Description*   A communication error has been detected between the electronic control board and the motor control board.

Solution   Check all wiring connections and wiring. Check for worn or failed components in the drive mechanism. Check the electronic control board, motor control board, and the motor for proper operation.

F/13   *Description*   The electronic control board has detected a problem in the dispenser circuitry.

Solution   Check the dispenser motor and the mechanical linkage from the motor to the dispenser. Check the electronic control board for proper operation.

F/14   *Description*   The electronic control board has detected an EEPROM error from the microprocessor chip on the board.

Solution   Disconnect the electricity to the washer for a few minutes and try to reprogram the electronic control. There might have been a power surge to the washer. Check the electronic control board for proper operation if the error code returns.

F/15   *Description*   The electronic control board detected a problem with the motor control board.

Solution   Check the wiring connections and wiring from the electronic control board, motor control board, and the motor. Check the following components: drive belt, motor control board, electronic control board, and motor for proper operation.

Sud   *Description*   A suds lock error was detected.

Solution   The motor control board has detected a suds lock condition by analyzing the current draw on the motor. Check the pump operation. In addition, check for a heavy wash load, too much detergent, or over-sudsing condition.

F/dt   *Description*   A door lock error occurred after six attempts to lock the door.

Solution   Check the door lock switch and assembly. Check the wiring connections and wiring. Check the electronic control board. Repair or replace the defective part.

F/du   *Description*   The electronic control board has detected a door unlock error after six attempts to open the door lock.

Solution   Check the door switch/lock assembly for proper operation and for foreign objects. Check the wiring connections and wiring from the door lock assembly to the electronic control board. Check the electronic control board.

Whirlpool Front-Load Dryer Error/Fault or Function Codes

The error/fault or function codes listed here will vary from model to model. The codes
on the model you are servicing might be different from what is listed. If this is the case, check
the technical data sheet, the use and care guide, the manufacturer's Web site, or call the
manufacturer for the description of the error/fault code you are experiencing. Sometimes, by
disconnecting the electricity to the product and waiting for two minutes and then turning
on the electricity, this may reset the error/fault or function code, allowing you to reprogram
and test the operation of the appliance.

For your safety and the safety of others, all testing and physical inspections of the product
must be performed with the electricity turned off. On occasion, you will have to test a
component with the electricity on. Extreme care must be taken to prevent electrical mishaps.

PF   *Description*   The electronic control board has detected a power failure.

      Solution   Press start and the dryer will continue with the cycle, or press cancel to
reprogram the dryer.

E1   *Description*   The electronic control has detected a faulty thermistor.

      Solution   Test the thermistor and the electronic control board, and check the wiring
connections and wiring.

E2   *Description*   The electronic control has detected a short in the thermistor.

      Solution   Test the thermistor and the electronic control board, and check the wiring
connections and wiring.

E3   *Description*   The electronic control has detected a communication or software error,
a defective electronic control board, or a defective component in the control panel.

      Solution   Check the electronic control board and check all of the components in the
control panel. Check all wiring connections and wiring.

Gas Appliance Parts

This chapter explains how to identify, locate, and understand the operation of gas appliance parts. In addition to electrical parts, the gas components play an important role in the proper operation and safety of gas appliances. Figures 10-1, 10-2, and 10-3 will help you to identify and locate the parts in a gas range, gas dryer, and gas water heater, respectively. Gas parts are divided into the following groups:

- Control parts: Manual and automatic controls used in gas appliances to turn the gas supply on or off or to regulate the flow of gas in the appliance.
- Safety parts: Gas controls that prevent a hazardous condition.
- Combination parts: Gas controls that act as both control parts and safety parts.
- Sensing parts: Sensing devices that are used to activate or deactivate a control.
- Ignition parts: Gas appliances require an ignition source to ignite the burners.

Gas appliance parts are factory-set upon installation and manufacture of the product. These settings should not be tampered with, unless it is determined that the setting was improperly set. It is recommended that you adjust the factory setting according to the manufacturer's recommendation.

Controls

Manual and automatic controls are the two types of controls used in major appliances. Manual controls are operated by the consumer and are adjusted by eyesight. For example, a consumer will manually turn on a gas burner and adjust the flame height with the burner knob. Automatic controls require three elements to control the gas flow:

- A device to sense the operating conditions
- A device to regulate the flow of gas
- A means to actuate the control

Over the years, controls have evolved from simple controls to complex electronic systems using microprocessors that provide integrated control over all of the components in an appliance.

Figure 10-1 Typical gas range parts identification.

Temperature selector switch

Start switch

Timer

Control console

Electric heater for electric models only

Thermostat

Drive belt

Door switch

Front drum seal

Lint screen

Drum

Exhaust duct

Idler

Thermostats

Drum slide

Blower wheel

Gas burner assembly for gas models only

Blower housing

Motor

Motor pulley

Leveling foot

FIGURE 10-2 Typical gas dryer parts identification.

Pressure Regulator Controls

The pressure regulator (Figures 10-4, 10-5, and 10-6) is either a mechanical or electric control that regulates and maintains gas flow. This device reduces the incoming gas pressure to a level that is desired for a particular application. It is recommended that a main shutoff valve be installed between the pressure regulator valve and the main gas supply entering the appliance (Figure 10-7). With the shutoff valve located near the appliance, the technician will be able to shutoff the gas supply to the product before beginning repairs.

PART IV

FIGURE 10-3
Typical gas water
heater parts
identification.

1	Vent pipe	5	Outlet	9	Ground joint union	13	Outer door	17	Name tag
2	Drafthood	6	Insulation	10	Sediment trap	14	Drain valve	18	Flue baffle
3	Anode	7	Gas supply	11	Air intake screen	15	Thermostat	19	TPR valve
4	Hot water outlet	8	Gas shutoff valve	12	Inner door	16	Gas igniter	20	Drain pan

FIGURE 10-4 A gas pressure
regulator valve.

FIGURE 10-5
A water heater
combination control.
This control includes a
gas pressure regulator
valve, thermostat, and
ignitor.

Figure 10-6   Two types of dryer gas valves.

Figure 10-7   The main gas supply line to an appliance should include a manual gas shutoff valve. If you intend to use a flexible gas line, you must check local building codes first.

Pressure regulator

Fitting

Flexible gas line

Fitting

Manual gas shutoff valve

Main gas supply

PART IV

FIGURE 10-8
(a) An illustration of a gas pressure regulator valve in the closed position. (b) An illustration of a gas pressure regulator valve in the open position.

As the gas enters the pressure regulator valve (see Figure 10-4 and Figure 10-8a), the gas pressure pushes against the spring-loaded diaphragm, forcing the valve to close and shutting off the supply of gas to the appliance. When the consumer turns on the appliance, or when the appliance itself is calling for more gas, the pressure within the pressure regulator valve (Figure 10-8b) decreases, allowing the spring tension to push down on the diaphragm and forcing the valve to open, allowing more gas to the burner(s). The design of the tapered plug and diaphragm allows for metering and maintaining a constant pressure of gas to the burner(s). Another feature incorporated into the pressure regulator is an air vent in the upper chamber. The main purpose of this air vent is to allow air to enter and leave the upper chamber during the operation of the pressure regulator. As a secondary feature, the vent will allow gas to escape at a predetermined amount if the diaphragm ever ruptures.

Dryer gas valves (see Figure 10-6) contain a pressure regulator and two solenoid-operated gas valves. During normal operation, both solenoid valves are energized simultaneously to allow gas to flow to the burner.

When diagnosing a pressure regulator failure, common causes to consider include:

- The valve portion within the regulator may have worn out or it may be broken.
- Accumulation of dirt and debris around the valve seat can cause erratic operation or a complete shutdown of the regulator valve.
- The air vent might be plugged or restricted.
- The diaphragm has ruptured and gas is venting into the atmosphere.
- With LP gas, corrosion can occur within the regulator valve if water enters the gas supply.
- An electrical component may have failed.

Pressure regulator valves and dryer gas valves are not serviceable and should be replaced with a duplicate of the original if they fail.

Water Heater Thermostat/Regulator Combination Control

Water heaters use a combination control that incorporates a thermostat and a gas pressure regulator in one control (see Figure 10-9). In addition, the control has a gas cutoff device incorporated into the control in the event that the thermostat fails to shutoff the gas supply. The combination valve is activated by a thermocouple that opens the gas inlet to the pressure regulator. The temperature probe will actuate a lever from within the pressure regulator to open or close the valve to the main burner. On top of the control is a knob that you will depress or turn, depending on the type of control, to begin the process to light the pilot light. If any part of this control fails, it is not serviceable and should be replaced with a duplicate of the original.

FIGURE 10-9
A water heater combination control valve.

Figure 10-10
(a) The single gas safety valve in a standing pilot system in the open position. (b) The single gas safety valve in a standing pilot system in the closed position.

Safety Valve

Ovens with a standing pilot-light ignition system have a safety valve (Figure 10-10) that controls the gas flow. The safety valve's main function is to allow the gas coming from the thermostat to enter the oven burner. In Figure 10-10a, as the pilot flame heats up the safety valve sensor, the mercury-filled sensor expands and forces the switch to open the safety valve, allowing the gas to enter the oven burner. When the temperature in the oven is satisfied, the sensor begins to cool down, closing the safety valve (Figure 10-10b), stopping the gas flow to the oven burner. The safety valve and the oven thermostat must work together to operate the oven burner correctly. Further discussion on the thermostat and safety valve operation will be covered in a later chapter.

Ovens that have a glo-bar ignition system use a bimetal-operated safety valve (Figure 10-11). This type of valve has one gas inlet and one gas outlet. It is used for the bake burner and the broil burner combination. At the outlet end of the safety valve, there is an electrically operated bimetal strip with a rubber seat that covers the outlet, preventing the flow of gas at room temperature. When current is applied to the bimetal strip, it will warp, allowing the safety valve to open. Gas ranges with the self-cleaning feature in a single oven cavity have a dual safety valve (Figure 10-12). This valve will allow the gas to flow to the bake and broil burners separately when needed. It will not operate both burners at the same time. The operation of the dual safety valve is similar to the single safety valve.

Gas outlet
(to oven burner)

120 volt
supply terminal

Safety valve

Gas inlet
(from regulator)

Current warps bimetal
& valve opens

(a)

Gas outlet
(to oven burner)

120 volt
supply terminal

Safety valve

Gas inlet
(from regulator)

Bimetal & heater coil
no current flow valve is closed

(b)

Pressure
regulator

Safety valve

Manual gas
shutoff valve

(c)

FIGURE 10-11   (a) The single gas safety valve in a glo-bar ignition system in the open position. (b) The single gas safety valve in a glo-bar ignition system in the closed position. (c) A bimetal single safety gas valve and gas regulator connected together in an automatic ignition system.

Figure 10-12   (a) A dual safety gas valve. (b) An internal view of a dual safety gas valve.

When diagnosing a safety valve failure, common causes to consider include:

- A broken capillary tube
- Loss of voltage to the safety valve
- Bimetal and heater coil failure from within the safety valve
- Debris buildup around orifice
- Mechanical failure

Safety valves are not serviceable and should be replaced with a duplicate of the original if they fail.

Dryer Gas Valve

The dryer gas valve in Figure 10-13 is a combination control, consisting of a pressure regulator and dual shutoff valves, housed in one body to regulate the gas flow when the thermostats call for more heat. The solenoid coils (Figure 10-14) will activate by means of electrical power and open the gas valves by electromagnetism, allowing gas to flow to the burner. When the temperature is satisfied, the electrical power is turned off and the solenoid coils deactivate, allowing the internal spring pressure to close the valve. Dryer gas valves are not serviceable; only the solenoid coils are serviceable. The gas valve body should be replaced with a duplicate of the original if it fails.

Sensing Devices

A sensing device can be a temperature-responsive or pressure-responsive device that transmits a signal or motion to activate or deactivate a control device. In electrical control circuits, resistive coils, resistance temperature detectors (RTD), and thermistors are used in a circuit to activate or deactivate the controls. The electrical resistance of these devices varies by temperature change to control current flow.

FIGURE 10-13
Two different designs
of a dryer gas valve
combination.

Gas valve
solenoid coils

Gas valve
regulator

(a)

Gas valve
solenoid coils

Gas valve
regulator

(b)

FIGURE 10-14
(a) A de-energized
solenoid coil in a
dryer gas valve
assembly indicating
no gas flow. (b) An
energized solenoid
coil in a dryer gas
valve assembly
indicating the flow
of gas.

Hood ass'y

Spring

Solenoid

Valve disc

Gas
inlet

Gas
flow

(a)

(b)

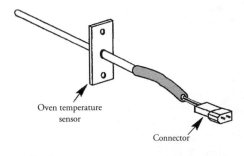

Oven temperature
sensor

Connector

FIGURE 10-15   A resistance temperature
detector (RTD).

Resistance Temperature Detector (RTD)

The RTD sensor operates on the principle that as the temperature increases, the resistance in the metal increases. With a constant voltage, the current through the metal will drop off as the temperature increases. Ovens with electronic control circuits use an oven temperature sensor (Figure 10-15) to activate or deactivate the bake, broil, and self-clean functions. This sensor is a resistance temperature detector (RTD) composed of a stainless steel tube, coated with platinum at one end and two wires connected to a connector that plugs into the electronic circuitry. The location of the sensor is in the upper corners of the oven cavity. This device is neither adjustable nor repairable, and should be replaced with a duplicate of the original if it fails.

Thermocouple

A thermocouple (Figure 10-16) is a measuring device consisting of two dissimilar metals, which produces a low DC voltage when heated by a gas pilot flame that is measured in millivolts. This thermoelectric device is commonly used in gas appliances to power automatic-pilot safety devices. The average output voltage for a single thermocouple is between 20 to 30 millivolts. If the thermocouple voltage drops below 5 millivolts, which can vary in design from manufacturer to manufacturer, the pilot safety device will shut off the gas supply to the burner. This device is neither adjustable nor repairable, and should be replaced with a duplicate of the original if it fails.

Flame Sensors

Flame sensors (Figure 10-17) are used in gas appliances to detect the presence of a pilot flame or the main burner flame. For safety reasons, in an automatic ignition system, it is required that a flame sensor be installed to detect the presence of a flame in the gas pilot or

Flexible tube with
insulated wire inside tube

Thermocouple

FIGURE 10-16   A thermocouple.

the main gas burner. Before the gas valve can open in an automatic pilot system, the flame sensor must detect the presence of the pilot flame. In an electronic ignition system (pilotless ignition), the flame sensor must be mounted over a window cut out in the burner tube to ensure that the burner flame is present or it will not allow the gas valve to open. The switch will open within 15 to 90 seconds, when the switch senses the burner flame, and the ignitor temperature, which will be at 1800 to 2500 degrees Fahrenheit. This device is neither adjustable nor repairable, and should be replaced with a duplicate of the original if it fails.

FIGURE 10-17
A flame sensor is a thermostatically controlled single pole, single throw, normally closed switch.

Flame sensor mounting screw

Flame sensor

Burner tube

Burner assembly mounting bracket

Ignition

There are two ways to ignite a gas burner: using matches or using an automatic ignition source. Many appliances that are manufactured today have some type of automatic ignition source. This automatic system can be continuous, intermittent, interrupted, or a combination of these things.

Glow-Bar Ignitor

To achieve direct ignition, a silicon carbide glow-bar device (Figure 10-18) is positioned in the path of the burner flame. The reason for this positioning is to achieve the best performance for

FIGURE 10-18
Two types of a silicon carbide glow-bar device (ignitor) used in gas appliances.

ignition and flame sensing. Line voltage is applied to the ignitor. When it reaches a temperature between 1800 and 2500 degrees Fahrenheit, in about 15 to 100 seconds (depending on design), a signal is sent to open the gas valve, allowing gas to flow to the burner, and gas ignition occurs. When using the glow-bar as a sensor, if the

FIGURE **10-19**
Type of ignitor used to
light the gas burner(s)
on a gas range.

temperature of the glow-bar begins to drop
below the ignition temperature, the gas
valve will close, shutting off the gas supply
to the burner.

There are times when the glow-bar
ignitor will appear to glow properly and
be reddish in color, but the gas burner will
not light. In addition, there are times when
the ignitor will not light at all. If this
happens, you will need to perform a visual
inspection and test the ignitor with a clamp-on multimeter. This ignitor is neither adjustable
nor repairable, and should be replaced with a duplicate of the original if it fails.

Spark Electrode Ignitor

The spark electrode ignitor replaces the standing pilot flame system with electrodes and a
spark module. The ignitor (Figure 10-19) consists of a metal rod embedded into a ceramic
insulating body that is wired to a spark module that is located in the gas appliance. A spark
module will send a number of pulses to the spark electrode ignitor, which will begin to arc
between the metal rod and the grounding strap bracket. This device is neither adjustable
nor repairable, and should be replaced with a duplicate of the original if it fails.

Spark Module

The spark module (Figure 10-20) is an electronic device that delivers a high-voltage pulse to the
spark electrode ignitor. These pulses are delivered by a repeatable timing sequence from within
the module every few seconds (depending on design) and operate at very low amperage.
When a flame is detected, the spark module will stop transmitting pulses to the ignitor. Some
gas appliance models are designed with automatic flame recovery and/or automatic lockout of
the gas valve. This device is neither adjustable nor repairable, and should be replaced with a
duplicate of the original if it fails.

FIGURE **10-20** A spark module.

V
PART

Appliance Service, Installation, and Preventive Maintenance Procedures

Automatic Dishwashers

When a dishwasher is properly used, it will provide satisfactory results. There are times, however, when the dishwasher is blamed for poor performance. Perhaps the customer does not know how to load the dishwasher properly or the wrong amount of detergent was used—in some cases, the dishwasher might not run at all. Whatever the case might be, it is up to the technician to either repair the dishwasher or instruct the owner in its proper usage.

This chapter provides the technician with the basic skills needed to diagnose and repair automatic dishwashers. The actual construction and features might vary, depending on what brand and model you are servicing.

Principles of Operation

After placing the dishes properly in the dishwasher, the detergent is placed in the dispenser, and the rinse conditioner is checked for the proper level. The door is closed, and the type of wash cycle is selected. The door latch holds the door closed and activates the door latch switch. This will complete the electrical circuit for the dishwasher to operate. If the door is opened during the cycle, this will cause all operations to cease.

The timer will energize the water inlet valve, and water will begin to enter the tub. The dishwasher does not fill with water like a washing machine. It is designed so that the tub does not have more than two gallons of water in it at any one time. Should the timer switch contacts fail to open during the fill cycle, a float switch assembly, located inside the tub, will open the electrical circuit to the water inlet valve at a preset level.

The fill safety switch is part of the float assembly. Should the timer fail to open its switch contacts, water will keep entering the tub until the float, located inside the tub, rises and engages the float switch to shutoff the water. *Note:* The float switch will not protect against a mechanical failure of the water inlet valve.

During the wash and rinse portion of a cycle, the heater element heats the water (on some models) to at least 140 degrees Fahrenheit. This feature is built into the dishwasher and is designed to save the customer money on the operating cost. Also, the customer does not have to raise the water temperature of the water heater to 140 degrees Fahrenheit. He or she only needs to set the water heater temperature at 120 degrees Fahrenheit. This will prevent any member of the household from getting burned.

FIGURE 11-1
An outside view of
inside a dishwasher
tub with the water
circulating.

Front view of
dishwasher with
water circulating
in wash cycle

Side view of
dishwasher with
water circulating
in wash cycle

The water is repeatedly pumped through the lower and upper spray arms and on to the
dishes (Figure 11-1). As the water runs off the dishes and back to the pump, it flows through
a filtering system. On some models, the filter is designed to separate most food particles
from the water so that they aren't sprayed back on to the dishes (Figure 11-2).

At the end of a wash or rinse cycle, the water is pumped out of the dishwasher, flushing
the filter of any small food particles. The larger pieces of food are trapped on the pump
guard, which must be cleaned out before the next use. On some models, the pump screen

Screw (4)

Accumulator cover

Screw

Foreign object plate

Accumulator bottom

Filter assembly

Inlet protector

Motor sheld

Screw

Chopper assembly

Sump check valve

Motor sub assembly

Sump

Separation screen

Pressure switch

Shield-capacitor

Drain pump

Capacitor

Nut

Drain loop
(Check value
molded inside)

Sump seal
joint

Face seal-head carbon

Seal

Motor

Impeller-weld assembly

Volute back

Face seal

Figure 11-2 An exploded view of a dishwasher pump, motor, and filter assembly.

removes food particles from the water, stores them, and then grinds them up as they are washed down the drain. During grinding, some sounds will be heard.

At the end of the cycle, the heater element (Figure 11-3) comes on (if selected) and helps dry the dishes. Certain models have a fan that circulates the air to speed up the drying cycle, thus making sure that the dishes dry evenly. Some models have a cool-dry cycle. This allows the dishes to be dried without the heater element operating. Combined with the heated air within the tub (from the wash cycle) and the dishwasher door opened a little bit, it will cause the water that remains on the dishes to condense and roll off them. Figure 11-3 illustrates component locations within the automatic dishwasher (under-the-counter and portable).

FIGURE 11-3a
Dishwasher
component locations.
(*continued*)

Upper
dishrack

Lower spray
arm and
motor/pump
assembly

Lower
dishrack

Overfill
protector

Silverware
basket

Heater

Vent

Detergent and rinse
conditioner dispensers

Safety First

Any person who cannot use basic tools or follow written instructions should *not* attempt to install, maintain, or repair an automatic dishwasher. Any improper installation, preventive maintenance, or repairs could create a risk of personal injury or property damage.

If you do not fully understand the installation, the preventive maintenance, or the repair procedures in this chapter, or if you doubt your ability to successfully complete the task on the automatic dishwasher, call your service manager.

Before continuing, take a moment to refresh your memory on the safety procedures in Chapter 2.

Dishwashers in General

Much of the troubleshooting information in this chapter covers the various types of dishwashers in general, rather than specific models, in order to present a broad overview of service techniques. The pictures and illustrations that are used in this chapter are for demonstration

FIGURE 11-3b
Dishwasher component locations.

Vent

Door latch

Timer and controls

Door springs

Water valve

Wheels

Pump and motor

purposes only—they clarify the description of how to service an appliance, and they in no way reflect a particular brand's reliability.

Location and Installation of a Dishwasher

Locate the dishwasher where there is easy access to existing drain, water, and electrical lines (Figure 11-4). Be sure to observe all local codes and ordinances for electrical and plumbing connections. It is strongly recommended that all electrical and plumbing work be done by qualified personnel. The best location for the dishwasher is on either side of the sink.

For proper operation and appearance of the dishwasher, the cabinet opening should be square and have the dimensions as shown in Figure 11-5. If the dishwasher is to be installed in a corner, there must be sufficient clearance to open the door (Figure 11-6).

FIGURE 11-4
FIGURE 11-4
Typical dishwasher
installation for a left
or right sink
application.

FIGURE 11-5
Undercounter
dishwasher cut-out
dimensions.

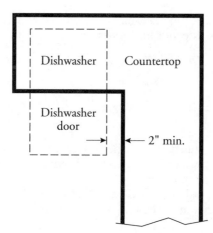

FIGURE 11-6
Corner dishwasher
installation
dimensions.

Take the time to read over the installation instructions and the use and care manual that comes with every new dishwasher. These booklets will provide you with very important information, such as:

- Safety
- Tools needed for the installation
- How to remove the panels
- How to change the color of the panels
- Locating drain, water, and electrical supplies
- Positioning, aligning, and leveling the dishwasher
- Drain hose connection
- Water line connection
- Connecting the dishwasher to the electrical supply
- Securing the dishwasher in the opening you have selected
- Proper operation of the dishwasher

Most important, read the warranty information that is supplied with the dishwasher.

Functions and Cycles

Dishwashers are similar to automatic clothes washers. They apply three kinds of energy on the things to be washed. These forces are:

- **Mechanical energy:** Water that is sprayed on to the dishes to remove the food particles.
- **Heat energy:** Using hot water to liquefy the fats and greases on dirty dishes.
- **Chemical energy:** Detergent dissolves the fat and grease off the dishes.

Dishwashers perform four basic functions that are modified and put together in different ways to create the various cycles. The four functions are:

- Fill
- Wash/rinse
- Drain
- Dry

As with clothes washers, the only difference between the wash and rinse cycles is the presence of detergent in the wash water. The mechanical activities that make up a wash and a rinse cycle are basically the same function.

Unlike clothes washers, most dishwashers fill and begin to wash (or rinse) at the same time. The functions are put together in various ways to make up different cycles. For example:

- **Normal wash cycle:** A single or double wash with several rinses and a dry.
- **Heavy wash cycle:** Adds a wash to the normal wash cycle.
- **Pots and pans:** Similar to the heavy wash cycle, but this cycle heats the water in the wash cycles and (on some models) also heats the rinse cycles. On certain models, the timer will not advance until the water temperature is 140 degrees Fahrenheit. This will extend the total time of operation. Depending on the model, the cycle time can increase from 15 minutes to 2 hours. (Check the use and care manual for exact details.)
- **Light wash cycle:** This cycle is like the normal wash cycle, but minus a wash cycle.
- **Rinse and hold:** Two rinses for holding dishes to wash later.

Water Temperature

The temperature of the incoming water is critical to the operation of a dishwasher. Most dishwashers have heaters, and some have delay periods that extend the time during which water is heated to a specified point, but this does not fully compensate for low temperature of the water supply. You can check the temperature of hot water at the sink nearest to the dishwasher with a thermometer. Open the hot water faucet. Let the water run until it is as hot as possible, and then insert the thermometer into the stream of water. On some models, if the thermometer reading is below 140 degrees Fahrenheit, you will have to raise the water heater thermostat setting. On other models, the dishwasher was designed to operate with water temperatures as low as 120 degrees Fahrenheit. These models have longer detergent wash periods that allow 120-degree Fahrenheit supply water to be heated up to a temperature that gives good washability results. The dishwasher delay periods occur in only one, two, or three of the water fills, and do little for the remaining rinses. Except during delay periods, the water is not in the dishwasher long enough to be heated adequately.

Water Temperature above 150 Degrees Fahrenheit

It is not recommended to have the water temperature above 150 degrees Fahrenheit in a domestic dishwasher; above this temperature, certain components in the dishwasher might be adversely affected.

Water Temperature of 150 Degrees Fahrenheit

A water temperature of 150 degrees Fahrenheit is the ideal temperature for a mechanical dishwasher. Detergent action and the dissolving of grease are at the maximum at this temperature. Drying of most materials in the dishwasher will be satisfactory.

Water Temperature of 140 Degrees Fahrenheit

Water temperature of 140 degrees Fahrenheit is the minimum temperature recommended by most dishwasher and detergent manufacturers. At this temperature, detergent is still quite active, and most fats are dissolved so that they can be emulsified in the water by the detergent and washed down the drain. Drying will be fair to poor, as water temperature in the last rinse is lowered toward 140 degrees Fahrenheit in some models. Some improvement in drying is possible if a liquid wetting agent is added to the dispenser.

Water Temperature between 130 and 140 Degrees F

Water temperature between 130 and 140 degrees Fahrenheit is outside the range for most dishwasher operations, and users will have to exert special care if they are to obtain satisfactory results. The cleaning action of detergents and the dissolving of fats are gradually reduced as temperatures drop below 140 degrees Fahrenheit, so the dishes will probably have to be well rinsed before putting them into the dishwasher. Satisfactory drying becomes less likely as water temperature becomes lower. Some dishwashers have an optional feature which will increase the temperature of the water in the tub at different points during the cycle.

Water Temperature between 120 and 130 Degrees Fahrenheit

Water temperatures between 120 and 130 degrees Fahrenheit will aggravate all of the conditions and problems mentioned for the 130- to 140-degree range. Very few fats will dissolve, so the greasy buildup in the lower areas of the tub will be accelerated. Sudsing and foaming are more likely to increase. Detergent action is further reduced, so pre-rinsing of the dishes becomes even more important. The water heat feature will most likely have to be used.

Water Temperature below 120 Degrees Fahrenheit

Because of poor washing, grease buildup, poor drying, and foaming, it is unlikely that the dishwasher will perform to the user's satisfaction if the water supply is less than 120 degrees Fahrenheit.

The Detergent

The kind and amount of dishwasher detergent that is used is an important part of getting the dishes clean. Different brands of dishwasher detergent contain different amounts of phosphorous, which works to soften water and prevent water spots. If the water is hard, you will have to instruct the customer to use a detergent with a higher phosphorous content—above 12 percent. If the water is soft, the customer can use a low-phosphorous dishwasher detergent. Some areas restrict the phosphate content to 8 percent or less. This means that the customer will have to increase the amount of detergent used in those areas where the water is hard. This is done by adding 1 teaspoon of dishwasher detergent manually in the main wash cycle for each grain of water hardness above 12 grains (water hardness is measured in grains):

- 0 to 3 grains for soft water
- 4 to 9 grains for medium-hard water
- 10 to 15 grains for hard water
- Over 15 grains for very hard water

If the hardness of the water supply is unknown, contact the local water department.

Always instruct the user to use automatic dishwasher detergent only. The use of soap, hand dishwashing detergent, or laundry detergent will produce excessive suds and will cause flooding and damage to the dishwasher.

Step-by-Step Troubleshooting by Symptom Diagnosis

In the course of servicing an appliance, you might overlook the simple things that could be causing the problem. Step-by-step troubleshooting by symptom diagnosis is based upon diagnosing the malfunctions, with possible causes arranged into categories relating to the operation of the dishwasher. This section is intended only to serve as a checklist to aid in diagnosing a problem. Look at the symptom that best describes the problem you are experiencing with the dishwasher, and then proceed to correct the problem.

Before testing any electrical component for continuity, disconnect the electrical supply to the appliance.

No Water to Dishwasher

1. Is the water turned on?
2. Is there voltage to the water inlet valve solenoid?
3. Is the water inlet valve solenoid defective? Disconnect the electrical supply, and check the solenoid coil with an ohmmeter.
4. Is the water valve plunger stuck? Disassemble the water valve and check.
5. Is the water valve inlet screen blocked? Turn off the water supply and remove the water supply line to inspect the screen.
6. Is the water fill line that goes to the water valve kinked? Check visually.
7. Is the water siphoning out of the dishwasher while it is filling? Check the drain hose installation.
8. On electronic models, run the diagnostic test mode.

Low Water Charge

1. Is there adequate water pressure to the water inlet valve? Pressure should be between 15 and 120 pounds per square inch.
2. Is the water volume adequate? Take a quart container and fill it at the tap. The water must fill the container in seven seconds or less.

3. Is the water valve inlet screen clean? Turn off the water supply and remove the water supply line to inspect the screen.

4. On portable models, check the aerating snap adapter on the faucet.

5. Is the timer defective? Disconnect the electric supply and check the switch contacts for continuity.

6. Is the electronic control functioning properly? Run the test mode.

7. Is the float switch improperly positioned? Is the switch defective? Disconnect the electric supply and check the switch contacts for continuity.

8. Is the water fill line that goes to the water valve kinked? Check visually.

Poor Washability on the Upper Rack

1. Is the upper spray arm turning?
   a. Are the holes in the spray arm plugged?
   b. Check to see if the spray arm is split.
   c. Is there uneven loading of the dishes?
   d. Check the filter assembly for blockage.
   e. Check the lower impeller to see if it is defective or blocked with debris.
   f. Are any objects protruding down that might prevent the upper spray arm from rotating?

2. Is the water charge okay?

3. What is the temperature of the water entering the tub? Is the temperature at 140 degrees Fahrenheit?

4. Is the user using the proper amount of detergent?

5. Is the detergent dispenser functioning properly?

6. Are the dishes loaded properly? Ask the user to load the dishwasher so that you can observe whether he or she is loading the dishwasher properly.

Poor Washability in the Lower Rack

1. Is the lower spray arm turning?
   a. Are the holes in the spray arm plugged?
   b. Check to see if the spray arm is split.
   c. Is there uneven loading of the dishes?
   d. Is the spray arm binding on the housing?
   e. Are any objects protruding down that might prevent the lower spray arm from rotating?

2. Is the water charge okay?

3. What is the temperature of the water entering the tub? Is the temperature at 140 degrees Fahrenheit?

4. Is the consumer using the proper amount of detergent?

5. Is the detergent dispenser functioning properly?

6. Are the dishes loaded properly? Ask the user to load the dishwasher so that you can observe whether he or she is loading the dishwasher properly.

Poor Drying

1. Is the water hot enough? Check the water supply temperature. On certain models, check to see if the cycle extender is functioning properly.

2. Is the heater working?

   a. Check the wiring.

   b. Check the timer or the electronic control board.

   c. Check for the correct voltage.

   d. Check the heater itself.

   e. Check the heater fan assembly.

3. Is a wetting agent being used?

4. Check the door baffle.

5. Check the vent system.

6. Suggest that the user open the door a little after the dry cycle has been completed.

Water Leaks at the Front of the Dishwasher

1. Is the spray arm turning? Is it split? Are any of the holes blocked with debris? If so, this can cause the water to spill out of the tub.

2. Is the tub gasket in place? Check the corners especially. Worn tub or corner gaskets will cause water to leak out of the corners.

3. Did you check the vent baffle gasket?

4. Check to see if the tub is not overcharged with water. Too much water will spill over the front tub flange.

5. Does the door close properly and tightly where the latch fits the strike?

6. On some models, are the gasket clips in place? These clips secure the gasket to the tub.

7. Is the dishwasher loaded properly? Ask the user to load the dishwasher so that you can observe whether he or she is loading the dishwasher properly.

8. Check to see if the dishwasher is draining properly. Inspect the drain hose where it enters the drain.

9. Most important, is the dishwasher level?

10. Check the door for any cracks.

11. Check to see if the wetting agent dispenser is working properly. Too much wetting agent will cause sudsing. Also check the gasket between the door and the dispenser.

Water Leaks at Sides, Top, or Bottom of Dishwasher

1. Check the side inlet tube on the side of the dishwasher.
2. Check to see if water is leaking out from the pump assembly.
3. Check the motor seals for water leakage.
4. Does the tub have a hole in it? Is the tub rusted out?
5. Check the nut on the water inlet port. Is it tight?
6. Are the heater element nuts tight?
7. Check to be sure that all screws are tight (motor and pump assembly, upper rack screws, etc.).

The Dishwasher Cycle Will Not Advance (but the Lights Will Come on)

1. Is there voltage to the timer motor or electronic control board?
2. Are the cams in the timer rotating?
3. Is the dishwasher wired correctly?
4. Check to see if the timer is jammed.
5. Check to see if the selector switch is defective.
6. On electronic models, run the test mode.

The Dishwasher Cycle Will Not Advance Past Start

1. Is the thermostat functioning properly? The thermostat must be flush with the underside of the tub.
2. Check for correct timer settings.

The Detergent Cup Won't Open

1. Is there voltage to the solenoid?
2. Is the detergent mechanism properly adjusted?
3. Check to see if there is any binding at any point.
4. Check to see if any dishes are preventing the detergent cup door from opening.

The Detergent Cup Won't Close

1. Check the detergent cup actuator and cams.
2. Check to see if there is any binding at any point.
3. Is the detergent mechanism adjusted properly?

The Main Motor Won't Operate

1. Check for continuity at the start and run windings. Check the motor for a shorted or grounded winding.
2. Is there voltage to the motor?

3. Is the motor jammed? Check for foreign debris in the pump assembly.

4. Check to see if there are any loose wiring terminals or burned wires.

5. Check to see if the motor assembly is wired correctly.

The Motor Runs but Goes into Overload

1. Check the relay.

2. Is there any binding? Check the pump assembly for broken pieces of glass.

3. Be sure you have the correct voltage and the correct polarity.

4. Check to see if the motor windings are shorted.

Door Liner Hits Side of Tub in Undercounter Models

1. Check to see if the tub is square.

2. Check the installation, and correct as needed.

Dishwasher Won't Start

1. Is there voltage to the dishwasher? Check the plug, circuit breaker, or fuse box. Also check the wires in the junction box, located behind the lower front panel.

2. Check the door switch for continuity of the switch contacts.

3. Check the selector switch for continuity of the switch contacts.

4. Check to see if the timer and selector switch have been wired correctly.

5. On electronic models, check the electronic control board.

6. Check for any wires that might have come off the timer or switches.

The Dishwasher Repeats the Cycle

1. Check the timer contacts; they should be open in the "off" position.

2. Check to see if the timer motor wires are shorted.

Wetting Agent Assembly Leaks

1. Open the door assembly, and check for leaks in the holding tank.

2. Check the wetting agent assembly for proper operation.

Water Siphons Out Through the Drain Pipe While the Dishwasher Is Trying to Fill

1. Is the drain line properly installed?

2. Check the installation instructions for proper installation.

Common Washability Problems

If there are no mechanical problems with the dishwasher's operation, and the complaints are that the dishwasher will not clean the dishes properly or that the glassware is cloudy, etc., the next step will be to look at the best possible cause for the problem that the customer is experiencing with the dishwasher. Then proceed to correct the problem. If necessary, instruct the user on how to get better results from the dishwasher.

Poorly Cleaned Dishes

On occasion, there might be some food particles left on the dishes at the end of the cycle.

Possible cause: Water temperature might be too low. Remember, the water temperature should be 140 to 150 degrees Fahrenheit as it enters the dishwasher tub.

Solution: Check the water temperature at the closest faucet. Let the water from the hot water tap run before starting the dishwasher in order to clear the water line of any cold water.

Possible cause: Not enough detergent for the degree of water hardness or for the amount of dirty dishes to be cleaned.

Solution: Use 1 teaspoon of detergent for each grain of hardness, with 3 teaspoons of detergent, at a minimum, in soft water. The dishwasher will require extra detergent for greasy pans.

Possible cause: The detergent was placed in the wrong side of the dispenser cup.

Solution: Instruct the user on how to fill the dispenser, and have the user reread the use and care manual so that the detergent is placed in the correct dispenser for the cycle that is selected.

Possible cause: Improper loading of the dishes into the dishwasher.

1. Blocking the spray nozzle. If a large bowl or pot is placed over the center of the lower rack, this will block the spray nozzle washing action when the lower rack is pushed in.

2. Larger items that shield smaller items from the washing action.

3. Observe to see if there is a nesting of the bowls or silverware so that the water cannot reach all surfaces.

4. The spray arms are blocked from turning, for example, tall items or an item that fell through the racks.

5. If an item blocks the detergent dispenser from opening, this will not allow the detergent to mix with the water.

Solution: Instruct the user to reread the use and care manual for the proper instructions on how to load the dishes for proper cleaning.

Possible cause: Improper filling of water in the tub. Water pressure must be between 15 and 120 pounds per square inch. After the fill has stopped, check the water level in the tub. On most models, it should be even with the heating element in the bottom of the dishwasher.

Solution: If the water pressure is low, be sure that no other faucets are in use while the dishwasher is operating.

Possible cause: Not enough hot water.

Solution: Instruct the user to use the dishwasher when the hot water is not being used for laundry, baths, or showers.

Possible cause: If the dishwasher detergent is old and caked, it will not dissolve completely.

Solution: Instruct the user to always purchase fresh detergent and store it in a dry place.

Etching

Etching occurs when the glass is pitted or eroded. It appears as a permanent film on the glass. The beginning stages of etching can be identified by an iridescent look—shades of blue, purple, brown, or pink when the glass is held at an angle to the light. In the advanced stages of etching, the glass surface appears frosted or cloudy.

Possible cause: Certain types of glass will etch in any dishwasher with the combination of soft water, the alkalinity of dishwasher detergents, and heat.

Solutions: There is no way to remove the filmy appearance caused by etching; the damage is permanent. There is no way to predict what glassware might be affected—it is not related to the cost or quality of the glass. To prevent etching from recurring:

- Adjust the amount of detergent according to the water hardness.
- Adjust the water temperature so that it enters the dishwasher at approximately 140 degrees Fahrenheit.
- Recommend to the customer that the Energy Saver dry cycle be used.
- Instruct the customer not to manually pre-rinse the dishes before loading them into the dishwasher.

Discoloration

Discoloration (red, black, or brown) might be present on the dishwasher interior or dishes.

Possible cause: If iron or manganese is in the water, the dishes and/or the interior of the dishwasher might turn red, black, or brown.

Solution: A rust remover can be used to remove the discoloration from the dishwasher interior. With the dishwasher empty, turn to the rinse and hold cycle, and start the dishwasher. During the fill, open the door and add 1/2 cup of rust remover to the water. Allow the dishwasher to complete the cycle. Then start the dishwasher on the normal wash cycle, with detergent but without the dishes. To keep this condition from returning, the customer might have to install special filters to remove the iron and manganese. Use a rust remover, as mentioned previously, according to the manufacturer's recommendations to remove the discoloration from the dishes and the glassware.

Lime Deposits on the Dishwasher Interior

Possible cause: If there is a lot of calcium in the water, a lime film or deposit might eventually build up on the interior surfaces of the dishwasher.

Solution: You can try one of the following methods:

- Use a mild scouring powder and a damp cloth to clean away the lime deposit.
- With the dishwasher empty, turn the timer to the rinse and hold cycle. During the fill portion, open the door and add 1/2 cup of white vinegar to the water. Let the dishwasher complete the remaining cycle. Do not use detergent. After the cycle is completed, run the dishwasher with a regular load.
- Use a product that removes lime deposits. Follow the manufacturer's directions.

Suds or Foam in Dishwasher

Possible causes:

- Sudsing in the dishwasher can be caused by protein foods (milk, eggs, etc.) and an insufficient amount of detergent.
- The water in the dishwasher is not hot enough to activate the defoaming agents in the dishwasher detergent.
- The user has used a non-automatic dishwasher detergent.

Solution: Increase the amount of detergent to reduce sudsing. Dishwasher detergents contain defoaming agents to break down the suds in the dishwasher water.

Check that the water temperature is between 140 and 150 degrees Fahrenheit. Only use detergents that are made for automatic dishwashers.

Darkened Aluminum

Possible cause: A combination of water, heat, and alkaline foods will darken or stain aluminum products.

Solution: To remove this discoloration, instruct the customer to use an aluminum cleaner and to clean the item by hand. Never allow undissolved dishwasher detergent to come in direct contact with the metal. Avoid placing aluminum items in the lower rack, right in front of the detergent dispenser.

Discoloration of Copper

Possible cause: Some copper items will discolor when washed by hand, as well as when washed in the dishwasher, because of the heat and detergent alkalinity.

Solution: Instruct the customer to use a copper cleaner to restore the copper color.

Cracking (Crazing) of China

Possible cause: Crazing is the appearance of tiny cracks that appear over the entire surface of the china. It can occur when porous earthenware, good china that is very old, or lower-quality china is exposed to heat and moisture.

Solution: Once the glaze is cracked or crazed, the damage is permanent. This characteristic is inherent in some clayware; and this type of damage can occur during use, handwashing, or automatic dishwashing.

PART V

Chipping of China and Crystal

Possible cause: Chipping usually occurs during normal use and handling, and simply might not be noticed until the dishes are removed from the dishwasher. When the dishes are loaded into the dishwasher according to the manufacturer's instructions, there is nothing in the dishwasher that can chip the dishes. The dishes should only come in contact with the cushioned vinyl-coated racks.

Solution: Instruct the user to follow the manufacturer's instructions in the use and care manual for loading the dishwasher properly.

Metal Marks on Dishes and Glassware

Possible cause: If a metal item, especially aluminum, touches a dish in the dishwasher, a metal mark might result. This symptom appears as small black or gray marks or streaks on dishes or glasses. However, most metal marks occur during normal use, when the dishes come in contact with the flatware.

Solution: Instruct the user to load the dishwasher carefully in order to prevent metal items from touching other dishes. There are products on the market that will remove these marks. Have the user read over the use and care manual or check with the manufacturer for this information.

Staining of Melamine Dinnerware

Possible cause: Stains on melamine dinnerware can result from contact with coffee, tea, and some fruit juices. If the surface is worn, it will stain more readily.

Solution: Some specialty products on the market are recommended for removing these stains. Instruct the customer to read over the use and care manual or check with the manufacturer for this information.

Melting or Warping of Plastic Items

Possible cause: Some plastic items cannot be exposed to the temperatures usually found in dishwashers without changing shape.

Solution: Once the plastic item has distorted, it cannot be returned to its original shape. In order to minimize or prevent plastic items from warping or melting, instruct the user to do one of the following:

- Choose the air-dry cycle to dry the dishes.
- Place the plastic items on the top rack.
- Purchase and use plastic items that are labeled "dishwasher-proof."

Discoloration of Silverplate

Possible cause: When silverplate takes on a copper- or bronze-colored appearance, the silverplate has worn thin and the base metal is showing through. The combination of dishwasher detergent and the lack of hand-toweling might result in discoloration of this base metal.

Solution: This discoloration can usually be removed by polishing the item with a silver polish or by soaking the item in vinegar for about 10 minutes. This is only a temporary solution, however. Only a replating with silver by a jeweler will correct the problem.

Tarnishing of Silverware (Sterling or Silverplate)

Possible cause: Sulfur in the water supply might be the cause. This effect might be accelerated by the automatic dishwasher because of the higher water temperature and because the usual hand-drying with a towel has been eliminated.

Solution: Because sulfur cannot be readily removed from the water supply, frequent polishing is the only answer.

Possible cause: Silver will tarnish easily if it is left in contact with foods such as mayonnaise and eggs.

Solution: If silverware has been in contact with such foods, instruct the user to rinse the item thoroughly as soon as possible after its use.

Bluish Discoloration of Stainless Steel

Possible cause: A bluish discoloration of some types of stainless steel is caused by heat and the alkalinity of the automatic dishwasher detergent.

Solution: This discoloration can be removed by using a paste of baking soda and water or a stainless steel cleaner.

Corrosion or Rusting of Stainless Steel

Possible cause: When the protective oxide film on the surface of the steel is removed, corrosion will take place, as with ordinary steels. Certain foods will remove the oxide film, such as table salt, vinegar, salad dressings, milk and milk products, fruits and juices, tomatoes and tomato products, and butter. However, if the stainless steel is washed, rinsed, and dried thoroughly, the oxygen of the air will heal the breaks in the oxide film and return the stainless property to the steel. But if food is not washed off promptly, the air cannot heal the break, and corrosion will occur.

Solution: There is no permanent solution. To minimize rusting, instruct the user to rinse or wash the flatware as soon as possible after use.

Dishwasher Maintenance

The dishwasher's interior is normally self-cleaning. However, there are times when the customer will have to remove food particles or broken glass in the bottom of the tub. Instruct the customer to clean the bottom edge of the dishwasher tub, which is sealed off by the gasket when the door is closed. Food and liquids drip on to this area when the dishwasher is loaded. The control panel should be cleaned with a soft damp cloth. Tell the customer not to use any abrasive powders or cleaning pads. Also, advise the customer to read the use and care manual for proper maintenance procedures for his or her brand of dishwasher. Dishwashers are designed to flush away all normal food soils that have been removed from the dishes. However, on occasion, certain foreign objects, such as fruit pits, bottle caps, etc., might collect in the openings of the pump. These items should be removed periodically to avoid clogging the drain system. Also, on occasion, some of these foreign objects can get caught in the spray arm openings and will have to be cleaned out. Check the racks carefully to see if there are any nicks or cuts in the vinyl. These nicks and cuts can be repaired. A liquid vinyl-repair material is available through the manufacturer or at any appliance supply store.

Repair Procedure

Each of the following repair procedures is a complete inspection and repair process for a single dishwasher component, containing the information you need to test a component that might be faulty and then replace it, if necessary.

Water Valve

The typical complaints associated with water valve failure (Figure 11-7) are:

- The dishwasher will run, but no water will enter dishwasher.
- The dishwasher will overfill and leak onto the floor.
- When the dishwasher is off, water still enters the tub.
- The dishes are not clean or not enough water enters the tub.

When dealing with these complaints, perform the following steps:

1. **Verify the complaint.** Verify the complaint by operating the dishwasher through its cycles. Listen carefully, and you will hear if the water is entering the dishwasher.

2. **Check for external factors.** You must check for external factors not associated with the appliance. Is the appliance installed properly? Is the voltage supply correct for the dishwasher?

3. **Disconnect the electricity.** Before working on the dishwasher, disconnect the electricity. This can be done by pulling the plug out of the wall receptacle. Be sure that you only remove the dishwasher plug. Double-check to ensure that the electrical supply has been disconnected before removing any service panels. Another way to disconnect the electricity is at the fuse panel or the circuit breaker panel. Turn off the electricity.

Figure 11-7
A typical dishwasher water valve.

Gasket   Screen

Bottom screws

Adjustable toeplate

Figure 11-8 Removing the bottom panel.

4. **Remove the bottom panel.** In order to gain access to the water valve, the bottom panel must be removed (Figure 11-8). The bottom panel is held on with either two or four screws, depending on the model. Remove the screws, and remove the panel.

5. **Remove the wire leads.** In order to check the solenoid coil on the water valve, remove the wire leads that connect to the solenoid coil from the wire harness (Figure 11-9). These are slide-on terminal connectors attached to the ends of the wire—just pull them off.

6. **Test the water valve.** Using the ohmmeter, set the range on R × 1000, and place the probes on the solenoid coil terminals (Figure 11-10). The meter should read between 700- and 900-ohms resistance. If not, replace the water valve.

7. **Inspect the inlet screen.** If you determine that the water valve is good, but there is little water flow through the valve, inspect the inlet screen. If this screen is filled with debris, it must be cleaned out. To accomplish this, use a small flat-blade screwdriver and pry out the screen (Figure 11-11). Then wash out the screen, making sure that all of the debris is removed. Reinstall the screen. Turn on the water supply. Plug in

Figure 11-9
Removing the wire
leads from the
solenoid coil.

Inlet valve solenoid terminals

Wire leads from water
valve solenoid coil

FIGURE 11-10
Connect meter probes
to the water valve
solenoid coil.

the dishwasher. Allow the water to enter the tub to check the flow rate of the water valve. The tub must be empty, because this will allow you to check the flow rate properly. On a normal fill, the water line should be over the heating element within two minutes (Figure 11-12). When you turn the dishwasher on and energize the water valve, if no water enters the dishwasher tub, replace the water valve. If the water valve checks out okay, then check the timer and the wiring harness.

Screen

8. **Remove the water valve.** Before removing the water valve, turn off the water supply to the dishwasher. Then disconnect the water supply line from the inlet end of the water valve and remove the fill hose from the outlet side (Figure 11-13). Next, remove the screws that hold the water valve to the chassis of the dishwasher. To install the new water valve, just reverse the disassembly procedure and reassemble. Check for water leaks. If none are found, reinstall the bottom panel and restore the electricity to the dishwasher.

9. **Operate the dishwasher.** Set the timer and control settings to operate the dishwasher through its cycle.

FIGURE 11-11   Removing water valve inlet screen. Be careful not to distort the screen.

FIGURE 11-12 Diagram indicates water level in dishwasher tub.

Motor and Pump Assembly

The typical complaints associated with pump and motor failure are:

- The water will not drain out of the dishwasher.
- Poor washability of the dishes.
- When the motor runs, there are loud noises.
- The fuses or circuit breaker will trip when the dishwasher is started.
- Water is leaking from the bottom of the dishwasher or leaking on the motor/pump assembly.

FIGURE 11-13
Disconnect the water supply line.

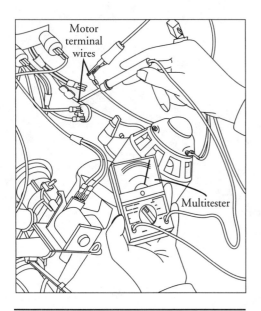

FIGURE 11-14   Check the motor windings for continuity.

When dealing with these complaints, perform the following steps:

1. **Verify the complaint.** Verify the complaint by operating the dishwasher through its cycles. Listen carefully, and you will hear if there are any unusual noises or if the circuit breaker trips.

2. **Check for external factors.** You must check for external factors not associated with the appliance. Is the appliance installed properly? Does the appliance have the correct voltage?

3. **Disconnect the electricity.** Before working on the dishwasher, disconnect the electricity. This can be done by pulling the plug out of the wall receptacle. Be sure that you only remove the dishwasher plug. Or disconnect the electricity at the fuse panel or at the circuit breaker panel. Turn off the electricity.

4. **Remove the bottom panel.** In order to gain access to the pump and motor assembly, the bottom panel must be removed (see Figure 11-8). The bottom panel is held on with either two or four screws, depending upon the model. Remove the screws and remove the panel.

5. **Disconnect the motor wire leads.** Disconnect the motor wire leads from the wiring harness. Check the motor windings for continuity (Figure 11-14). Check for resistance from the common wire lead to the run winding (Figure 11-15). Check from the common to the wash winding (Figure 11-16).

FIGURE 11-15
Check from the common to the drain winding.

Blue - run winding
White - common
Red - drain winding
Black - wash winding

FIGURE 11-16
Check from the common to the wash winding.

To check for a grounded winding in the motor, take the ohmmeter probes and check from each motor wire lead terminal to the motor housing (Figure 11-17). The ohmmeter will indicate continuity if the windings are grounded (Figure 11-18).

If the motor/pump assembly shows signs of water leaking, replace the motor/pump assembly as a complete unit. Most part manufacturers give a one-year warranty on the motor/pump assembly. There is no advantage in tearing down the motor/pump assembly to replace only one or two parts and receiving only a partial warranty. Replacing the motor/pump assembly as a complete assembly will save you time and money in the long run. If the motor/pump assembly checks out okay, then check the timer and the motor relay (if the model you are repairing has one) and check for a kinked or plugged drain line.

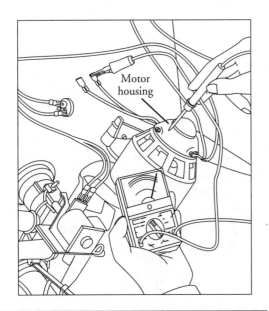

FIGURE 11-17   Check for the grounded motor.

6. **Remove motor and pump assembly.**
Remove the lower dishwasher rack
from the dishwasher. As shown in
Figure 11-19, remove the wash tower
and spray arm assembly. If there is a
filter, remove it also. Remove the motor
wiring leads from the wiring harness.
Then remove the drain line from the
pump assembly. Reach underneath the
tub and rotate the four pump hold-
downs 90 degrees inward (Figure 11-20).
You are now ready to remove the motor/
pump assembly. Lift the motor/pump
assembly out from the inside of the tub
(Figure 11-21). Keep the work area dry
to help prevent electrical shocks.

FIGURE 11-18   The grounded motor.

7. **Reinstall motor and pump assembly.**
To reinstall the motor/pump assembly,
just reverse the order of step 6. Before
restoring the electricity to the dishwasher, pour a gallon of water into the tub and check for leaks
underneath the tub. Restore the electricity to the dishwasher, and run the dishwasher through a
cycle. Check for leaks again. If no leaks are found, reinstall the bottom panel.

FIGURE 11-19   Remove the wash tower and the spray arm assembly.

PART V

Figure 11-20
An exploded view of a dishwasher motor/pump assembly.

Upper spray arm

Water column

Nozzle cap

Washer

Lower spray arm

Sump guard

Power clean module and motor

Motor hold-downs

Figure 11-21
Remove the motor/pump assembly. After removal, inspect the tub for rust.

Dishwasher Timer

The typical complaints associated with dishwasher timer failure are:

- The cycle will not advance.
- The dishwasher won't run at all.
- The dishwasher will not fill.
- The dishwasher will not pump the water out.

When dealing with these complaints, perform the following steps:

1. **Verify the complaint.** Verify the complaint by operating the dishwasher through its cycles. Before you change the timer, check the other components controlled by the timer.

2. **Check for other factors.** You must check for other factors that could affect the operation of the appliance. Is the appliance installed properly? Does the appliance have the correct voltage?

3. **Disconnect the electricity.** Before working on the dishwasher, disconnect the electricity. This can be done by pulling the plug out of the wall receptacle. Be sure that you only remove the dishwasher plug. Or disconnect the electricity at the fuse panel or circuit breaker panel. Turn off the electricity.

4. **Remove the console to gain access.** Begin by removing the four screws from the console to access the timer (Figure 11-22). Turn the timer knob counterclockwise to remove it from the timer shaft, and slide the indicator dial off the shaft. Remove the console panel from the dishwasher. On some models, the latch handle knob will also have to be removed.

5. **Test the timer.** Disconnect the timer motor wire leads from the timer assembly. Using the ohmmeter, set the range on R × 1000, and place the probes on the timer motor terminals (Figure 11-23). The meter should indicate some resistance. If not, replace the timer. If the motor checks out, then check the door latch switch, cycle extender relay, and the float switch assembly.

Remove screw

Remove screw

Console

Remove screw

Remove screw

6. **Remove the timer.** To remove the timer, remove the timer mounting screws (Figure 11-24). Remove the wire lead terminals from the timer. Mark the wires as to their location on the timer. Some timers have a disconnect terminal block instead of individual wires, which makes it easier to remove the timer wires.

7. **Install the new timer.** To install a new timer, just reverse the disassembly procedure, and reassemble. Reinstall the console panel, and restore the electricity to the dishwasher. Test the dishwasher operation.

FIGURE 11-22   Remove the screws to access the control panel.

FIGURE 11-23
Attach the meter probes and test for continuity.

Timer motor terminals

Timer

The Dishwasher Door Is Hard to Close

The typical complaints associated with a dishwasher door being hard to close are:

- The door is hard to close.
- The door won't latch.
- The dishwasher won't run.

Timer plug

Timer

Timer mounting screws

FIGURE 11-24 Removing the timer.

When dealing with these complaints, perform the following steps:

1. **Verify the complaint.** Verify the complaint by trying to close and latch the door.

2. **Check for external factors.** You must check for external factors not associated with the appliance. Is the appliance installed properly? Be sure that the dishwasher door is not binding against the side cabinet.

3. **Run the dishwasher.** By pushing on the door, you finally latch the door closed. Next, run the hot water faucet closest to the dishwasher to flush out the supply line. Then turn the timer to the normal wash cycle with heated dry,

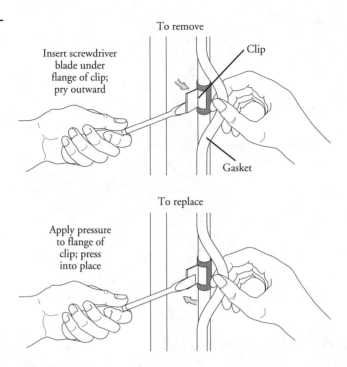

FIGURE 11-25
Removing dishwasher
door gasket.

To remove

Insert screwdriver
blade under
flange of clip;
pry outward

Clip

Gasket

To replace

Apply pressure
to flange of
clip; press
into place

and let it run through the entire cycle. By doing this procedure, the gasket will soften enough to form itself to the tub while it is compressed. This should make it easier to latch the door. If not, adjust the latch assembly. If it is still too hard to close, replace the door gasket.

4. **Replace the door gasket.** To remove the old gasket, remove the screws or clips that hold the old gasket in place (Figure 11-25). On some models, the gasket is pressed into the inner door assembly. Remove the inner panel from the dishwasher, and pull off the gasket from the panel (Figure 11-26). Soak the new gasket in warm water to make it more flexible.

5. **Test a new gasket.** After you have installed the new door gasket, close the dishwasher door, and test to see if the door latches without pushing hard against it.

6. **Check for water leaks.** Run the dishwasher through another cycle to check for water leaking out the door. Most models have a baffle, or tub, gasket. If the dishwasher that you're working on has one, check to make sure it is not defective. The tub gasket is located either on the tub or behind the inner door panel.

Dishwasher Heating Element

The typical complaints associated with the heating element are:

- Dishes are still wet at the end of the cycle.
- The dishwasher stalls in the middle of the cycle.

When dealing with these complaints, perform the following steps:

1. **Verify the complaint.** Verify the complaint by operating the dishwasher, starting at the dry cycle.

2. **Check for external factors.** You must check for external factors not associated with the appliance. For example, check the Energy Saver switch. Is it set for heat-dry? Is the appliance installed properly? Does the appliance have the correct voltage?

3. **Disconnect the electricity.** Before working on the dishwasher, disconnect the electricity. This can be done by pulling the plug out of the wall receptacle. Be sure that you only remove the dishwasher plug. Or disconnect the electricity at the fuse panel or the circuit breaker panel. Turn off the electricity.

4. **Remove the bottom panel.** In order to gain access to the heater wire terminals, the bottom panel must be removed (see Figure 11-8). The bottom panel is held on with either two or four screws, depending on the model. Remove the screws and remove the panel.

5. **Test the heating element.** To test the heating element (Figure 11-27), remove the wires from the heating element terminals (Figure 11-28). These are slide-on terminal connectors attached to the ends of the wire. Use the ohmmeter to check for continuity between the two element terminals (Figure 11-29). If the meter indicates no continuity between the terminal ends, replace the heater.

   To check for a shorted-out heating element, take one end of the ohmmeter probe and touch the element terminal; then, with the other probe, touch the sheath (outer cover of element), shown in Figure 11-30. If the meter indicates continuity, the element is shorted out and should be replaced.

FIGURE 11-27 The dishwasher heating element.

FIGURE 11-28 Remove the wire terminals from the heater element.

6. **Remove the heating element.** To remove the heating element (with the wires already removed from the heater terminals), unscrew the locknuts that hold the element in place. From inside the tub, remove the heating element (Figure 11-31).

7. **Install a new heating element.** To install a new element, just reverse the disassembly procedure and reassemble. Then test the new element by repeating step 1.

FIGURE 11-29 Set the meter on the ohms scale. Connect the probes to the heater terminals.

FIGURE 11-30 To check for a shorted-out element, attach the meter probe to one terminal and attach the other probe to the sheath.

FIGURE 11-31
Removing the heating element.

Locknut

Cycle Selector Switch

The typical complaints associated with the cycle selector switch are:

- Inability to select a different cycle.
- The consumer inadvertently selected the wrong cycle.

When dealing with these complaints, perform the following steps:

1. **Verify the complaint.** As shown in Figure 11-32, verify the complaint by trying to select different cycles.

FIGURE 11-32 Selecting a cycle.

2. **Check for external factors.** You must check for external factors not associated with the appliance. For example, is there any physical damage to the component?

3. **Disconnect the electricity.** Before working on the dishwasher, disconnect the electricity. This can be done by pulling the plug out of the wall receptacle. Be sure that you only remove the dishwasher plug. Or disconnect the electricity at the fuse panel or circuit breaker panel. Turn off the electricity.

4. **Remove the control panel.** To remove the control panel, remove the screws along the top inside edge of the door (see Figure 11-22). On some models, you might have to remove the door latch knob in order to completely remove the control panel.

5. **Test the cycle selector switch.** To test the cycle selector switch, remove all wires from the switch. Just remember—you will have to identify the wires according to the wiring diagram in order to reinstall them back on the cycle selector switch properly. Take your ohmmeter, and check for continuity on the switch contacts; press the switch that coincides with the terminals that are being checked (Figure 11-33). At this point, you have to use the wiring diagram to identify the switch contacts.

6. **Remove the cycle selector switch.** To remove the cycle selector switch, remove the screws that hold the component to the control panel (Figure 11-34).

7. **Reinstall the cycle selector switch.** To reinstall the cycle selector switch, just reverse the disassembly procedure and reassemble. Remember—you will have to identify the wires according to the wiring diagram in order to reinstall them back on the cycle selector switch properly.

FIGURE 11-33
Testing the selector switch contacts.

Multitester   Switch terminals

Control panel

FIGURE 11-34
Removing the selector switch.

Selector switch

Dishwasher Door Switch

Sometimes, the dishwasher door switch malfunctions and the door will not close properly. When dealing with these complaints, perform the following steps:

1. **Verify the complaint.** As shown in Figure 11-35, verify the complaint by closing the dishwasher door and turning the timer dial to start the wash cycle.

2. **Check for external factors.** You must check for external factors not associated with the appliance. Is the appliance installed properly? Does the appliance have the correct voltage? Is there any physical damage to the component?

3. **Disconnect the electricity.** Before working on the dishwasher, disconnect the electricity. This can be done by pulling the plug out of the wall receptacle. Be sure that you only remove the dishwasher plug. Or disconnect the electricity at the fuse panel or circuit breaker panel. Turn off the electricity.

4. **Remove the control panel.** To remove the control panel, remove the screws along the top inside edge of the door (see Figure 11-22). On some models, you might have to remove the door latch knob in order to completely remove the control panel.

FIGURE 11-35 Closing the door and running the dishwasher.

5. **Test the door switch.** To test the door switch, remove the two wires from the switch (Figure 11-36). Close and latch the door. With your ohmmeter, check for continuity between the two terminals on the switch. Then open the door latch and check for no continuity between the terminals. If the switch fails these tests, replace the switch.

6. **Remove the door switch.** Remove the screws that secure the switch in place. Remove the switch.

7. **Install new door switch.** To install the new door switch, just reverse the disassembly procedure and reassemble. Then reconnect the electricity and test the dishwasher.

Float and Float Switch

Sometimes, the dishwasher float and the float switch malfunction and water will not enter the tub, or water overfills and spills onto the floor. When dealing with these complaints, perform the following steps:

1. **Verify the complaint.** Verify the complaint by closing the dishwasher door and turning the timer dial to start the wash cycle.

2. **Check for external factors.** You must check for external factors not associated with the appliance, as was done in the previous sections.

FIGURE 11-37
Inspecting the float
assembly. Check for
soap buildup around
the stem and inside of
the float.

3. **Disconnect the electricity.** Before working on the dishwasher, disconnect the electricity. This can be done by pulling the plug out of the wall receptacle. Be sure that you only remove the dishwasher plug. Or disconnect the electricity at the fuse panel or the circuit breaker panel. Turn off the electricity.

4. **Check the float.** Check the float from the inside of the tub (Figure 11-37). Be sure that the float moves freely up and down. If there is any soap buildup around the float, clean it off.

5. **Remove the bottom panel.** In order to gain access to the float switch terminals, the bottom panel must be removed (see Figure 11-8). The bottom panel is held on with either two or four screws, depending on the model. Remove the screws, and remove the panel.

6. **Test the float switch.** The float switch is located under the float, underneath the tub (Figure 11-38). Remove the wires from the terminals, and test for continuity. Lift the float—there should be no continuity. Let the float rest, and you should have continuity. If the test fails, replace the float switch.

7. **Remove the float switch.** To remove the float switch, you will have to remove the screws that hold the switch in place (Figure 11-39).

8. **Install a new float switch.** To install the new float switch, just reverse the disassembly procedure and reassemble. Then reconnect the electricity and test a dishwasher cycle that fills the dishwasher.

PART V

Figure 11-38
Testing the float
switch contacts.

Mounting
bracket

Continuity tester

Float
switch
terminals

Figure 11-39
An exploded view of
the float switch
assembly.

Float

Screw

Collar

"O" Ring

Screw

Float assembly

Switch

Wetting Agent and Detergent Dispensers

The typical complaints associated with the wetting agent and detergent dispensers are:

- The dishes are dirty.
- The dishes have spots on them.
- Detergent remains in the dispenser.
- Wetting agent is leaking onto the floor.

When dealing with these complaints, perform the following steps:

1. **Verify the complaint.** Verify the complaint by operating the dishwasher through its cycles. Make sure that the detergent dispenser door is not blocked by dishes. Also check the wetting agent level in the dispenser.

2. **Check for external factors.** You must check for external factors not associated with the appliance. For example, is there any physical damage to the component? Is the water in the dishwasher at the correct level? Did the consumer fill the wetting agent dispenser with wetting agent?

3. **Disconnect the electricity.** Before working on the dishwasher, disconnect the electricity. This can be done by pulling the plug out of the wall receptacle. Be sure that you only remove the dishwasher plug. Or disconnect the electricity at the fuse panel or circuit breaker panel. Turn off the electricity.

4. **Remove the control panel.** To remove the control panel, remove the screws along the top inside edge of the door (see Figure 11-22). On some models, you might have to remove the door latch knob in order to completely remove the control panel. Then reconnect the electricity, latch the dishwasher door, and test a dishwasher cycle. A cam follower next to the timer will actuate the detergent dispenser when the wash cycle is in the second wash. The cam follower will engage the draw bar (Figure 11-40) and pull up on the detergent lid latch, releasing the detergent lid and allowing the detergent to enter into the wash cycle. In the final rinse cycle, the same operation occurs to release the wetting agent from the wetting agent dispenser.

FIGURE 11-40 A side view of the detergent dispenser.

5. **Remove the wetting agent dispenser.** As illustrated in Figure 11-41a, remove the drip cover and drawbar spring. Now begin to remove the upper and lower drawbars by aligning the lower drawbar holes with the locating tabs. In Figure 11-41b, you will note that the locating tabs will align with their respective holes while removing the drawbars. Remember where they reattach when it comes to reassembly later. Begin to slide the drawbars apart, as shown in Figure 11-41c. Next, you will have to remove the wetting agent cap by turning the cap and seal assembly counterclockwise, as in Figure 11-41d. Using a ¾-inch socket wrench, begin to insert it into the wetting agent dispenser locking tabs, shown in Figure 11-41d, to release the wetting agent dispenser from the inner door liner. Now you are ready to remove the wetting agent dispenser actuator. Spread apart the two locking tabs, as shown in Figure 11-41e, and remove the upper and lower slides from the dispenser actuator. Next remove the diaphragm spring and diaphragm (Figure 11-41e). To remove the detergent door and actuator assembly, use a small flat-blade screwdriver to depress and release the lower tabs, noting the position of the door latch level through the hole in the lower slide (Figure 11-41f). Finally, remove the upper slide, slide spring, and the lower slide (Figure 11-41g).

(a)

(b)

FIGURE 11-41a Remove the drip cover and drawbar spring.

FIGURE 11-41b Removal of the drawbars. Take note of the alignment of the locating tabs.

(c)

FIGURE 11-41c The upper and lower drawbars. To remove, follow the direction of the arrows.

(d)

FIGURE 11-41d Removal of the wetting agent dispenser.

(e)

FIGURE 11-41e Removal of the actuator, diaphragm spring, and the diaphragm from the wetting agent dispenser.

(f)

FIGURE 11-41f Removal of the detergent door actuator assembly and detergent door latch. Remember the tab alignment when dissembling and reassembling.

PART V

(g)

FIGURE **11-41g** The upper slide, slide spring, lower slide, detergent door actuator, and spring will be removed last in disassembly; it will be the first step in reassembly.

6. **Install new wetting agent and detergent dispenser.** To install the new dispenser, just reverse the disassembly procedure and reassemble. Then reconnect the electricity and test the dishwasher.

Water Heater Operating Thermostat

The typical complaints associated with the water heater operating thermostat are:

- Poor washability
- The dishwasher timer will not advance.
- Low inlet water temperature.
- Dishwasher water line hooked up to the cold water line.

When dealing with these complaints, perform the following steps:

1. **Verify the complaint.** Verify the complaint by operating the dishwasher through its cycles. Take the temperature of the water in the dishwasher tub when the dishwasher is in the wash cycle. Also check the hot water temperature at the sink. It should be between 120 to 140 degrees Fahrenheit.

2. **Check for external factors.** You must check for external factors not associated with the appliance. Test the hot water temperature at the sink. The water temperature should be around 120 to 140 degrees Fahrenheit. The purpose of this thermostat is to delay the wash cycle until the water temperature reaches between 135 and 145 degrees Fahrenheit. The rule of thumb is it takes 20 minutes to raise the water temperature two degrees. The thermostat will then close and complete the timer motor circuit, allowing the dishwasher cycle to continue.

3. **Disconnect the electricity.** Before working on the dishwasher, disconnect the electricity. This can be done by pulling the plug out of the wall receptacle. Be sure that you only remove the dishwasher plug. Or disconnect the electricity at the fuse panel or circuit breaker panel. Turn off the electricity.

FIGURE 11-42
The heater operating thermostat will allow the water to heat up to 145 degrees Fahrenheit.

Thermostat located underneath the dishwasher tub

Water heater operating thermostat

4. **Remove the bottom panel.** In order to gain access to the water heater operating thermostat terminals, the bottom panel must be removed (see Figure 11-8). The bottom panel is held on with either two or four screws, depending on the model. Remove the screws, and remove the panel. Inspect the wiring connections first.

5. **Remove and test the water heater operating thermostat.** Locate the thermostat (Figure 11-42) underneath the tub, and remove the wires from the terminals. Remove the thermostat from the dishwasher tub. Take a metal pot of water, and heat it up on the range until the temperature reaches between 150 and 160 degrees Fahrenheit. Next, set your multimeter to the ohm scale. Test the thermostat before placing it against the metal pot—it should read infinite resistance. Now place the thermostat against the metal pot of hot water—the thermostat should close and the multimeter will read continuity. If not, replace the thermostat.

6. **Install new water heater operating thermostat.** To install the new thermostat, just reverse the disassembly procedure and reassemble. Then reconnect the electricity and test the dishwasher.

High-Limit Heater Thermostat

The typical complaints associated with the high-limit heater thermostat are:

- The dishes are extremely hot at the end of the cycle.
- It smells as if something is burning.
- Excessive steam is coming from the vent.

When dealing with these complaints, perform the following steps:

1. **Verify the complaint.** Verify the complaint by operating the dishwasher through its cycles. The purpose of this thermostat is to act as a safety switch to disconnect the heater in the event the temperature rises above 185 degrees Fahrenheit.

2. **Check for external factors.** You must check for external factors not associated with the appliance.

3. **Disconnect the electricity.** Before working on the dishwasher, disconnect the electricity. This can be done by pulling the plug out of the wall receptacle. Be sure that you only remove the dishwasher plug. Or disconnect the electricity at the fuse panel or circuit breaker panel. Turn off the electricity.

PART V

FIGURE 11-43
The high-limit heater
thermostat location.

4. **Remove the bottom panel.** In order to gain access to the high-limit heater
   thermostat terminals, the bottom panel must be removed (see Figure 11-8). The
   bottom panel is held on with either two or four screws, depending on the model.
   Remove the screws, and remove the panel. Inspect the wiring connections first.

5. **Remove and test the high-limit heater thermostat.** Locate the high-limit thermostat
   (Figure 11-43) underneath the tub, and remove the wires from the terminals.
   Remove the thermostat from the dishwasher tub. Take a metal pot of water, and
   heat it up on the range until the temperature reaches 185 degrees Fahrenheit. Next,
   set your multimeter to the ohm scale. Test the thermostat before placing it against
   the metal pot of hot water—the multimeter will read continuity. Now place the
   thermostat bimetal section against the metal pot of hot water—the thermostat
   should open and the multimeter will read infinite resistance. If not, replace the
   thermostat.

6. **Install a new high-limit heater thermostat.** To install the new thermostat, just
   reverse the disassembly procedure and reassemble. Then reconnect the electricity
   and test the dishwasher.

Dishwasher Motor Start Relay

The typical complaints associated with the motor start relay are:

1. The dishwasher motor will not start.

2. The dishwasher motor hums and shuts off on overload.

When dealing with these complaints, perform the following steps:

1. **Verify the complaint.** Verify the complaint by operating the dishwasher. If the motor does not start, turn the dishwasher off immediately.

2. **Check for external factors.** You must check for external factors not associated with the appliance. Is there electricity to the dishwasher? The purpose of the motor start relay is to protect the dishwasher motor. This current-sensitive relay will close the contacts when the current increases for about one or two seconds when the motor begins to start. When the dishwasher motor starts and runs, the current will decrease in the motor start relay, thus opening the contacts.

3. **Disconnect the electricity.** Before working on the dishwasher, disconnect the electricity. This can be done by pulling the plug out of the wall receptacle. Be sure that you only remove the dishwasher plug. Or disconnect the electricity at the fuse panel or circuit breaker panel. Turn off the electricity.

4. **Remove the bottom panel.** In order to gain access to the water heater operating thermostat terminals, the bottom panel must be removed (see Figure 11-8). The bottom panel is held on with either two or four screws, depending on the model. Remove the screws, and remove the panel. Inspect the wiring connections first.

5. **Remove and test the motor start relay.** Remove the motor start relay from the dishwasher (Figure 11-44) by removing the screw that holds the relay in place. Disconnect the wires from the relay. Set your ohmmeter to the R x 1 scale, and measure the coil's resistance. The coil will read less than 1 ohm if it is good. Next, attach your ohmmeter probes to the relay contacts; you will check the relay contacts for continuity by turning the relay upside down. If you do not know which way is up or down on the relay, there is an arrow and the word "up" stamped on the relay. The motor start relay contacts should be closed, and the ohmmeter will read zero resistance. With the ohmmeter probes still attached to the relay, turn the relay right-side-up to check the continuity of the contacts—the relay contacts will be open, and the ohmmeter will read infinite resistance.

6. **Install a new dishwasher motor start relay.** To install the new motor start relay, just reverse the disassembly procedure and reassemble. Then reconnect the electricity and test the dishwasher.

FIGURE 11-44
The dishwasher motor start relay is located behind the lower front panel.

Screw

Electronic Control Board, Display Board, and Touch Pad Panel

The typical complaints associated with the electronic control board, display board, and touch pad panel are:

- Dishwasher won't run or power up.
- Unable to program the dishwasher.
- The display board will not display anything.
- One or more key pads will not accept commands.
- Unusual display readouts.

When dealing with these complaints, perform the following steps:

NOTE *To prevent electrostatic discharge (ESD) from damaging expensive electronic components, follow the steps in Chapter 9.*

1. **Verify the complaint.** Verify the complaint by operating the dishwasher. Turn off the electricity to the appliance, and wait for two minutes before turning it back on. If a fault code appears, look up the code. If the dishwasher will not power up, locate the technical data sheet behind the control panel or behind the lower panel for diagnostics information.

2. **Check for external factors.** You must check for external factors not associated with the appliance. Is there electricity to the dishwasher? Check for a blown thermal fuse in the circuit.

3. **Disconnect the electricity.** Before working on the dishwasher, disconnect the electricity. This can be done by pulling the plug out of the wall receptacle. Be sure that you only remove the dishwasher plug. Or disconnect the electricity at the fuse panel or circuit breaker panel. Turn off the electricity.

WARNING *Some diagnostic tests will require you to test the components with the power turned on. When you disassemble the control panel, you can position the panel against the door so that the wiring will not make contact with the door. This act will allow you to test the components without electrical mishaps.*

4. **Remove the control panel.** To remove the control panel, remove the screws along the top inside edge of the door (see Figure 11-22). On some models, you might have to remove the door latch knob in order to completely remove the control panel.

5. **Test the electronic control board, display board, or touch pad panel.** If you are able to run the dishwasher diagnostic test mode, check the different functions of the dishwasher. Use the technical data sheet for the model you are servicing to locate the test points from the wiring schematic. Check all wiring connections and wiring. Using the technical data sheet, test the key pad matrix, display board LEDs, input voltages, and output voltages on the electronic control board.

FIGURE 11-45
An exploded view of the electronic components in a dishwasher.

6. **Remove the electronic control board, display board, or touch pad panel.** To remove the defective component, remove the covers that shield the control, and then remove the screws that secure the boards to the control panel (Figure 11-45). Disconnect the connectors from the electronic control board, display board, or the touch pad panel. On some models, the touch pad panel might be glued to the control panel. Once you remove the defective part, you will have to replace it with a new one.

7. **Install the new component.** To install a new electronic control board, display board, or touch pad panel, just reverse the disassembly procedure and reassemble. Reinstall the console panel, and restore the electricity to the dishwasher. Test the dishwasher operation.

Diagnostic Charts

The following diagnostic charts will help you to pinpoint the likely causes of the various dishwasher problems (Figures 11-46 through 11-52). Schematics are included in Figures 11-53 through 11-56.

Figure 11-46
Dishwasher diagnostic
flow chart: Dishwasher
will not run at all.

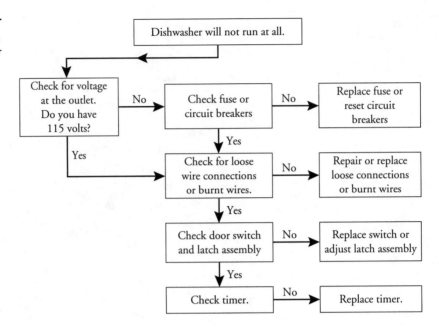

Figure 11-47
Dishwasher diagnostic
flow chart: Water not
entering the
dishwasher.

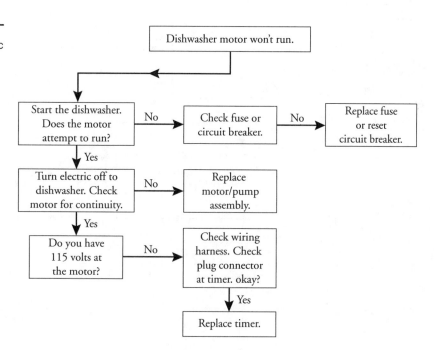

PART V

FIGURE 11-50
Dishwasher diagnostic
flow chart:
Dishwasher won't dry.

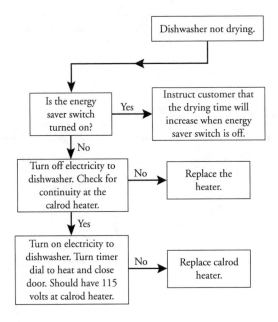

Dishwasher not drying.

Is the energy saver switch turned on? → Yes → Instruct customer that the drying time will increase when energy saver switch is off.

No ↓

Turn off electricity to dishwasher. Check for continuity at the calrod heater. → No → Replace the heater.

Yes ↓

Turn on electricity to dishwasher. Turn timer dial to heat and close door. Should have 115 volts at calrod heater. → No → Replace calrod heater.

FIGURE 11-51
Dishwasher diagnostic
flow chart: Water in
the tub at the end of
the cycle.

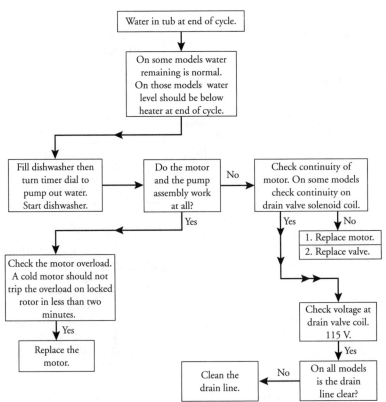

Water in tub at end of cycle.

On some models water remaining is normal. On those models water level should be below heater at end of cycle.

Fill dishwasher then turn timer dial to pump out water. Start dishwasher. → Do the motor and the pump assembly work at all? → No → Check continuity of motor. On some models check continuity on drain valve solenoid coil.

Yes

Check the motor overload. A cold motor should not trip the overload on locked rotor in less than two minutes.

Yes ↓

Replace the motor.

Yes / No

1. Replace motor.
2. Replace valve.

Check voltage at drain valve coil. 115 V.

Yes ↓

Clean the drain line. ← No ← On all models is the drain line clear?

Water valve noise chart

Complaint				
Noise	Description	When noise occurs	Which water valve parts	External factors
Whistle	Makes a "hissing" to a shrill whistle.	While the water valve is open.	Outlet assembly and hose to fill funnel, flow-washer.	* Close or open the supply water valve to change the pressure band. * Water heater temperature too high. Set the water heater thermostat to 140 degrees Fahrenheit.
Flutter	Low-frequency "rumble."	When the water valve is about to close.	Flow-washer.	
Chatter	Low-frequency pulsing.	When the water valve closes.	Spring, diaphragm.	* Long runs of unsupported pipes. You must secure the pipes to stop any vibrations. * Water supply line to the water valve is undersized. Must have 3/8" I.P.S. or 1/4" I.D. * Water heater temperature is too high. Set the water heater thermostat at 150 degrees Fahrenheit maximum.
Water hammer	The water valve will make a single "thud."	At closing.	Armature spring, diaphragm.	

FIGURE 11-52   Water valve noise chart.

FIGURE 11-53 The typical dishwasher electrical schematic: The wash cycle has already been selected, and the pushbutton switch contacts are closed.

FIGURE 11-54 This illustration shows the active circuits while the dishwasher tub is filling up with water.

FIGURE 11-55 This illustration shows the active circuits while the dishwasher is draining the water out of the tub.

FIGURE 11-56 A typical wiring schematic for a dishwasher with electronic components. (*continued*)

Turbo zone activation   Used only on some models, otherwise open, pump is also washing during sensing periods (see wash/rinse).

Drain

Drain

Vent   Used only on some models, otherwise open.

Heat dry

FIGURE 11-56   A typical wiring schematic for a dishwasher with electronic components.

Garbage Disposers

The garbage disposer provides a convenient and sanitary way to dispose of food waste (Figure 12-1). No sorting or separating of the food waste is necessary. It is designed to handle all types of food waste. If, however, the garbage disposer drains into a septic tank, organic wastes, such as egg shells, lobster, crab, and shrimp shells should be kept at a minimum. Items such as tin cans, glass, china, bottle caps, metal, etc., should never be placed in the garbage disposer because they might damage the appliance and plug the drain.

Types

There are two types of operating methods available in garbage disposers: continuous feed and batch feed.

Continuous Feed Disposers

A continuous feed disposer requires an on-off electrical switch that is remote from the disposer. Before food is placed into the hopper, the user must turn on the cold water faucet and turn on the electrical switch. Then the food waste is placed into the hopper. The food waste will be ground up and expelled into the drain.

Batch Feed Disposers

In a batch feed disposer, the on-off operation of the disposer is controlled by the stopper. The on-off switch is built into the hopper. The stopper is the component that completes the circuit that allows the disposer motor to run. The hopper will hold the food waste until the user is ready to dispose of it. The food waste will be ground up and expelled into the drain.

Principles of Operation

When the food is placed into the hopper and the disposer is running, the centrifugal force throws the food waste outward against the cutting edges of the shredder ring (Figure 12-2). The pivoting impeller arms, attached to the flywheel, push the food waste around and into the teeth of the shredder ring (Figure 12-3). As the food waste is pushed against and cut by the shredder ring, water running into the hopper flushes the ground food waste between the flywheel and the shredder ring, washing it into the drain housing assembly, where it is expelled into the drain.

FIGURE 12-1   Exploded view of a batch feed garbage disposer.

FIGURE 12-2   Food waste is forced outward against the shredder ring.

FIGURE 12-3   Impeller directing food waste into the teeth of the shredder ring.

Safety First

Any person who cannot use basic tools or follow written instructions should not attempt to install, maintain, or repair any garbage disposers. Any improper installation, preventive maintenance, or repairs could create a risk of personal injury or property damage.

If you do not fully understand the installation, preventive maintenance, or repair procedures in this chapter, or if you doubt your ability to complete the task on the garbage disposer, call your service manager.

These precautions should also be followed:

- Do not put your fingers or hands into the garbage disposer.
- When removing foreign objects, always use a long-handled pair of tongs or pliers.
- To reduce the risk of flying debris that can be expelled by the disposer, always be sure that the splash guard is properly installed.
- When replacing the garbage disposer, always be sure that it is properly grounded.
- Before attempting to free a jam in the garbage disposer, always disconnect the electric supply to the disposer.

Garbage Disposers in General

Much of the troubleshooting information in this chapter covers the various types of garbage disposers in general, rather than specific models, in order to present a broad overview of service techniques. The pictures and illustrations that are used in this chapter are for demonstration purposes only, to clarify the description of how to service these units, and in no way reflect on a particular brand's reliability.

To Free Jams from Foreign Objects

When foreign material falls into the disposer and the motor jams, it will trip the motor overload protector—otherwise known as the reset button (see Figure 12-1). When this happens, follow these steps to free the jam:

1. Turn off the electricity to the disposer, and shut off the water.
2. Insert the service wrench into the center hole at the bottom of the disposer (Figure 12-4), and work the wrench back and forth until the motor assembly is turning freely.
3. To free a tight jam, insert a prying tool alongside the grinding protrusion near the outside edge of the flywheel. Be sure that you place the prying tool on the proper side of the protrusion so that when pressure is applied, the flywheel will move (Figure 12-5).
4. Remove the foreign object with tongs.
5. Wait for the motor to cool for a few minutes before you press the reset button.
6. Reconnect the electric supply, turn on the cold water, and test the disposer.

FIGURE 12-4 Insert service wrench, turn wrench, and reset the overload protector button.

Remove splash guard

Wooden broom or mop handle

Turn flywheel

FIGURE 12-5 Insert prying tool into the hopper.

Installation of Garbage Disposer

Figure 12-6 shows some typical installations of garbage disposers. Every newly purchased garbage disposer comes with installation instructions, a use and care manual, and a warranty. The steps taken to install a new disposer are as follows:

1. Read the use and care manual.

2. Clean out the sink's drain line.

3. Disconnect the electrical supply:

   a. Continuous feed disposers need a wall switch and a receptacle.

   b. Batch feed disposers need a receptacle or must be wired directly.

4. Be sure to observe all local codes and ordinances for electrical and plumbing connections when installing or repairing disposers.

5. Follow the manufacturer's installation instructions.

6. After completing the installation, check for water leaks.

7. Turn on the electricity, and test the operation of the garbage disposer.

Garbage Disposer Maintenance

The garbage disposer is permanently lubricated; thus, it never has to be oiled. When the disposer is used properly, it cleans itself. If there is an odor coming from the inside of the disposer, you can deodorize it. To do this, take some orange or lemon rinds and grind them up in the disposer. This will dispel unpleasant odors and leave the sink with a sweet smell. Another way to deodorize the disposer is to take about a dozen or so ice cubes sprinkled with a generous amount of household scouring power and grind them up in the disposer

FIGURE 12-6
Typical garbage
disposer installations.

Double bowl center outlet

Double bowl end outlet

Single bowl

without running the water. Flush the disposer for one minute. This will allow any debris to be expelled into the drain.

Drain Blockage

To avoid drain blockage when using the garbage disposer, allow the cold water to flow for a sufficient time after grinding the food waste to be sure that all of the waste is flushed away. The ground waste and water mixture flows at the rate of two seconds per foot in a horizontal drain line. It is recommended that the user allow the water to flow for a minimum of 15 to 30 seconds after grinding the food waste.

The use of cold water in the garbage disposer will congeal and harden the grease, making its disposal easier. Never use chemicals or solvent drain compounds, because they can cause serious damage to the disposer.

The disposer should be used daily to flush the lines. If the dishwasher drain hose is connected to the disposer, it, too, should be used daily to prevent the dishwasher drain hose from becoming clogged.

Step-by-Step Troubleshooting by Symptom Diagnosis

In the course of servicing an appliance, don't tend to overlook the simple things that might be causing the problem. Step-by-step troubleshooting by symptom diagnosis is based on diagnosing malfunctions, with their possible causes arranged into categories relating to the operation of the garbage disposer. This section is intended only to serve as a checklist to aid you in diagnosing a problem. Look at the symptom that best describes the problem you are experiencing with the garbage disposer, and then proceed to correct the problem.

Overload Protector Trips

1. Check for voltage at the garbage disposer.

2. Does the motor hum when the disposer is turned on?

3. Was the foreign material removed from the hopper? Make sure that the flywheel turns easily.

4. Does the motor stay in the start winding? Use the ammeter to perform this check. Look at the name plate for the correct amperage rating.

5. Did you check the overload protector?

6. Is there a shorted or open winding in the motor?

7. Is the start relay okay?

Erratic Operation

1. Are there any loose connections? For example: wiring, switch, motor, etc.

2. Is the stopper worn or broken?

3. Is the wall switch defective?

4. Is the garbage disposer wired correctly?

Garbage Disposer Won't Run

1. Did the overload protector trip?

2. Is the fuse or circuit breaker blown or tripped?

3. Is the wall switch okay?

4. Are the motor windings burned out?

5. Check for open or shorted wiring.

6. Is the relay defective?

Garbage Disposer Won't Stop

1. Is the wall switch defective?

2. Check for a short in the wiring.

3. Check for incorrect wiring.

Slow Water Flow from Garbage Disposer

1. Check to see if the drain is partially clogged.
2. Check the shredder teeth, and see if they are clogged with food waste. This is caused by insufficient water flow.

Garbage Disposer Is Grinding Slowly

1. Check for undisposable matter in the hopper.
2. Is the flywheel damaged?
3. Is there sufficient water flow?
4. Is the shredder ring worn or broken?

Abnormal Noise in the Garbage Disposer

1. Check for undisposable matter in the hopper.
2. Are the mounting screws loose?
3. Is the flywheel broken?
4. Check the motor bearings—they might be damaged or worn.

Garbage Disposer Leaks

1. Disconnect the electricity. Then locate the water leak.
2. Check for loose mounting screws.
3. Check the sink flange. There might be insufficient putty around the flange.
4. Check for leaks around the tail pipe gasket.
5. Check for holes in the hopper.
6. Is the dishwasher/disposer connector cracked and leaking?
7. Is water leaking through the motor assembly? The seals might be faulty.

Repair Procedure

The repair procedure in this chapter is a complete inspection and repair process for the garbage disposer components. It contains the information you need to test a component that might be faulty and to replace it, if necessary.

Garbage Disposer Assembly

The typical complaints associated with garbage disposer failure are:

- When the motor runs, there are loud noises.
- Water is leaking from the bottom of the disposer.
- The fuses or circuit breaker will trip when the disposer is started.

To handle this problem, perform the following steps:

1. **Verify the complaint.** Verify the complaint by operating the disposer. Listen carefully, and you will hear if there are any unusual noises or if the overload protector trips.

2. **Check for external factors.** You must check for external factors not associated with the garbage disposer. Is the disposer installed properly? Does the disposer have the correct voltage?

3. **Disconnect the electricity.** Before working on the garbage disposer, disconnect the electricity. This can be done by pulling the plug out of the wall receptacle. Be sure that you only remove the disposer plug. Or disconnect the electricity at the fuse panel or at the circuit breaker panel. Turn off the electricity.

4. **Remove the garbage disposer.** To access the disposer assembly, the disposer must be removed. Start by disconnecting the drain line from the disposer discharge tube. Next, remove the dishwasher drain hose connection, if so equipped (Figure 12-7a). Insert the service wrench into the right side of the flange body. *Caution:* Place one hand under the disposer to keep it from falling.

   Now, turn the flange to the left to free the disposer from its mounting (Figure 12-7b). Turn the disposer upside down, and place it on a protective surface. This will protect the floor in the cabinet from damage. Remove the terminal plate (Figure 12-7c). Remove the ground wire and wire nuts from the service cord (Figures 12-7d and 12-7f). Separate the service cord wires from the motor wires. On models that are wired directly, loosen the cable clamp screws and remove the cable from the disposer (Figure 12-7e).

5. **Disassemble the garbage disposer.** Place the garbage disposer upside down on a protective surface or on a workbench. Some models have an insulated outer shell that must be removed first. Most disposers will separate into four or five sections, depending on the model you are servicing.

   Disposers that separate into four sections consist of:

   - Top container or hopper body
   - Upper-end bell or drain, housing assembly, cutting elements, and rotor and shaft assembly
   - Stator
   - Lower-end bell assembly

   Disposers that separate into five sections consist of:

   - Top container or hopper assembly
   - Stationary shredder
   - Upper-end bell or drain, housing assembly, rotor shredder, and rotor and shaft assembly
   - Stator
   - Lower-end bell assembly

   Before removing any screws from the disposer, you must place scribe marks on the lower-end bell and stator, the upper-end bell and stator, and the upper-end bell and hopper. These scribe marks will be used to align the sections properly when the

Drain line

Dishwasher
drain hose

(a)

(b)

Terminal
plate

Wire nuts

Green ground
screw

Cable clamp

(c)

(d)

Wire nuts

Wire nuts

Green ground
screw

Cable clamp

(e)

G

Ground screw

(f)

FIGURE 12-7   The steps taken to remove the garbage disposer.

garbage disposer is reassembled. Remove the four through-bolts from the lower-end bell assembly, and separate the sections of the disposer. Label the wires from the stator to the lower-end bell assembly. Disconnect them. Be cautious if the garbage disposer has a capacitor.

WARNING *Do not touch capacitor terminals. A charged capacitor can cause severe shock when both terminals are touched. A charged capacitor will hold a charge until it is discharged.*

6. **Inspect and test the garbage disposer components.** Inspect the hopper assembly for cracks, holes, and (on some models) corrosion around the dishwasher connection. Also inspect the gasket area for nicks or rough spots. Inspect the gasket or "O" ring for cuts, breaks, or worn areas.

Replace any defective parts. Inspect the upper-end bell or drain housing assembly for defective parts. Test the centrifugal switch for binding by manually operating it. It should slide up and down freely. Inspect the rotor and thrust washer for cracks, breaks, or wear. If the upper drain housing and rotor assembly fail inspection, replace the garbage disposer as a complete unit.

Before testing the capacitor, it must be discharged. Use a screwdriver with an insulated handle. Discharge the capacitor by shorting across both terminals. Remove the wire leads, one at a time, with a pair of needle nose pliers. Set the ohmmeter on the highest scale, place one probe on one terminal, and then attach the other probe to the other terminal (Figure 12-8).

Observe the meter action. While the capacitor is charging, the ohmmeter will dip toward zero ohms for a short period of time. Then the ohmmeter reading will

FIGURE 12-8 Place ohmmeter test leads on the capacitor terminals.

return to infinity. If the ohmmeter reading deflects to zero and does not return to infinity, the capacitor is shorted and should be replaced.

If the ohmmeter reading remains at infinity, the capacitor is open and should be replaced.

Inspect the stator windings for nicks, cuts, or burned spots. Set the ohmmeter on R × 1, and test the stator windings for open, grounded, or shorted windings. Test both the start and run windings. If the start or windings fail the test, replace the garbage disposer as a complete unit. Set the ohmmeter on R × 1, and test the start switch. Place the probes on the wires leading to the switch. To ensure an accurate meter reading, isolate the start switch wires from the remainder of the circuitry. The meter reading will indicate an open circuit. If the test fails, replace the start switch. Next, test the overload protector. Isolate the overload protector wires from the remainder of the circuitry. Then place the ohmmeter probes on the overload protector wires. Push the red reset button on the overload protector. The ohmmeter reading will indicate a closed circuit. If this test fails, replace the overload protector.

FIGURE 12-9 Illustrating the steps taken to reinstall the garbage disposer.

7. **Reassemble and reinstall the garbage disposer.** To reassemble the disposer, just reverse the disassembly procedure, and reassemble. Check to be sure that all wiring is correct. Remember to use the scribe marks to align the sections properly before inserting the through-bolts. Also, be careful not to damage any of the gaskets or "O" rings when assembling the garbage disposer. Reconnect the service cord wires to the motor wires, and tighten the cable clamp screws (see Figure 12-7e). Next, reconnect the ground wire to the green ground screw (see Figures 12-7d and 12-7f). Reinstall the terminal plate and screw (Figure 12-9a).

FIGURE 12-10   Before installing a new garbage disposer, remove the dishwasher connection plug if a dishwasher drain hose will be connected to the disposer.

Lift the disposer, and position it so that its three mounting ears are lined up under the ends of the sink mounting assembly screws (Figures 12-9b and 12-9c). The disposer will now hang by itself. After the plumbing is reconnected, you will tighten the ring using the service wrench. Rotate the disposer to align the discharge tube with the drain trap (Figure 12-9d). Reconnect them. Next, reconnect the dishwasher drain hose connection, if so equipped (Figure 12-9e). Finally, place the service wrench into the left side of one of the disposer mounting lugs, located on the top of the disposer. Then turn the wrench to the right until it is firmly secured, engaging the locking notch (Figure 12-9f).

Before you reconnect the electricity, check for leaks by running the water. Inspect all connections. Inspect the disposer for water leaks also. If no leaks are present, turn on the electricity and run the disposer; reinspect for any water leaks.

Before installing a new garbage disposer, the dishwasher connection plug must be removed if the defective disposer was equipped with a dishwasher drain hose connection. The dishwasher connection plug is removed by knocking the plug out with a hammer and some blunt instrument, such as a dowel or a punch (Figure 12-10). Do not use a screwdriver or any sharp tool. This can be driven through the plug, and make it difficult to remove. Remove the plug from the hopper before installing or operating the disposer.

Diagnostic Charts

The following diagnostic flow charts and wiring diagrams will help you to pinpoint the likely cause of the problem (Figures 12-11, 12-12, and 12-13).

FIGURE 12-11
Garbage disposer flow
chart: overload
protector trips.

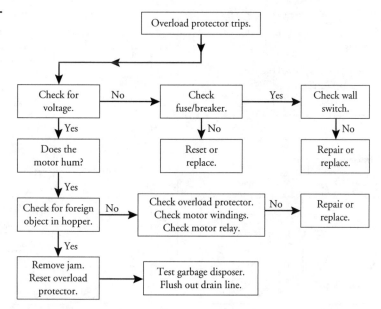

FIGURE 12-12
Garbage disposer
flow chart: Garbage
disposer motor won't
run.

Figure 12-13 Typical household garbage disposer wiring diagrams.

Electric Water Heaters

Electric water heaters are heat-producing appliances. They normally have two heating elements (Figure 13-1), with each element controlled by a thermostat. The thermostats are mounted on the outer wall of the water heater tank just above the elements, from where they sense the temperature of the water through the outer wall of the water heater tank. A temperature and pressure-relief valve, which is mounted on the outside of the tank, is a device applied to a water heater, which will open to pass water or steam if excessive pressure or temperature occurs in the water heater tank.

Electric water heaters are available in many different heights, widths, and capacities.

Principles of Operation

The water heater tank is full of water, the electricity is on, and the upper thermostat (see Figure 13-1) senses that the water is cold. This condition energizes the upper element. At the same time, the lower element is not activated. The upper heating element will heat approximately one-quarter of the tank's capacity. When the temperature of the water reaches the thermostat setting, the upper thermostat will shut off. Then the lower thermostat becomes energized and heats the remainder of the water in the tank. When the temperature of the water reaches the thermostat setting, the lower thermostat will deactivate the lower heating element. The tank is now filled with hot water. As the consumer uses the hot water, it is drawn from the top of the tank and is replaced with cold water through a dip tube (see Figure 13-1) located near the bottom of the tank. When the lower thermostat senses the cold water, the thermostat activates the lower heating element, heating the water. Figure 13-2 identifies one of the many types of thermostats. The figure also shows a schematic drawing of the thermostat switch contacts.

Safety First

Any person who cannot use basic tools or follow written instructions should *not* attempt to maintain or repair any electric water heaters. Any improper installation, preventive maintenance, or repairs could create a risk of personal injury or property damage.

FIGURE **13-1**
Exploded view of the
electric water heater.

FIGURE **13-1**
Exploded view of the
electric water heater.

1. Top pan
2. Anode
3. T & P relief valve
4. Jacket
5. Insulation
6. Tank
7. Bottom pan
8. Draincock
9. Draincock panel
10. Front panel
11. Nipple
12. Dip tube
13. Junction bracket
14. Thermostat-upper
15. Thermostat-lower
16. Thermostat-bracket
17. Element-upper
18. Element-lower
19. Element gasket

If you do not fully understand the preventive maintenance or repair procedures in this chapter, or if you doubt your ability to complete the task on your electric water heater, please call your service manager.

These precautions should also be followed:

- Before checking the water heater, turn off the electricity.

- Never restore the electricity to the water heater if the tank is empty; do so only after it is full of water.

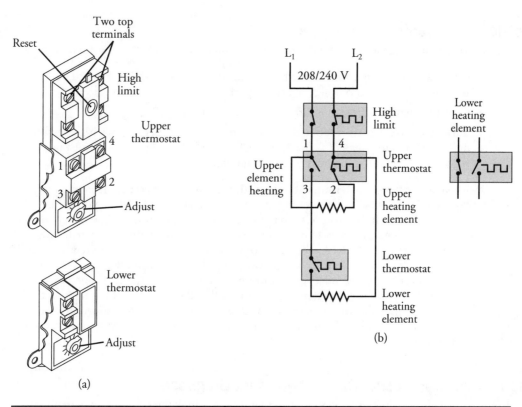

FIGURE 13-2   (a) Typical upper and lower water heater thermostats. (b) Wiring schematic for upper and lower thermostats.

- Never remove the heating elements with the tank full of water.
- If a water heater needs to be replaced, it is strongly recommended that all electrical, plumbing, and placement of the tank be done by qualified personnel. Observe all local codes and ordinances for electrical, plumbing, and installation procedures.

Before continuing, take a moment to refresh your memory concerning the safety procedures in Chapter 2.

Electric Water Heaters in General

Much of the troubleshooting information in this chapter covers electric water heaters in general, rather than specific models, in order to present a broad overview of service techniques. The pictures and illustrations that are used in this chapter are for demonstration purposes only, to clarify the description of how to service these heaters. In no way do they reflect on a particular brand's reliability.

Electric Water Heater Maintenance

Every so often, inspect the water heater tank for possible water leaks. Check the following:

- Check all pipe connections to the tank. If corroded, they must be repaired before there is serious water damage to the property.
- Check the temperature/pressure-relief valve. Is it leaking?
- Turn off the electricity; remove the access panels; and inspect the wiring, elements, and insulation for signs of water leakage. If water is leaking, the leak must be repaired or the unit must be replaced immediately.

Once a year, the water heater tank should be drained and flushed out. This will increase the life expectancy of the tank by removing any unwanted sediment. Before you begin draining the tank, turn off the electricity to the water heater. Then turn off the water supply to the tank. Connect a garden hose to the draincock, and open the valve. To increase the water flow draining out of the tank, open the hot water faucet. When you are ready to refill the tank, close the draincock and turn on the water supply to the water heater.

Leave the closest hot water faucet open during refilling; the air that is trapped in the tank will escape through the hot water faucet. When all of the air is out of the tank and water lines, turn off the hot water faucet. Go to all of the other hot water faucets and open them to remove air in the lines. Turn off all water faucets. Now you are ready to turn on the electricity to the water heater.

Step-by-Step Troubleshooting by Symptom Diagnosis

In the course of servicing an appliance, don't tend to overlook the simple things that may be causing the problem. Step-by-step troubleshooting by symptom diagnosis is based upon diagnosing malfunctions, with their possible causes arranged into categories relating to the operation of the water heater. This section is intended only to serve as a checklist to aid you in diagnosing a problem. Look at the symptom that best describes the problem you are experiencing with the water heater, and then proceed to correct the problem.

No Hot Water

1. Do you have the correct voltage at the water heater?
2. Check for loose wiring.
3. Make sure that the reset button did not trip.
4. With the electricity off, check the thermostats for continuity.
5. With the electricity off, check the heating elements for continuity. Also check for a grounded heating element.

Not Enough Hot Water

1. Is the water heater undersized for the usage?
2. Check the lower heating element for continuity. Also check for a grounded heating element.

3. Check for leaking faucets and pipes.

4. Check to see if anyone is wasting water.

5. Drain the tank and remove the lower element. See if any sediment is in the bottom of the tank or around the heating element.

6. Is there any lime formation on the elements?

7. Are the thermostats operating properly? Run a cycle test.

8. Check the thermostat settings. Verify if they are set too low.

Water too Hot

1. Check to see if the thermostat is snug against the tank.

2. Is the thermostat set too high?

3. Are the thermostats operating properly? Run the cycle test.

4. Check for a grounded element.

Water Heater Element Failure

1. Check for loose or burned wiring connections.

2. Check for the correct voltage to the elements.

3. Is the element shorted or grounded? Test the elements for continuity.

Discolored or Rusty Water

1. Drain the tank and remove the lower element. See if there is any sediment in the bottom of the tank.

2. Is the water supply to the water heater rusty in color?

3. Check the water for softness.

4. Check to see if there are excessive mineral deposits.

5. Inspect the anode rod to note its condition. If it has dissipated, replace the anode.

Repair Procedures

Each repair procedure is a complete inspection and repair process for a single water heater component, containing the information you need to test a component that may be faulty and to replace it, if necessary.

Upper Thermostat

The typical complaints associated with the upper thermostat are:

- No hot water
- Burned wires
- Water heater runs continuously
- The fuse blows or the circuit breaker trips

To handle this problem, perform the following steps:

1. **Verify the complaint.** To verify the complaint, turn on the hot water. Does the water get hot?

2. **Check for external factors.** You must check for external factors not associated with the water heater. Is the water heater installed properly? Does the water heater have the correct voltage?

3. **Disconnect the electricity.** Before working on the water heater, disconnect the electricity. This can be done by pulling the plug out of the electrical outlet. Or disconnect the electricity at the fuse panel or at the circuit breaker panel. Many installations have a disconnect switch/breaker box near the heater. Turn off the electricity.

4. **Remove the access panel.** To gain access to the upper thermostat, the access panel must be removed (see Figure 13-1). The access panel is held on with two screws on most models. Peel back the insulation and then remove the thermostat's protective cover.

5. **Test the thermostat.** To test the upper thermostat, use your voltmeter. Set the range on the 300-volt scale. Touch the probes to terminals 1 and 3 above the reset button (Figure 13-3). Turn on the electricity. If the meter reads no voltage, have the customer call an electrician to find out where the power failure is. If the meter reads the proper rated voltage, as stated on the name plate, the electricity is okay. Next, touch the probes to terminals 2 and 4 below the reset button (Figure 13-4). If the meter reads the proper rated voltage, the electricity is okay. If the meter reading is

FIGURE 13-3 Place the voltmeter probes on terminals 1 and 3 to check the voltage.

FIGURE 13-4 Place the voltmeter probes on terminals 2 and 4 to check the voltage.

FIGURE 13-5   Remove the wires from the element. Place the ohmmeter probes on terminals 1 and 2. When the thermostat is calling for the element to come on, the ohmmeter scale will read continuity.

FIGURE 13-6   Place the ohmmeter probes on terminals 1 and 4. If water in the tank is cold, the ohmmeter scale will read no continuity.

"0" volts, press the reset button on the thermostat. If there is still no voltage reading on the meter scale, either the water temperature in the tank is too hot or the thermostat is inoperative. At this stage, turn off the electricity to the tank again. Remove the wires from the upper heating element (Figure 13-5). Be absolutely sure that the electricity is off before any further tests are made using your ohmmeter.

Use your ohmmeter to test terminals 1 and 2 on the upper thermostat (see Figure 13-5). If the thermostat is calling for heat, the ohmmeter needle will swing to the right, showing continuity. Now touch the probes to terminals 1 and 4 on the thermostat (Figure 13-6). If the water in the tank is cold, you will not get a reading on the ohmmeter scale. If the water in the tank is hot, as called for by the thermostat, the needle will swing to the right, thus showing continuity.

6. **Cycle the upper thermostat.** Reconnect the wires to the heating element. Turn on the electricity to the tank. Take your ammeter and clamp the jaws around the wire that goes from terminal 4 to the upper element (Figure 13-7). With cold water in the tank and the thermostat calling for heat, you should get a reading on the meter in amps. The reading you get will depend on the rating (in watts) of the heating element, divided by the voltage supplied (see the formula in Figure 13-8). After the temperature is reached, the needle on the ammeter scale will read "0."

   Next, place the ammeter jaws on the terminal 4 wire that goes from the upper thermostat (just below the temperature gauge setting) to the lower element (Figure 13-9). The ammeter should show a reading, indicating that the lower element is working.

7. **Replace the upper thermostat.** Turn off the electricity to the water heater. To replace the upper thermostat, remove all of the wires from the thermostat. Be sure that you mark the wires so that you can replace them exactly the way you took them off. Next, pry back the thermostat bracket with your fingers far enough to slide the thermostat up and out; remove the thermostat. To install the new thermostat, just reverse the disassembly procedure, and reassemble. Be sure that the thermostat is snugly mounted against the tank so that it can sense the temperature better.

8. **Test the new upper thermostat.** After you have completed the new installation, turn on the electricity to the water heater. Repeat this testing process as described, and verify that the new unit is working correctly.

FIGURE 13-7   The ammeter jaws encircling the terminal 4 wire go to the element.

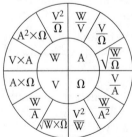

Conversion chart for determining amperes, ohms, volts, or watts (Amperes = A, Ohms = Ω, Volts = V, Watts = W)

FIGURE 13-8   Ohm's Law Equation Wheel: The conversion chart for determining amperes, ohms, volts, or watts.

FIGURE 13-9   Place the ammeter jaws around terminal 4. The ammeter reading is 0.

Lower Thermostat

Testing the lower thermostat is similar to testing the upper thermostat, with the following exception. With the wires removed from the element, place the ohmmeter's probes on each terminal. Take a screwdriver, place it on the set screw of the thermostat, and turn it clockwise. The needle on your ohmmeter will swing to the right, indicating that there is continuity. If the needle does not move, the thermostat needs to be replaced. Replace this thermostat as per the upper thermostat replacement procedure previously described. Then test the thermostat by placing the jaws of the ammeter around the wire that connects to the lower thermostat. Check to be sure that the thermostat cycles on and off.

Heating Element

The typical complaints associated with the heating element are:

- No hot water
- Burned wires
- The fuse blows or the circuit breaker trips

To handle this problem, perform the following steps:

1. **Verify the complaint.** To verify the complaint, turn on the hot water. Does the water get hot?
2. **Check for external factors.** You must check for external factors not associated with the water heater. Is the water heater installed properly? Does the water heater have the correct voltage?
3. **Disconnect the electricity.** Before working on the water heater, disconnect the electricity. This can be done by pulling the plug out of the electrical outlet. Or disconnect the electricity at the fuse panel, the circuit breaker panel, or the disconnect switch. Turn off the electricity.
4. **Remove the access panel.** In order to gain access to the heating element, the access panel must be removed (see Figure 13-1). The access panel is held on with two screws on most models. Peel back the insulation and then remove the wires from the heating element.
5. **Test the heating element.** Use the ohmmeter to test the heating element. Set the range to R × 1. Touch the probes to the element screws (Figure 13-10). If the element is good, the ohmmeter scale will show continuity. If the element is bad, the ohmmeter needle will not move, showing an open element. Now, set the ohmmeter on R × 100. Take one probe and place it on either terminal of the element. Take the other

Figure 13-10 Testing the heater element. Place the ohmmeter probes on the screw terminals of the element.

probe and touch the element head (Figure 13-11). If you get any reading, the element is grounded and should be replaced. Repeat this procedure for the other terminal on the element. Both terminals must be measured for grounding.

6. **Remove the element.** Turn off the electricity to the water heater. Next, turn off the water supply. Connect a garden hose to the draincock, and open the draincock. At the same time, open a nearby faucet to allow the water to drain faster. After the tank is empty, disconnect the wires from the element and remove it.

7. **Install the new element.** To install the new element, just reverse the disassembly procedure, and reassemble. When installing the new element, always replace the gasket. Before you insert the new element into the tank, clean the flange of debris.

8. **Test the new element.** When you are done installing the replacement element, close the draincock, and open the water supply. Close the nearby faucet after all of the air in the tank has

FIGURE 13-11   Testing the heater element for ground.

dissipated. Next, check for water leaks. Turn on the electricity to the water heater. Use the ammeter to check the element. The ammeter should show a reading.

FIGURE 13-12   Pictorial wiring diagram for a standard two-wire 208/240 volt circuit, single element.

FIGURE 13-13   Pictorial wiring diagram for a standard two-wire 208/120 volt circuit, single element.

Diagnostic Charts

The following diagnostic flow charts, wiring diagrams, and tables will help you to pinpoint the likely causes of the problem (Figures 13-12 through 13-19 and Tables 13-1 and 13-2).

FIGURE 13-14   Pictorial wiring diagram for a standard two-wire interlocking 208/240/277 volt circuit, double element.

FIGURE 13-15   Pictorial wiring diagram for a four-wire, off-peak double element, non-simultaneous operation.

FIGURE 13-16
Water heater flow
chart: Circuit breaker
out or fuse blown.

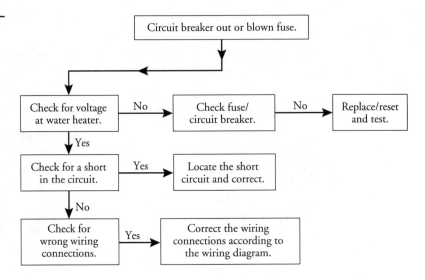

FIGURE 13-17
Water heater flow
chart: Water heater
runs continuously.

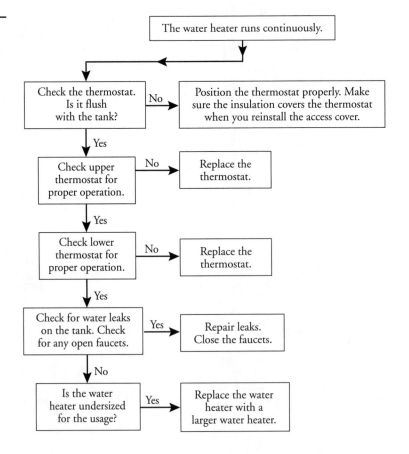

Figure 13-18
Water heater flow
chart: Relief valve
opens.

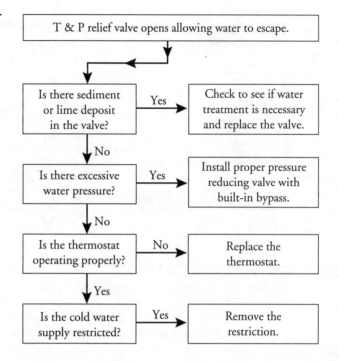

Figure 13-19
Water heater flow
chart: Water heater
tank bulged.

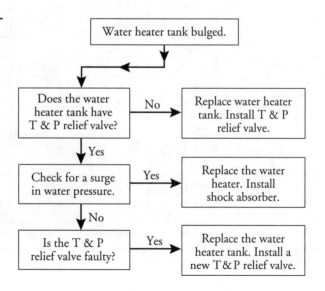

PART V

Activity	Gallons
Handwashing	0.9
Shower	5–7
Tub bath	7–12
Shaving	2
One meal per person	2–5
Automatic washer	11–34
Conventional washing	30
Dishwasher	8
Average Bathroom Use Per Day	
Adult	12
Child	24

TABLE 13-1   Typical Hot Water Usage Chart.

GPH Recovery at Indicated Temperature Rise					
Heating Element Wattage	**Degrees Fahrenheit**				
	60	70	80	90	100
750	5.1	4.4	3.8	3.4	3.1
1000	6.8	5.8	5.1	4.5	4.1
1250	8.5	7.3	6.4	5.7	5.1
1500	10.2	8.8	7.7	6.8	6.1
2000	13.7	11.7	10.2	9.1	8.2
2500	17.1	14.6	12.8	11.4	10.2
3000	20.5	17.5	15.4	13.6	12.3
3500	23.9	20.5	17.9	15.9	14.3
4000	27.3	23.4	20.5	18.2	16.4
4500	30.7	26.3	23.0	20.5	18.4
5000	34.1	29.2	25.6	22.7	20.5
5500	37.6	32.2	28.2	25.0	22.5
6000	41.0	35.1	30.7	27.3	24.6

TABLE 13-2   Electric Water Heater Recovery Chart.

Help on the Internet

To obtain additional technical help or information on selecting an electric water heater, you can go to the following Web sites:

http://www.aceee.org/consumerguide/topwater.htm

http://www.eere.energy.gov/consumer/your_home/water_heating/index.cfm/mytopic=12770

http://www.americanwaterheater.com/

http://www.rheem.com/

http://www.hotwater.com/

http://www.tankless-water-heater.com/

http://www.bradfordwhite.com/

http://www.marathonheaters.com/

http://www.statewaterheaters.com/

Gas Water Heaters

Gas water heaters are similar to electric water heaters, with the exception of how the water is heated. They use liquefied petroleum gas (LP or LPG) or natural gas to heat the water. The tank is constructed in such a way that it allows for a vent (chimney) up the middle of the tank, along with a gas burner on the bottom of the tank to heat the water (Figure 14-1). The thermostat that controls the water temperature and gas flow is mounted on the outside of the tank, with the temperature probes inserted into the tank to sense the water temperature. A temperature and pressure-relief valve, which is mounted on the outside of the tank, is a device applied to a water heater that will open to pass water or steam if excessive pressure or temperature occurs in the water heater tank.

Gas water heaters are available in many different sizes. When purchasing a gas water heater, it is important to specify the type of gas fuel that will be used to heat the water.

Principles of Operation

The gas water heater fills up in the same way as an electric water heater. Water enters the cold-water inlet port through a dip tube, filling the tank from the bottom up. Once the tank is full of water, the gas supply is turned on and the consumer or technician can begin the process of lighting the burner according to the manufacturer's use and care manual. Once the pilot light is lit, the thermocouple, which is near or in the gas flow, will begin to heat up and send a signal to the gas combination valve to open the gas inlet to the pressure regulator section of the combination control valve. At the same time, the thermostat senses the cold water and sends a signal to the combination control to actuate a lever from within the pressure regulator to open the gas valve, allowing the gas to enter the burner for ignition. While the gas flames are heating the water, the thermostat begins monitoring the water temperature. When the water temperature reaches a preset temperature, a signal will be sent to the combination valve to Shutoff the gas flow.

Safety First

Any person who cannot use basic tools or follow written instructions should *not* attempt to maintain or repair any gas water heaters. Any improper installation, preventive maintenance, or repairs could create a risk of personal injury, death, or property damage.

FIGURE 14-1
Exploded view of a gas water heater.

1.	Thermostat
2.	Pilot supply tube
3.	Thermocouple
4.	Burner supply tube
5.	Burner orifice
6.	Burner
7.	Pilot burner
8.	Drain valve shroud
9.	Drain valve
10.	Jacket door
11.	Inner door
12.	Flue baffle
13.	Anode rod
14.	Draft hood
15.	T&P relief valve

If you do not fully understand the preventive maintenance or repair procedures in this chapter, or if you doubt your ability to complete the task on your gas water heater, please call your service manager.

These precautions should also be followed:

- Before servicing the water heater, turn off the gas supply.

- Never turn on the gas to the water heater if the tank is empty; do so only after it is full of water.

- At least once a year perform maintenance on the water heater according to the manufacturer's recommendations.

- If a water heater needs to be replaced, it is strongly recommended that all gas connections, plumbing, ventilation, and placement of the tank be done by qualified personnel. Observe all local codes and ordinances for gas fuels, venting, plumbing, and installation procedures.

Before continuing, take a moment to refresh your memory concerning the safety procedures in Chapter 2. In addition, please follow the manufacturer's recommendations for safety as outlined in the use and care manual.

Do not store any flammable materials near a gas water heater, and avoid ignition sources in the event you smell gas leaking. Make sure you have proper ventilation when servicing the water heater to prevent carbon monoxide (CO) poisoning.

Gas Water Heaters in General

Much of the troubleshooting information in this chapter covers gas water heaters in general, rather than specific models, in order to present a broad overview of service techniques. The pictures and illustrations that are used in this chapter are for demonstration purposes only, to clarify the description of how to service these heaters. In no way do they reflect on a particular brand's reliability.

Gas Water Heater Maintenance

Maintenance on gas water heaters *must* be performed annually. The following checklist will assist you in performing the maintenance procedures:

1. A visual inspection must be made to the following components:

   - **The venting system:** A properly working ventilation system will prevent toxic fumes from entering the home. Check the venting system for leakage, corrosion, obstructions, sooting, and rust forming on top of the water heater. If any of these conditions are present, you must take action to correct these problem areas to protect the health and lives of the people living in the home. Most important, check the fitting (vent cap) on the end of the vent pipe for obstructions.

     A properly designed and working vent system will convey the combustion gases from the water heater environment to the atmosphere outside of the home. It will also protect the home from fire hazards due to overheating of walls and other surfaces, prevent condensation of hot combustion gases, and provide clean air circulation and adequate oxygen supply for the water heater.

     Check for adequate air supply to the area that the water heater is located in. A gas water heater will consume about 12.5 cubic feet of air (at sea level) for every 1000 BTUs of heat generated.

   - **The burner components:** Inspect the burner assembly and the pilot assembly. If your inspection reveals that the burner, pilot, or base assemblies need cleaning, turn off the gas supply and remove the burner and pilot assemblies for cleaning. Some models have a removable screen at the bottom of the tank that keeps dust, lint, and oils from entering the combustion chamber. This filter screen must also be cleaned.

   - **Surrounding area:** Clean the area surrounding the water heater of dust, lint, and debris. Make sure that combustible materials are removed completely from the area.

2. Inspect the temperature and pressure-relief valve (T&P) for leaks. If it is leaking water, you must replace it. Make sure that the valve is connected to a drain line that terminates to a drain or to the outside of the residence. Next, you will manually operate the T&P valve and check for proper operation. Be careful—the discharged water is very hot. Lift the T&P lever gently and then release it—the lever should return to its normal closed position. If the water is still leaking out of the valve after you release the lever, replace the valve.

3. Once a year, the water heater tank should be drained and flushed out. This procedure will increase the life expectancy of the tank by removing the unwanted sediment. Before you begin draining the tank, turn off the gas supply to the water heater. To drain the water heater tank, turn off the water supply to the tank. Connect a garden hose to the draincock, and open the valve. To increase the water flow draining out of the tank, open the hot water faucet—this will increase the flow of water draining out of the tank. When you are ready to refill the tank, close the draincock, turn on the water supply to the water heater, and begin to refill the tank.

Leave the closest hot water faucet open during refilling; the air that is trapped in the tank will escape through the hot water faucet. When all of the air is out of the tank and water lines, close the hot water faucet. Go to all of the other hot water faucets, and open them to remove the air in the lines. Close all water faucets. Now you are ready to turn on the gas supply to the water heater. Light the pilot light according to the manufacturer's use and care guide for the model you are servicing.

Step-by-Step Troubleshooting by Symptom Diagnosis

In the course of servicing an appliance, don't tend to overlook the simple things that may be causing the problem. Step-by-step troubleshooting by symptom diagnosis is based upon diagnosing malfunctions, with their possible causes arranged into categories relating to the operation of the water heater. This section is intended only to serve as a checklist to aid you in diagnosing a problem. Look at the symptom that best describes the problem that you are experiencing with the water heater and then proceed to correct the problem.

Gas Odors

1. Check the pilot light.
2. Check for a gas leak in the supply line.
3. Check the ventilation system.

The Burner Will Not Ignite

1. Make sure that the gas supply to the water heater is turned on.
2. Check the gas line for debris. Also, check the pilot line for debris.
3. Check the line to the burner for a blockage. If blocked, locate the source and correct.
4. Check the thermostat setting. It may be set too low.
5. Test the value of the thermocouple; replace if necessary.

6. Test the thermostat; replace if necessary.

7. Check the installation and fresh air supply.

The Pilot Will Not Light or Stay Lit

1. Make sure that the gas supply to the water heater is turned on.

2. Check the gas pressure to the water heater.

3. Check for air or debris in the lines. Purge the gas line.

4. Check the pilot line and orifice for debris.

5. Test the thermocouple, thermostat, and ignitor. Replace if necessary.

6. Check for drafts that can extinguish the pilot.

7. Check the intake air for obstruction. This is the air that is needed for combustion.

8. Check for flammable vapors.

Not Enough Hot Water

1. Check the thermostat; it might be set too low.

2. Check the installation for correct piping connections. Is the tank adequate for the size of the household and usage?

3. Check the home for hot water leaking from the faucets. Did you notice any wasting of hot water? If so, speak to the consumer.

4. Where are the water pipes located? Are the pipes exposed to the elements? Are they insulated? Check on the length of the piping.

5. Check the tank for sediment or lime. You may have to flush out the tank.

Slow Hot Water Recovery

1. Check for insufficient secondary air supply. Check the venting system and the burner.

2. Check the gas pressure. Is the type of gas correct for the water heater?

3. Check the thermostat setting. Is it out of calibration?

4. If this is a new installation, check the piping connections.

5. Check for wasting of water.

6. Is the water heater the correct size for the household?

The Burner Flame Is too High

1. Is the orifice too large? You can change the orifice and retest the burner.

2. Check for sufficient secondary air supply. Check the venting system and the burner.

PART V

The Flame Burns at the Orifice

1. Check the thermostat. Replace if necessary.
2. Check the gas pressure to the water heater.

The Pilot Flame Is too Small

1. Check the pilot line and orifice for debris.
2. Check the gas pressure to the water heater.

The Burner Flame Floats and Lifts Off the Ports

1. Is the orifice too large? You can change the orifice and retest the burner.
2. Check for high gas pressure.
3. Check the flue and burner. Locate the source and correct the problem.
4. Check for drafts that can blow the flame off the burner.

A Yellow-Lazy Burner Flame

1. Check for sufficient secondary air supply.
2. Check the gas pressure.
3. Check the venting system. Is the flue clogged?
4. Check the burner line for debris.
5. Check the water heater installation. Is the tank installed in a confined space?
6. Check the burner orifice for obstruction.

The Formation of Soot (Smoke and Carbon)

1. Check for sufficient secondary air supply.
2. Check the flue way, flue baffle, and burner.
3. Test the draft hood for proper ventilation.
4. Check for a yellow-lazy burner flame.
5. Check for low gas pressure to the water heater.
6. Check the thermostat. Replace if necessary.
7. Is the water heater installed in a confined space?

Condensation

Check the thermostat setting. Is it set too low?

The Water Has an Odor

Check for sulfides in the water. If necessary, replace the anode.

The Water Heater Jacket Is Discolored Above the Combustion Area

1. Check that the burner orifice is the proper size.
2. Is the flue clogged?
3. Check for high gas pressure.

Combustion Odors

1. Check for sufficient secondary air. Check the flue way, flue baffle, and the burner.
2. Is the flue clogged?
3. Is the water heater installed in a confined space?

Repair Procedures

Each repair procedure is a complete inspection and repair process for a single water heater component. It contains the information you need to test a component that may be faulty and to replace it, if necessary.

Burner/Manifold Assembly

The typical complaints associated with the burner/manifold assembly are:

- No gas to the burner.
- Yellow-lazy burner flame.
- The flames lift off the burner ports.
- The burner will not ignite.
- High temperatures. Some models have a thermal cutoff device installed to prevent the gas flow if the temperature in the combustion chamber exceeds its limit.
- Poor maintenance. It's time to clean the burner/manifold assembly.

To handle this problem, perform the following steps:

1. **Verify the complaint.** To verify the complaint, run the hot water. Does the burner come on? Check the thermal cutoff device. What color is the burner flame? Does the burner/manifold assembly need cleaning?
2. **Check for external factors.** You must check for external factors not associated with the water heater. Is the gas supply turned on? Is the water heater installed properly? Is there sufficient air supply? Is the surrounding area clear of debris?
3. **Shutoff the gas supply.** Before you begin servicing any gas components, Shutoff the supply of gas to the water heater. Turn the gas control knob on the combination gas control to the off position (Figure 14-2a). See the use and care guide for the model you are servicing.
4. **Remove the burner/manifold.** Remove the screws that secure the manifold door to the skirt (see Figure 14-2a). At the combination gas control, disconnect the following: thermocouple, ignitor wire from the ignitor button, pilot tube, and the

FIGURE 14-2a
Component identification and location on a gas water heater.

Piezo ignitor button

Gas valve/thermostat

Thermocouple

Pilot tube

Manifold tube

Viewport

Manifold door

Two piece wire connector

Manifold screw (2)

Ignitor button

Thermocouple

Ignitor

Pilot tube

Manifold tube

FIGURE 14-2b The combination control valve. The ignitor, thermocouple, manifold tube, pilot tube, and ignitor will have to be disconnected in order to remove the valve assembly.

manifold tube (Figure 14-2b). Gently push down on the manifold tube to free the manifold, thermocouple, and pilot tube. Next, carefully remove the burner/manifold assembly from the burner compartment. Try not to damage any of the internal parts.

5. **Clean the burner.** Now that the burner/manifold assembly is removed from the water heater, you can inspect the burner and clean it. Disassemble the burner from the manifold assembly by removing the two screws that secure it to the manifold (Figure 14-3). To clean the burner, use soap and hot water, rinse, and let it dry. Clean the remainder of the manifold assembly with a brush, and blow off any debris from the flame trap. Inspect the pilot tube and the manifold tube for a blockage. Before reinstalling the burner/manifold assembly, clean out the combustion chamber (Figure 14-4) with a shop vacuum, making sure that you remove all of the loose debris. Some models are manufactured with a flame arrestor to prevent flames from escaping the combustion chamber. This component must also be cleaned with a vacuum to remove debris.

FIGURE 14-3
The top and side view
of the burner/
manifold assembly.

Burner

Two piece wire
connector

Manifold door
assembly

Ignitor wire

Pilot tube

Manifold
tube

Manifold
door
gasket

Thermocouple

Pilot bracket

Viewport

Thermocouple
retainer clip

Pilot assembly

Pilot tube

Thermocouple

Manifold door
assembly

Burner

Pilot assembly

Manifold
tube

Screws

FIGURE 14-4
The burner/manifold
assembly located in
the combustion
chamber.

Tab

Flame-trap

Bracket

6. **Reinstall the burner/manifold assembly.** To reinstall the burner/manifold assembly, just reverse the disassembly procedure. Before you get started, inspect the gaskets for wear and tear, and replace them if necessary. Reconnect the pilot tube, manifold tube, thermocouple, and ignitor wire. Next, turn on the gas supply and light the burner according to the manufacturer's recommendations for the model you are servicing. Check for gas leaks by using a chloride-free soap solution. *Do not use a flame to check for gas leaks*. The soap solution will begin to bubble if a leak is present. If no leaks are found, you are ready to reinstall the outer door.

7. **Test the water heater.** Once assembled, run the water heater for a number of cycles by letting the hot water run for a while and then shutting it off. The water heater will begin cycling. Inspect the flame—it should be a blue flame. Adjust the air shutter as needed. On some models, you will have to remove the burner/manifold assembly to adjust the air shutter.

Combination Control Valve

The typical complaints associated with the combination control valve are:

- The burner will not ignite.
- Unable to control the water temperature.
- Slow hot water recovery.
- Smoke and carbon buildup (sooting).
- The combination control will not Shutoff the gas to the burner.
- Some models have a thermal cutoff device installed to prevent gas flow if the temperature in the combustion chamber exceeds its limit.

To handle this problem, perform the following steps:

1. **Verify the complaint.** To verify the complaint, check the burner operation. Test the water temperature and the thermostat setting. Does the burner cycle at the predetermined temperature? Check the thermal cutoff device. Check for sooting. Can you Shutoff the burner with the thermostat control knob?

2. **Check for external factors.** You must check for external factors not associated with the water heater. Is the water heater installed properly? Is there sufficient air supply? Is the surrounding area clear of debris? Is the ventilation system clear of debris?

3. **Shutoff the gas supply.** Before you begin servicing any gas components, Shutoff the supply of gas to the water heater. Turn the gas control knob on the combination gas control to the "off" position (see Figure 14-2a). See the use and care guide for the model you are servicing.

4. **Remove the combination gas control valve.** Turn off the combination gas control valve according to the use and care guide for the model you are servicing. Turn off the main gas supply to the water heater. Before you can remove the control valve, you first have to drain the water heater tank. To drain the water heater tank, turn off the water supply to the tank. Connect a garden hose to the draincock, and open

the valve. To increase the water flow draining out of the tank, open the hot water faucet—this will increase the flow of water draining out of the tank. While the tank is draining, disconnect the ignitor wire from the ignitor, and remove the ignitor button and bracket. Remove the manifold tube and pilot tube, and disconnect the thermocouple (see Figures 14-2a and 14-2b). Next, disconnect the ground joint union located between the gas shutoff valve and the combination valve. Disconnect the remaining pipe from the combination gas control valve. When the water heater tank is empty, you are ready to remove the control valve. Turn the valve counterclockwise and remove it from the tank.

5. **Install the combination gas control valve.** To install a combination gas control valve, just reverse the order of disassembly. Make sure that you use approved pipe joint compound or Teflon tape on the pipe connections (according to local codes) and on the threads of the control valve before you install the valve into the water heater tank. This will prevent water from leaking out of the tank. Close the draincock and turn on the water supply to the tank, allowing it to fill up. Keep a faucet open near the tank to allow air to escape from it. As for the gas lines, you will perform a leak check using a chloride-free soap and water solution when the gas supply is turned back on. Finally, check for any water leaks before lighting the burner.

Thermocouple

The typical complaints associated with the thermocouple are:

- The burner will not ignite.
- The pilot light will not remain lit.

To handle this problem, perform the following steps:

1. **Verify the complaint.** To verify the complaint, check the burner operation. Does the burner come on? Check the pilot light. Is it lit?

2. **Check for external factors.** You must check for external factors not associated with the water heater. Is the water heater installed properly? Check the gas supply line. Is it turned on? Is there sufficient air supply? Is the surrounding area clear of debris?

3. **Remove the burner/manifold.** Remove the screws that secure the manifold door to the skirt (see Figure 14-2a). At the combination gas control, disconnect the following: thermocouple, ignitor wire from the ignitor button, pilot tube, and the manifold tube (see Figure 14-2b). Gently push down on the manifold tube to free the manifold, thermocouple, and pilot tube. Next, carefully remove the burner/manifold assembly from the burner compartment. Try not to damage any of the internal parts. Remove the thermocouple two-piece connector by removing the retainer clip that secures the thermocouple connector to the manifold door (Figure 14-5). Disconnect the thermocouple from the pilot bracket (Figure 14-6).

4. **Install the new thermocouple.** To install a new thermocouple, just reverse the order of disassembly. Insert the new thermocouple into the bracket holes until it clicks into place. Now you are ready to reinstall the burner/manifold assembly and test.

Ignitor wire

Retainer clip

Manifold door

Thermocouple

Two piece wire connector

Notice: Thermocouple must not be bent in this location.

Ignitor

The typical complaint associated with the ignitor is that the burner will not light or remain lit. To handle this problem, perform the following steps:

1. **Verify the complaint.** To verify the complaint, check the burner operation. Does the burner come on? Check the pilot light. Is it lit? Turn off the gas supply and observe the ignitor electrode when activating the ignitor. Do you see a spark at the ignitor?

2. **Check for external factors.** You must check for external factors not associated with the water heater. Is the water heater installed properly? Check the wiring connections and wiring. Is the ignitor electrode broken?

Pilot bracket

Thermocouple

Pilot

0.125 in.

Tip   Electrode

3. **Remove the burner/manifold.** Remove the screws that secure the manifold door to the skirt (see Figure 14-2a). At the combination gas control, disconnect the following: thermocouple, ignitor wire from the ignitor button, pilot tube, and the manifold tube (see Figure 14-2b). Gently push down on the manifold tube to free the manifold, thermocouple, and pilot tube. Next, carefully remove the burner/manifold assembly from the burner compartment. Try not to damage any of the internal parts. Remove the ignitor electrode from the pilot bracket (Figure 14-7a and b).

4. **Install the new ignitor.** To install the new ignitor, just reverse the order of disassembly. When positioning the ignitor electrode in the pilot bracket, the spark gap should be 0.125 inches. Reinstall the burner/manifold assembly and test.

FIGURE 14-7
(a) The natural gas ignitor assembly.
(b) The LP gas ignitor assembly.

(a)

(b)

Water Heater Leakage Inspection Points

A technician must check the following areas for leakage when servicing the water heater (Figure 14-8):

A Check the draft hood for condensation. Water vapor that has formed from combustion products is an indication of a problem in the vent system. If this problem is not corrected, the top of the water heater will begin to rust.

B On humid days, condensation may appear on the pipes and connections.

C Check the anode rod fitting for leakage.

D Check for leakage at the temperature and pressure-relief valve and piping. The leakage could be from thermal expansion or high water pressure to the household. If there is a steady stream of water coming from the valve, the temperature and pressure-relief valve may be defective.

E Check for leakage from the temperature and pressure-relief valve fitting at the tank.

F The drain valve may be leaking. Check if the valve is slightly open.

G Check for leakage from the drain valve fitting at the tank.

H Check for leakage in the combustion chamber. Combustion products that contain water vapor could form droplets in that area. When cold water enters the tank, condensation may form droplets in that area, too. These droplets can fall on the burner/manifold area, causing damage to the component. In addition, the droplets can run onto the floor. This usually happens on new installation startups.

I Check the rim around the outside of the water heater tank for leakage, the water heater bottom, and on the floor. Check for leakage at all pipe connections and at the temperature and pressure-relief valve. Check and see if the temperature and pressure-relief valve discharge line is directed over a drain or runs to the outside of the residence. In some cases, I found the discharge line was draining on the floor and was not piped correctly.

FIGURE 14-8 Leakage inspection points on a gas water heater.

Other possibilities for water leakage around the water heater tank can come from other appliances (for example, an automatic washer or air conditioner), water lines, and ground water seepage.

Diagnostic Charts

The following diagnostic flow chart (Figure 14-9) will help you to pinpoint the likely causes of the problem. See also Figures 13-18 and 13-19 and Tables 13-1 and 13-2.

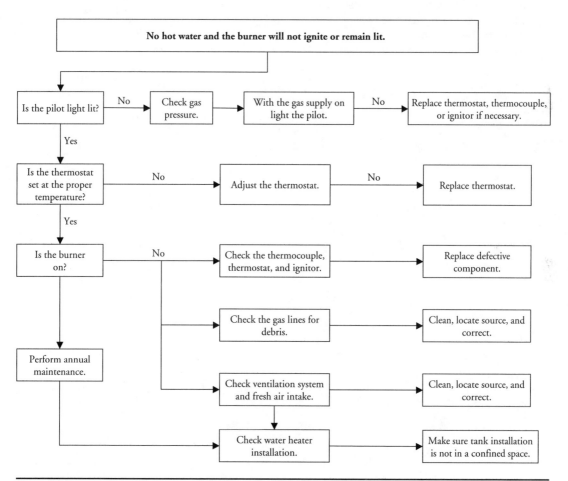

FIGURE 14-9   The diagnostic flow chart: No hot water and burner will not ignite or remain lit.

Help on the Internet

To obtain additional technical help or information on selecting a gas water heater, you can go to the following Web sites:

http://www.aceee.org/consumerguide/topwater.htm

http://www.eere.energy.gov/consumer/your_home/water_heating/index.cfm/mytopic=12770

http://www.americanwaterheater.com/

http://www.rheem.com/

http://www.hotwater.com/

http://www.tankless-water-heater.com/

http://www.bradfordwhite.com/

http://www.marathonheaters.com/

http://www.statewaterheaters.com/

Automatic Washers

The automatic washer is a complex electromechanical machine, and Figure 15-1 is used as an example only. The actual construction and features of the washer you are servicing might vary, depending on brand and model. The automatic washer performs various cycles to clean clothes. There are times when a washer fails to operate properly. Don't let its complexity intimidate you. This chapter will provide the basics needed to diagnose and repair the washing machine.

Principles of Operation

The clothes are placed evenly into the washer basket, making sure that the washer is not overloaded and that the proper cycle is selected. The user activates the washer through the timer. The internal switches of the timer distribute the electricity to activate the other components in the washing machine during a given time period, designated by the internal cam of the timer.

The water enters the tub through the water fill hoses, the water inlet valve, and water inlet hose. Hot, warm, or cold water is selected by the user via the water temperature selector switch, located on the console panel. On some models, the water temperature selection is controlled by the timer.

The amount of water that fills the tub is controlled by the water level control (pressure switch). The water level control offers a choice of water levels, depending on the amount of clothing being washed. As the water level rises in the tub, it forces air through the air dome and up the plastic tube to the water level control. The pressure that is exerted on the water level control's diaphragm will trip the water level switch from empty to full, supplying electricity to the washer drive motor and, thus, operating the transmission.

The transmission/gearbox is operated by the drive motor, either by belt drive or by direct drive, depending upon the model. Agitation is accomplished by the agitator, which is located in the center of the basket and which is driven by the transmission/gearbox. The agitator turns clockwise and counterclockwise, creating a water motion that moves the clothes within the basket.

When the washer goes into the drain mode, the agitator will stop agitating, and the water leaves the bottom of the tub through the water pump, to be pumped into the drain. The water pump may operate by belt drive or by direct motor drive.

Water fill valve and
flowmeter valve

Pressure switch

Electronic
control board

Display board and
touch pad panel

Controls and
cycle selection

Detergent
dispenser

Basket/drum

Drain pump
assembly

Motor control
board

Pedestal

The drive motor and
belt are serviceable
in the rear of washer.

FIGURE 15-1 Parts identification and location in a front-loading automatic washer.

A final deep rinse cycle (the tub fills to the selected water level and begins to agitate) will be introduced to wash off any remaining residue of soap or dirt.

The timer will now select the spin cycle, and the washer will then go into the spin mode. In the spin mode, the washer spins the clothes, removing most of the water from the clothing by centrifugal force. Some models use brief sprays of water to remove any residue of soap or dirt remaining on the clothes in the spin cycle.

Functions and Cycles

The removal of soil from clothing and fabrics is accomplished by a combination of mechanical and chemical processes:

- Mechanical process: Soil is removed by agitating and by forcing the detergent through the clothing.
- Chemical process: The detergent used will dissolve and loosen the soil in the clothing. As the washing machine operates through its cycles, it is aided by hot, soft water, which increases the chemical processes of the detergent being used.

Washing machines perform four basic functions, which are modified and put together in different ways to create the various cycles. The four functions are:

- Fill
- Agitate
- Drain
- Spin

Safety First

Any person who cannot use basic tools or follow written instructions should *not* attempt to install, maintain, or repair any automatic washers. Any improper installation, preventive maintenance, or repairs could create a risk of personal injury or property damage.

If you do not fully understand the installation, preventive maintenance, or repair procedures in this chapter, or if you doubt your ability to complete the task on the automatic washer, please call your service manager.

The following precautions should also be followed:

- Never bypass or disconnect any part or device (originally designed into the washer) as a temporary repair.
- Always reconnect all ground wires, and be sure that they are secure.
- Be careful of moving parts and sharp edges.

Before continuing, take a moment to refresh your memory of the safety procedures in Chapter 2.

Automatic Washers in General

Much of the troubleshooting information in this chapter covers automatic washers in general, rather than specific models, in order to present a broad overview of service techniques. The pictures and illustrations that are used in this chapter are for demonstration purposes only, to clarify the description of how to service washing machines and in no way reflect a particular brand's reliability.

Location and Installation of Automatic Washer

The following are some general principles that should be followed when performing the installation of a washing machine:

- Locate the washing machine where there is easy access to existing drain, water, and electrical lines.
- Be sure you observe all local codes and ordinances for the electrical and plumbing connections.
- The washing machine should be installed and leveled on a firm floor to minimize vibration during operation.
- Do not install the washing machine in an area where the temperature might be below freezing.
- To reduce the risk of a fire, never install a washing machine on any type of carpet.
- Always follow the installation instructions that are provided with every new washing machine model purchased.

Common Installation Problems

Automatic washer installations are not complicated. As a service technician, you will come across a washer that has not been installed according to the manufacturer's installation instructions. The following sections describe some of the problems that you might run into and how to solve them.

When you arrive at a service call and the consumer tells you that the washer is full of water and it will not drain out, check for the following:

- The drain hose is kinked.
- The drain hose has a blockage.
- The drainpipe might be too small, not allowing for proper venting. The drainpipe must be a minimum of 1½ inches in diameter.
- The drainpipe and the drain hose were installed over six feet above the floor.

If the consumer tells you the washer will not fill up with water or very little water is entering the washer, check the following:

- The water faucets were never turned on or they were turned on only a little bit.
- There is a blockage in the water inlet screen. The consumer did not flush out the water lines before installing the hoses on the washer.

- When the washer is filling with water, the water is siphoning out through the drain line. The drainpipe is too low or the wrong size diameter. You might have to install a siphon break kit at the end of the drain hose.

When you arrive at the service call, you notice water on the floor. Check the following:

- The water hoses to the water inlet valve
- The water faucet might be leaking
- The drain hose is not connected properly to the washer
- The drain hose comes out of the drainpipe
- The drainpipe might have a blockage
- The household drain cannot handle the capacity of the discharge water

When the washer goes into the spin cycle, it will begin to vibrate and walk across the floor. Check the following:

- The feet or leveling legs on the washer are not set properly, causing the washer to teeter-totter.
- Is the washer level?
- Is the floor level?
- Check and see if the packing and shipping straps have been removed. Read the installation instructions on removing the packing straps.

Water Supply

The water supply for an automatic washer should have a hot and cold faucet located within five to seven feet of the washer. The faucets should be a 3/4-inch threaded type to accept the fill hose connection.

The water pressure must be between 25 and 125 pounds per square inch for the washer to operate properly. The water coming out of the fill hoses should be equal in both pressure and in the volume of water to prevent unacceptable water temperature changes when entering and filling the washer.

The hot water supply to the washer should be between 130 and 150 degrees Fahrenheit. If the hot water temperature is below 70 degrees Fahrenheit, the clothes being washed will not clean properly and the detergent will not dissolve properly. You can check the temperature of the hot water by operating the washer in the fill mode with the water temperature setting on hot. Let the water run until it is as hot as possible, and then insert a thermometer into the stream of water. If the thermometer reading is below 130 degrees Fahrenheit, you will have to raise the water heater thermostat setting. On some front-loading models, the manufacturer has incorporated an electric heater in the washer to heat the water to 153 degrees Fahrenheit for select wash modes.

The cold water temperature should be between 70 and 100 degrees Fahrenheit. When the washer is in its rinse stage, the cold water will prevent wrinkles from setting into the fabrics. Some fabric manufacturers require that their fabrics be washed in cold water, both to prevent shrinkage and to eliminate the possibility of destroying the fabric. When the user selects the warm fill, the temperature of the water should be 100 degrees Fahrenheit.

It is recommended that the consumer read the use and care manual before performing a wash. Most use and care manuals have a water temperature guide to assist the user in the proper selection of the water temperature.

Drain Requirements

The drain to which the washer's drain hose is connected must be able to accept at least a 20- to 30-gallon-per-minute flow in order to remove the water from the tub. The standpipe should be at least 32 inches high and not exceed 60 inches in height. The internal diameter of the drain pipe should be a minimum of 1½ inches in order to provide an air gap around the drain hose and, thus, to prevent the suction from siphoning the water out of the tub during the wash cycle.

Detergent

The kind and amount of detergent that is used is an important part of getting clothes clean. Different brands of detergent contain different amounts of phosphorous, which works to soften the water and to boost the cleaning action. If the water is hard, you might need to recommend a detergent with a higher phosphorous content. If the water is soft, the user can use a low-phosphorous detergent.

Some areas restrict the phosphate content to 8 percent or less. This means that the user will have to increase the amount of detergent used in those areas where the water is hard. This is done by adding a certain extra amount of detergent manually to the wash cycle.

It is recommended that the consumer read the use and care manual before performing a wash. Most use and care manuals have a detergent guide to assist the user in the recommended amount of detergent to use.

Water hardness is measured in grains:

- 0 to 3 grains: soft water
- 4 to 9 grains: medium-hard water
- 10 to 15 grains: hard water
- Over 15 grains: very hard water

If you do not know the hardness of the water supply, contact the local water department.

Step-by-Step Troubleshooting by Symptom Diagnosis

In the course of servicing an appliance, don't overlook the simple things that might be causing the problem. Step-by-step troubleshooting by symptom diagnosis is based on diagnosing malfunctions, with their possible causes arranged into categories relating to the operation of the washer. This section is intended only to serve as a checklist to aid you in diagnosing a problem. Look at the symptom that best describes the problem that you are experiencing with the washer, and then proceed to correct the problem.

No Water Entering Washer

1. Is the washer plugged in?
2. Check for proper voltage at the wall receptacle.

3. Is the water supply turned on? The fill hoses should feel stiff.

4. Test the water temperature switch contacts for continuity.

5. Check for an open circuit in the timer contacts. On electronic models, run the test mode.

6. Check for loose wires to the water valve solenoid.

7. Test water valve solenoid coils for continuity.

8. Check the water valve inlet strainer screens. Remove the fill hoses to inspect these filters.

9. Test the water level control switch for continuity.

10. Are the water supply hoses kinked?

Motor Will Not Run

1. Check for proper voltage at the wall receptacle.

2. Is the washer plugged in?

3. Check for a blown fuse or a tripped circuit breaker.

4. Check for a faulty timer. On electronic models, run the test mode and check the electronic control board.

5. Are there any loose wires to the timer, motor, etc.?

6. Test the motor windings for continuity.

7. Test the thermal overload in the motor for continuity.

8. Test the water level control switch contacts for continuity.

9. Test the motor speed selector switch contacts for continuity.

10. Are there any open wires in the washer circuit?

11. Test the capacitor on the motor.

12. Check the centrifugal switch in the motor.

13. Check for continuity of the lid switch contacts.

Washer Will Not Agitate

1. Check for a broken or worn belt.

2. Check the motor to the transmission drive coupling (direct-drive models).

3. Test the timer contacts for continuity. On electronic models, run the test mode.

4. Are there any loose wires within the wiring harness?

5. Are there any loose wires to the timer, motor, etc.?

6. Test the continuity of the motor windings.

7. Test the agitator solenoid on the transmission.

8. Check for loose pulleys on the transmission and motor.

9. Check the water level control switch.

10. Test for continuity of the start relay.

11. Test the capacitor on the motor.

12. Check the centrifugal switch in the motor.

13. Test for continuity of the lid switch contacts.

14. Test for continuity of the speed selector switch.

15. Check the clutch assembly.

16. Check the transmission.

Water Will Not Drain

1. Check for a clogged drain connection.

2. Inspect the pump for obstructions.

3. Check the drain hose and be sure it is not kinked.

4. Check the belt that goes to the pump.

5. Check for suds lock. If this happens, just add cold water and flush the suds out of the pump. (Suds lock is caused by too much soap remaining in the tub, pump, and the connecting hoses. This condition will prevent water from draining effectively.)

6. Check the pump coupling. The pump and motor must be removed for a visual inspection of the coupling.

7. Check for air lock in the pump (air trapped inside the pump caused by debris).

8. Check to be sure that the motor is not running in the agitation direction.

9. Check the height of the drain.

10. Does the pump pulley turn freely?

Washer Will Not Spin

1. Check for a loose or broken belt.

2. Check for loose pulleys.

3. Check the clutch assembly.

4. Check for loose or broken wires in the washer circuit.

5. Test the continuity for a faulty lid switch assembly.

6. Test the continuity for a defective spin solenoid.

7. Check for a broken drive coupling (direct-drive models). The pump and motor must be removed for a visual inspection of the coupling.

8. Test the continuity of the water level control switch.

9. Test the continuity of the speed selector switch.

10. Test the continuity in the motor windings and the motor overload protector.

11. Test the timer contacts for continuity.

12. Check for clothing jammed between the inner basket and the outer tub.

13. Check all seals and mechanical linkages.

14. Check the transmission.

15. Check to be sure that the motor is running in the spin direction.

Common Washability Problems

If there are no mechanical problems with the washer's operation and the complaints are that the washer does not clean the clothes properly, you have a washability problem. The next step should be to look at the cause that best describes the problem that the customer is experiencing with the washer. Then proceed to correct the problem. If necessary, instruct the user how to get better results from the automatic washer.

Stains on the Clothing

Stains on clothing can be caused by a number of different things. As the servicer, you will have to determine if it is caused by the washer's components or from an external source. Many stains are blamed on leaking transmissions. This type of problem is related to the increasing use of synthetic fabrics and to the poor washing practices of the user. Many of these stains consist of cooking oil or grease and are not visible when they first occur during cooking or eating. The oil that is embedded in the clothing acts like glue, attracting dirt from the wash water. When the wash cycle is completed, the clothes come out dirty and spotted. If the transmission oil leaked into the wash water, there will be stains on all of the clothes in a random pattern. The color of transmission oil embedded into the clothing is usually a brownish-yellow stain. Transmission oil cannot be washed out of the clothes; a solvent is required to remove the stains.

The following are some stain-removal rules for clothing in general:

- Stains are easier to remove when they first appear on the clothing. If the stains are old, they might never come out of the clothing.

- Before attempting to remove any stain, you must know: what type of stain, what kind of fabric, and how old is the stain?

- Use only cold or warm water to remove stains. Hot water will set the stain permanently into the fabric.

- When bleach is recommended for the removal of the stain, use a bleach that is safe for the fabric. When using a chlorine bleach, always dilute it with water to prevent the bleach from destroying the fibers.

- Always test stain remover products on a hidden corner of the garment to see if the color remains in the fabric.

- When preparing to remove the stain from the fabric, face the stained area down on a paper towel or a white cloth. Then apply the stain remover to the back of the stain so that the stain will be forced off the fabric, instead of through the fabric.

- Some protein stains can be removed with an enzyme pre-soak or with meat tenderizer.

- When using dry-cleaning solvents, always use them in a well-ventilated room, away from flames and sources of ignition to prevent personal injury.

- Alcoholic beverage stains: Alcoholic beverage stains turn brown the longer they stay on the fabric. As soon as the stain appears on the fabric, start treating the stain immediately. Wash or soak the stain in cold water, and then wash the garment.

- Blood stains: To remove blood stains, rinse or soak the garment in cold water with an enzyme pre-soak product. You can use diluted chlorine bleach on white fabrics, if necessary. For colored fabrics, use a powdered oxygen-type bleach. Then wash the garment.

- Chewing gum stains: To remove chewing gum, use ice on the stain to make the chewing gum hard. Then scrape most of it off the fabric. Next, use a non-flammable dry-cleaning solvent with a sponge to remove the excess chewing gum. Wash the garment.

- Coffee or chocolate stains: To remove coffee or chocolate stains, soak the garment in warm or cold water. Next, make a paste of detergent mixed with hot water, and brush it on the stain. Wash the garment.

- Milk product stains: To remove a milk product stain, use a non-flammable dry-cleaning solvent with a sponge. Wash the garment.

- Antiperspirant and deodorant stains: To remove antiperspirant and deodorant stains, wash the garment with laundry detergent in the hottest water that is safe for the fabric. If the stain remains on the fabric, place the stain face down on a white towel, and treat the stain with a paste of ammonia and a powdered oxygen-type bleach. Let the paste stay on the stain for 30 minutes, and then wash the garment in the hottest water that is safe for the fabric.

- Fruit stains: To remove fruit stains, soak the stain in cool water. Do not use soap—it will set the stain. Wash the garment. If the stain remains, cover the stain with a paste made of a powdered oxygen-type bleach, a few drops of hot water, and a few drops of ammonia. Let the paste stay on the stain for about 15 to 30 minutes. Then wash the garment.

- Iron or rust stains: To remove iron or rust stains, apply some lemon juice mixed with salt. Then place the garment in the sun. Alternatively, a commercial rust-removing solution can be used. Wash the garment.

Yellowing in Fabric

Some causes of yellowing in fabrics are:

- Poor body soil removal
- Clothes washed in water treated with a water softener
- Hard water or minerals in the water, such as iron
- Body oils released into the garment
- The water supply might pick up the color of decaying vegetation

To remove body oils, the user will have to increase the amount of detergent and use 150-degree Fahrenheit wash water. The user must also increase the frequency of using bleach in the wash.

To remove the yellowing from garments that are washed in water treated with a water softener, the user will have to decrease the amount of detergent used, approximately to the point that the decreased amount will not affect the soil removal process. The user must also increase the frequency of using bleach in the wash.

Hard water and minerals in the water can be treated with a water-conditioning apparatus. The user might have to drain the water heater and flush the tank. Never use chlorine bleach to remove hard water stains or iron stains.

To remove body oils from the garment, use a paste made of detergent and water. Let it stay on the fabric for 15 to 30 minutes. Then wash the garment.

To remove the yellowing caused by decaying vegetation, increase the amount of detergent, and bleach more often. White fabrics typically respond quite well to bleaching.

Fabric Softener Stains

Fabric softener stains are becoming more prevalent because it is now being recommended that some fabric softeners be used in the wash cycle, instead of the rinse cycle. These types of stains show up on synthetics as well as cotton fabrics. They can be removed from the fabric by pretreating the stain with liquid detergent and following the washing procedures listed in the use and care manual.

Lint

Lint is cotton fiber that has broken away from the cotton garment. Lint likes to attach itself to synthetic fabrics. When this happens, the user often thinks that the washer is not performing properly. Therefore, to solve the problem of lint on synthetic fabrics, the user must sort the items before washing the clothes. For example:

- The user must separate cottons from permanent press and knits.
- The user must separate light colors from dark colors.

Another cause of lint on clothes is overwashing. This causes the clothes to wear out faster. To correct the overwashing problem, use only one minute of wash time per pound of dry laundry with normal soil. Any more time than this is a waste, and it usually does not get the laundry cleaner.

If the drain cycle is excessive, this, too, will cause lint to remain on the garments. Check for improper drain hose connections. For example:

- Drain hose is too long (over 10 feet)
- Drain hose is too high (over 5 feet)
- Drain hose is kinked

If excessive drain times still exist, check the following:

- Check the filter, located under the wash basket on most newer models.
- Check to be sure that the pump is operating properly.
- Check for any obstructions in the drain system.
- Check for any obstructions within the water circulatory system of the washer.

Automatic Washer Maintenance

The interior is normally self-cleaning. However, there are times when you might have to remove objects from the inner basket. Clean the control panel and outer cabinet with a soft damp cloth. Do not use any abrasive powders or cleaning pads. Clean and inspect the interior underneath the washer. Read the use and care manuals for the proper maintenance of the brands of washers you service.

Repair Procedures

Each repair procedure is a complete inspection and repair process for a single washer component, containing the information you need to test a component that might be faulty and to replace it, if necessary.

Washer Timer

The typical complaints associated with washing machine timer failure are:

1. The cycle will not advance.
2. The washer won't run at all.
3. The washer will not fill.
4. The washer will not pump the water out.
5. The washer will not shut off.

To handle these problems, perform the following steps:

1. **Verify the complaint.** Verify the complaint by operating the washer through its cycles. Before you change the timer, check the other components controlled by the timer.

2. **Check for external factors.** You must check for external factors not associated with the appliance. Is the appliance installed properly? Does the appliance have the correct voltage?

3. **Disconnect the electricity.** Before working on the washer, disconnect the electricity. This can be done by pulling the plug out from the electrical outlet. Be sure that you only remove the washer plug. Or disconnect the electricity at the fuse panel or the circuit breaker panel. Turn off the electricity.

4. **Remove the console panel to gain access.** Begin by removing the screws from the console panel to gain access to the timer (Figure 15-2). Roll the console toward you. On some models, the console will roll upwards.

5. **Test the timer.** Remove the timer motor leads from the timer assembly. Test the timer motor by connecting the ohmmeter probes to the timer motor leads (Figure 15-3). Set the range on the ohmmeter to R × 100. The meter should indicate between 2000 and 3000 ohms. Next, test the timer switch contacts using the wiring diagram configuration for the affected cycle. Place the meter probe on each terminal being tested, and turn the timer knob. If the switch contact is good, your meter will read continuity. If the timer

Figure 15-2   Remove the screws that hold the control console. Fold over the console to gain access.

motor measures suitably, then connect a 120-volt, fused service cord (Figure 15-4) to the timer motor leads. *Note:* Connect the ground (common) wire test lead to the console ground wire. Be cautious whenever you are working with "live" wires. Avoid any shock hazards.

If the motor does not operate, replace the timer. If the timer motor runs but does not advance the cams, then the timer has internal defects and should be replaced.

6. **Remove the timer.** To remove the timer, remove the timer mounting screws (Figure 15-5). Remove the wire lead terminals from the timer. Mark the wires as to their location on the timer. Some timers have a disconnect block instead of individual wires, which makes it easier to remove the timer wires.

Figure 15-3
Checking the washer timer motor.

FIGURE 15-4
120-volt fused service test cord.

Male plug with ground

6-foot cord

Ground lead - Also grounded to electrical box (green)

Neutral lead (white)

Electrical box

L1 - Hot lead (black)

On/Off switch

20-A resettable fuse

24-inch cord with three rubber insulated, NEC color-coded alligator clips

FIGURE 15-5
Removing the timer.

Disconnect harness plug

Remove timer screws

Timer

Turn the timer knob counterclockwise to remove it from the timer shaft, and slide the indicator dial off the shaft.

7. **Install a new timer.** To install a new timer, just reverse the disassembly procedure, and reassemble. Replace the wires on the timer. Reinstall the console panel, and restore the electricity to the washer. Test the washing machine cycles.

Electronic Control Board and User Interface Controls

The typical complaints associated with the electronic control board or the user interface controls are:

- The washer won't run or power up.
- Unable to program the washer.
- The display board will not display anything.
- One or more key pads will not accept commands.
- Unusual display readouts.

To prevent electrostatic discharge (ESD) from damaging expensive electronic components, follow the steps in Chapter 9.

To handle these problems, perform the following steps:

1. **Verify the complaint.** Verify the complaint by operating the washer. Turn off the electricity to the appliance and wait for two minutes before turning it back on. If a fault code appears, look up the code. If the washer will not power up, locate the technical data sheet behind the control panel for diagnostics information.

2. **Check for external factors.** You must check for external factors not associated with the appliance. Is there electricity to the washer?

3. **Disconnect the electricity.** Before working on the washer, disconnect the electricity. This can be done by pulling the plug out of the wall receptacle. Or disconnect the electricity at the fuse panel or circuit breaker panel. Turn off the electricity.

WARNING *Some diagnostic tests will require you to test the components with the power turned on. When you disassemble the control panel, you can position the panel in such a way so that the wiring will not make contact with metal. This act will allow you to test the components without electrical mishaps.*

4. **Remove the console panel to gain access.** Begin by removing the screws from the washer top (for front-loading models only) to gain access to the electronic control board. Remove the console panel screws to gain access to the user interface controls (Figure 15-6). On top-loading models, the console will roll upwards or toward you after removing the console screws.

5. **Test the electronic control board and user interface controls.** If you are able to run the washer diagnostic test mode, check the different functions of the washer. Use the technical data sheet for the model you are servicing to locate the test points from the wiring schematic. Check all wiring connections and wiring. Using the technical

FIGURE 15-6   An exploded view of the control panel parts for a front-load automatic washer.

data sheet, you can test the electronic control board or user interface controls and input voltages. On some models, fuses are soldered to the printed circuit board (PCB). These fuses must be tested first before condemning the component.

6. **Remove the electronic control board and user interface controls.** To remove the defective component, remove the screws that secure the board to the control panel or washer frame. Disconnect the connectors from the electronic control board or user interface control.

7. **Install the new component.** To install a new electronic control board or user interface control, just reverse the disassembly procedure, and reassemble. Reinstall the console panel, and restore the electricity to the washer. Test the washer operation.

Water Temperature Selector Switch

The typical complaints associated with the water temperature selector switch are:

- Inability to select a different water temperature.
- The consumer inadvertently selected the wrong water temperature.

To handle these problems, perform the following steps:

1. **Verify the complaint.** Verify the complaint by trying to select different water temperatures.

2. **Check for external factors.** You must check for external factors not associated with the appliance. Is the appliance installed properly? Is there any physical damage to the component? Are the fill hoses connected to the hot and cold water supply correctly? Be sure that both the hot and cold water faucets are turned on.

3. **Disconnect the electricity.** Before working on the washer, disconnect the electricity. This can be done by pulling the plug out of the electrical outlet. Be sure that you only remove the washer plug. Or disconnect the electricity at the fuse panel or at the circuit breaker panel. Turn off the electricity.

4. **Remove the console panel to gain access.** Begin by removing the console panel to gain access to the water temperature selector switch (see Figure 15-2).

5. **Test the water temperature selector switch.** To test the water temperature selector switch, remove all wires from the switch. Label the wires.

   Remember, you will have to identify the wires according to the wiring diagram in order to reinstall them onto the water temperature selector switch properly. Take your ohmmeter and check for continuity on the switch contacts. Press or turn on the switch that coincides with the terminals that are being checked (Figure 15-7). At this point, you have to use the wiring diagram to identify the switch contacts.

6. **Remove the water temperature selector switch.** To remove the water temperature selector switch, remove the screws that hold the component to the console panel (Figure 15-8).

FIGURE 15-7
Checking the water
temperature switch.

FIGURE 15-7
Checking the water
temperature switch.

Water temperature
switch

7. **Reinstall the water temperature selector switch.** To reinstall the water temperature selector switch, just reverse the disassembly procedure, and reassemble. *Note:* You will have to identify the wires according to the wiring diagram in order to reinstall them onto the water temperature selector switch properly. Reinstall the console panel, and restore the electricity to the washer. Test the washing machine water temperature cycles.

FIGURE 15-8
Removing the water
temperature switch.

Outlet

Metal screen

Cold

Hot

Figure 15-9 A typical water valve used in automatic washers.

Water Valve

The typical complaints associated with water valve failure (Figure 15-9) are:

- The washer will not fill with water.

- The washer overfills and leaks onto the floor.

- When the washer is off, water still enters the tub.

To handle these problems, perform the following steps:

1. **Verify the complaint.** Verify the complaint by operating the washer through its cycles. Listen carefully, and you will hear whether the water is entering the washer.

2. **Check for external factors.** You must check for external factors not associated with the appliance. Is the appliance installed properly? Does the appliance have the correct voltage? Is the water turned on?

3. **Disconnect the electricity.** Before working on the washer, disconnect the electricity. This can be done by pulling the plug from the electrical outlet. Be sure that you only remove the washer plug. Or disconnect the electricity at the fuse panel or at the circuit breaker panel. Turn off the electricity.

4. **Gain access to the water valve.** Turn off the water supply to the water valve. To access the water valve, the rear panel must be removed. Turn off the water supply to the washer. Disconnect the fill hoses from the inlet end of the water valve (Figure 15-10). Next, remove the screws that hold the water valve to the chassis of the washer (Figure 15-11).

Figure 15-10 Turn off the water supply, and remove the fill hoses.

Water inlet valve

Figure 15-11 Remove the wires from the solenoid coil.

PART V

Test leads

Inlet
valve

FIGURE 15-12 Attaching test leads to the solenoid coil on the water valve.

5. **Test the water valve.** In order to check the solenoid coils on the water valve, remove the wire leads (label them) that connect to the coils from the wire harness (see Figure 15-11). These are slide-on terminal connectors attached to the ends of the wire. Just pull them off. Set the ohmmeter on $R \times 100$, and attach the probes to the terminals of one of the solenoid coils (see Figure 15-12). The meter should read between 500 and 2000 ohms. Repeat this test for the second solenoid coil.

To test the fill rate of the water valve, just reverse the disassembly procedure, and reassemble the water valve. The rear panel does not have to be reinstalled for this test. Attach the 120-volt fused service cord—including the ground wire test lead to the cabinet ground (see Figure 15-4)—to the water valve solenoid coil (see Figure 15-12). Then energize the solenoid coil to allow water to enter the tub and to check the flow rate of the water valve (Table 15-1). This step is repeated for each solenoid coil. If, when you energize the water valve, no water enters the washer tub, replace the water valve. If the water valve checks correctly, check the timer and the wiring harness.

6. **Remove the water valve.** To remove the water valve, follow the instructions in step 4. Remove the water outlet hose from the water valve.

7. **Install a new water valve.** To install the new water valve, just reverse the disassembly procedure, and reassemble. Reconnect the wire leads to the solenoid coils. After the installation of the new valve, turn on the water supply and check for water leaks. If none are found, reinstall the rear panel and restore the electricity to the washer. Set the timer and the water temperature control settings to operate the washer through its cycles.

Water Input P.S.I.	Gallon Per Minute One Side of Valve	Gallon Per Minute Both Sides of Valve
20	3.7	4.5
30	4.6	5.5
40	5.3	6.2
50	5.6	6.3
60	5.8	6.7
80	6.6	7.0
100	7.0	7.2
120	7.2	7.4
140	7.0	7.3
160	6.8	7.0

TABLE 15-1 Water Fill Rate for a Typical Water Value.

Washer Motor (Older Models)

The typical complaints associated with motor failure are:

- Fuse is blown or the circuit breaker trips.
- Washer fills up with water but the motor will not run.

To handle these problems, perform the following steps:

1. **Verify the complaint.** Verify the complaint by operating the washer through its cycles. Listen carefully, and you will hear if there are any unusual noises or if the circuit breaker trips.

2. **Check for external factors.** You must check for external factors not associated with the appliance. Is the appliance installed properly? Does the appliance have the correct voltage?

3. **Disconnect the electricity.** Before working on the washer, disconnect the electricity. This can be done by pulling the plug from the electrical outlet. Be sure that you only remove the washer plug. Or disconnect the electricity at the fuse panel or at the circuit breaker panel. Turn off the electricity.

4. **Gain access to the motor.** To access the motor, the back panel must be removed (Figure 15-13). The back panel is held on with screws. Remove the screws and remove the panel.

5. **Disconnect the motor wire leads.** Disconnect the motor wire leads from the wiring harness. Set the ohmmeter on R × 1, and attach the probes to the motor lead wires (Figure 15-14). Refer to the wiring diagram for the common, start, and run motor winding leads identification. Test these for continuity, from the common wire lead

FIGURE 15-13
Removing the screws that hold the back panel.

FIGURE 15-14 Checking the motor windings for continuity.

FIGURE 15-15 Checking the motor for ground.

to the run winding. Then test for continuity from the common wire lead to the start winding. Next, test for continuity from the start winding to the run winding. To test for a grounded winding in the motor, take the ohmmeter probes and test from each motor wire lead to the motor housing (Figure 15-15). The ohmmeter will indicate continuity if the windings are grounded. If the motor has no continuity between the motor windings, replace the motor. If the motor checks out okay, check the timer and motor relay (if the model that you are repairing has one).

6. **Remove the motor.** To remove this type of motor, you must first loosen the two nuts that hold the motor support bracket (Figure 15-16). Then slide the assembly to

FIGURE 15-16
Removing the bolts that hold the motor on the bracket.

PART V

FIGURE 15-17
Removing the motor
and then removing the
pulley.

Start switch

Motor pulley

disengage the belt from the pulley. Next, remove the four nuts that hold the motor
to the motor support bracket (Figure 15-17). Remove the motor from the washer.
Remember to remove any remaining wires from the motor and label them. Remove
the pulley from the motor after loosening the set screw.

7. **Install the new motor.** To install the new motor, just reverse the disassembly
   procedure, and reassemble. To adjust the belt, refer to the "drive belt" section of this
   chapter (step 5). Restore the electricity to the washer, and test the motor. If the
   motor is working, reinstall the back panel.

Washer Motor (Direct-Drive Models)

The typical complaints associated with motor failure are:

- Fuse is blown or the circuit breaker trips.
- Washer fills up with water but the motor will not run.

To handle these problems, perform the following steps:

1. **Verify the complaint.** Verify the complaint by operating the washer through its
   cycles. Listen carefully, and you will hear if there are any unusual noises or if the
   circuit breaker trips.

2. **Check for external factors.** You must check for external factors not associated with
   the appliance. Is the appliance installed properly? Does the appliance have the
   correct voltage?

3. **Disconnect the electricity.** Before working on the washer, disconnect the electricity.
   This can be done by pulling the plug from the electrical outlet. Be sure that you only
   remove the washer plug. Or disconnect the electricity at the fuse panel or at the
   circuit breaker panel. Turn off the electricity.

4. **Gain access to the motor.** To access the motor, you must first remove the cabinet from the washing machine (Figure 15-18). Remove the two screws that secure the control console to the cabinet. Tilt the control panel upwards. Next, insert a flat-blade screwdriver in the cabinet retaining clip and push forward to remove it. Remove the wiring harness connector and ground wire. Finally, remove the cabinet as shown in Figure 15-18.

Selector switch and timer

Pressure switch

Clip

Wiring harness connector

Clip

Tilt the cabinet toward you and slide it off the base, and then pull it away from the washer.

Cabinet

FIGURE 15-18   Removing the outer cabinet on a direct-drive automatic washer.

FIGURE 15-19
The direct-drive water pump.

Water pump clamps

Direct-drive washer water pump

5. **Test the drive motor.** To test the drive motor, disconnect the wire connector from the motor relay. Refer to the wiring diagram for the motor protector; common; start, low, and high speed; and motor winding leads identification. Set your ohmmeter on the $R \times 1$ scale and place your ohmmeter probes on the start winding. The meter should read around 7 ohms. Next test the high and low speed windings—the resistance should be between 1 and 3 ohms. Now test the motor overload protector—the reading should be zero ohms. After testing the motor, spin the motor shaft; it should spin freely. To test for a grounded winding in the motor, take the ohmmeter probes and test from each motor wire lead to the motor housing. The ohmmeter will indicate continuity if the windings are grounded. If the motor has no continuity between the motor windings, replace the motor. If the motor checks out okay, check the timer and motor relay.

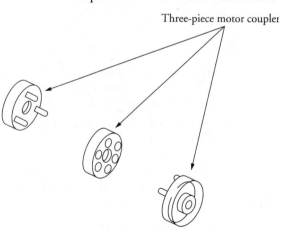

Three-piece motor coupler

6. **Remove the motor.** To remove the motor, you must first remove the water pump and the motor coupler. Before removing the pump, make sure that most of the water is removed from the tub. Use a pinch-off tool to restrict the water lines at the pump hoses. Next, proceed to remove the water pump by removing the clamps that hold the pump in place (Figure 15-19). Remove the three-piece motor coupler (Figure 15-20). Finally, remove the clamps that secure the motor to the gearcase (Figure 15-21). Pull the motor toward you to remove it from the washer.

PART V

FIGURE 15-20   The three-piece motor coupler.

FIGURE 15-21
The drive motor and
relay.

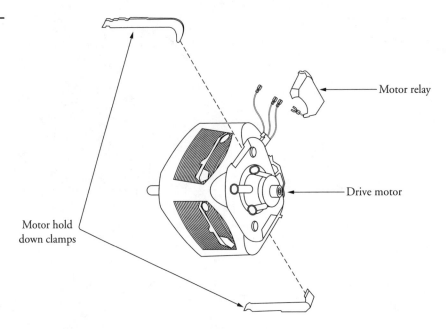

Motor relay

Drive motor

Motor hold
down clamps

7. **Install the new motor.** To install the new motor, just reverse the disassembly procedure, and reassemble. Restore the electricity to the washer, and test the motor. If the motor is working, reinstall the cabinet.

Capacitor

The typical complaints associated with capacitor failure are:

- Fuse is blown or the circuit breaker trips.
- Washer motor will not run.
- Motor has a burning smell.
- Motor will try to start and then shuts off.

To handle these problems, perform the following steps:

1. **Verify the complaint.** Verify the complaint by operating the washer. Listen carefully, and you will hear if there are any unusual noises or if the circuit breaker trips. If you smell something burning, immediately turn off the washer and pull the plug.

2. **Check for external factors.** You must check for external factors not associated with the appliance. Is the appliance installed properly? Does the appliance have the correct voltage?

3. **Disconnect the electricity.** Before working on the washer, disconnect the electricity. This can be done by pulling the plug from the electrical outlet. Be sure that you only remove the washer plug. Or disconnect the electricity at the fuse panel or at the circuit breaker panel. Turn off the electricity.

4. **Gain access to the capacitor.** Some models have the capacitor mounted on the motor, and some are mounted to the cabinet interior in the rear of the machine. Access might be achieved through the front or rear panel, depending on which model you are working on. Do not touch the capacitor until it's discharged.

WARNING *A capacitor will hold a charge indefinitely, even when it is not currently in use. A charged capacitor is extremely dangerous. Discharge all capacitors immediately any time that work is being conducted in their vicinity. Redischarge after repowering the equipment if further work must be done.*

5. **Test capacitors.** Before testing the capacitor, it must be discharged. Use a screwdriver with an insulated handle to discharge the capacitor by shorting it across both terminals. Remove the wire leads, one at a time, with needle nose pliers. Set the ohmmeter on the highest scale, and then place one probe on one terminal and the other probe on the other terminal (Figure 15-22). Observe the meter action. While the capacitor is charging, the ohmmeter will read nearly zero ohms for a short period of time. Then the ohmmeter reading will slowly begin to return toward infinity. If the ohmmeter reading deflects to zero and does not return to infinity, the capacitor is shorted and should be replaced. If the ohmmeter reading remains at infinity and does not dip toward zero, the capacitor is open and should be replaced.

6. **Remove the capacitor.** Remove the capacitor from its mounting bracket.

7. **Install a new capacitor.** To install the new capacitor, just reverse the disassembly procedure, and reassemble. *Note:* A capacitor is rated by its working voltage (WV or WVac) and by its storage capacity in microfarads (μF). Always replace a capacitor with one that has the same voltage rating and the same (or up to 10 percent greater) microfarad rating.

Drive Belt

The typical complaints associated with belt failure are:

- Washer will not agitate.
- Washer will not spin.
- Washer motor spins freely.
- There is a smell of something burning.

FIGURE 15-22
Placing ohmmeter test leads on the capacitor terminals.

To handle these problems, perform the following steps:

1. **Verify the complaint.** Verify the complaint by operating the washer in the spin cycle. Listen carefully, and you will hear and see if the inner basket is turning or if the circuit breaker trips.

2. **Check for external factors.** You must check for external factors not associated with the appliance. Is the appliance installed properly? Does the appliance have the correct voltage?

3. **Disconnect the electricity.** Before working on the washer, disconnect the electricity. This can be done by pulling the plug out from the electrical outlet. Be sure that you only remove the washer plug. Or disconnect the electricity at the fuse panel or at the circuit breaker panel. Turn off the electricity.

4. **Gain access to the belt.** You must gain access to the belt—whether by removing the back or the front panel, tilting the washer, or laying it onto its back—depending on which model washer you are working on. The back panel (or the front panel) is usually held on with two screws. Remove the screws, and remove the panel.

5. **Adjust the belt.** Before adjusting the belt, use your finger and press on the belt; it should only deflect about 1/4 inch. To adjust the belt (see Figure 15-16), loosen the motor bracket nut just enough to move the bracket. Take hold of the motor bracket, and pull against the belt just enough to take up the slack in it and to properly re-tension it. If you are unable to adjust the tension, or if the belt is worn, replace the belt. Some models have more than one belt: one is for the drive system that is attached to the motor and the transmission pulleys; the other belt is for the water pump. This belt is attached to the motor pulley and to the water pump pulley (Figure 15-23). To adjust the water pump belt in this type of washer, just loosen the pump mounting screws and adjust to obtain the correct tension, about 1/4-inch deflection. There are even some models that use a direct-drive system (which has no belts) to drive the motor, transmission, and pump (Figure 15-24). This type of washer has the motor and the water pump attached to the transmission with retaining clips.

Figure 15-23
View from underneath the washer base. Check belt deflection.

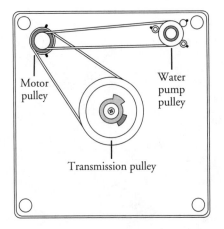

Motor pulley

Water pump pulley

Transmission pulley

Water pump belt

1/4"

Water level switch

Water temperature switch

Timer

Control console

Water inlet

Water valve

Basket

Agitator

Air dome

Transmission

Motor

Pump

FIGURE 15-24 The direct-drive washer has no belts.

Between the motor shaft and the transmission, there is a coupling. To replace this coupling, remove the retaining clips that secure the water pump and the motor. Install the new coupling, and reattach the motor and the water pump.

6. **Replace the drive belt.** To replace the drive belt on this type of washer, you must remove the flexible pump coupling (Figure 15-25). Next, loosen the motor bracket nuts, and slide the motor forward to take the tension off the belt. Then remove the belt from the motor pulley and the transmission pulley (Figure 15-26). Install the new belt on the transmission pulley and the clutch pulley. Be sure that the belt is in the pulley grooves. Next, adjust the belt tension, and tighten the motor bracket nuts. Reinstall the flexible pump coupling and clamps, making sure that the coupling is not twisted and that it is seated on the pump and clutch pulleys. The drive belt tension on this type of washer should be approximately 1/2 inch when deflected.

7. **Test the washer.** You are now ready to test the washer. Begin the wash cycle with a full load of laundry in the basket. Check the agitate and the spin cycles. If these check out okay, reinstall the outer panels. If not, readjust the belt tension.

Figure 15-25
Loosening the screw
on each clamp and
removing the flexible
coupling.

Figure 15-25
Loosening the screw
on each clamp and
removing the flexible
coupling.

Flexible
pump
coupling

Transmission pulley

Drive belt

Clutch pulley

Water Level Control

The typical complaints associated with the water level control (pressure switch) are:

- Water is flowing over the top of the tub.
- Tub does not fill to the proper level selected.
- Washer will not agitate.
- Washer will not spin.

Figure 15-26
Sliding the belt over
the motor pulley.

Clutch pulley

To handle these problems, perform the following steps:

1. **Verify the complaint.** Verify the complaint by trying to select different water levels when operating the washer through its cycles.

2. **Check for external factors.** You must check for external factors not associated with the appliance. Is the appliance installed properly? Does the appliance have the correct voltage? Is there any physical damage to the component? Is the plastic line connected to the water level control and air dome? Check to be sure that the water is turned on all the way.

3. **Disconnect the electricity.** Before working on the washer, disconnect the electricity. This can be done by pulling the plug from the electrical outlet. Be sure that you only remove the washer plug. Or disconnect the electricity at the fuse panel or at the circuit breaker panel. Turn off the electricity.

4. **Remove the console panel to gain access.** Begin by removing the console panel to gain access to the water level control (see Figure 15-2).

5. **Test the water level control.** To test the water level control switch, remove and label all of the wires from the switch (Figure 15-27). Remember, you will have to identify the wires according to the wiring diagram in order to reinstall them on the water level control switch properly. With the washer empty, use your ohmmeter (set on R × 1), and test it for continuity on the switch contacts numbered 1 and 2 (Figure 15-28). If you have continuity, this means that the water valve will be energized, allowing the water to enter the tub. Now test for continuity between contacts 1 and 3; the ohmmeter should read no continuity. Reconnect the wires to the water level switch.

   Plug in the washer, and start the wash cycle. Let the console rest on top of the washing machine for this test. Be careful not to touch any live wires or to short them to the washer chassis while the test is being performed. As the water level rises in the tub, it forces air through the air dome and up the plastic tube to the

FIGURE 15-27
Removing the wires from the water level control before checking continuity.

Water level switch

FIGURE 15-28
Adjustable water level control terminal identification for this model only.

Normally open terminal
Normally closed terminal
3 2
1
Common terminal

water level control. The pressure that is exerted on the water level control's diaphragm will trip the water level switch from empty to full, which will start the agitation cycle. When the agitation cycle begins, turn off the washer. Pull the power plug from the wall socket to ensure that power has been removed.

At this point, test for continuity again between the contacts numbered 1 and 2. The ohmmeter should read no continuity. Next, test for continuity between contacts numbered 1 and 3. The ohmmeter will read continuity. If the water level switch checks out, then check the plastic hose that goes from the air dome to the water level control. Ensure that there are no holes or cracks in the line. If the switch does not check out okay, replace it. Remember: Never blow into the water level control switch to activate it. Why? You might activate the switch, but it will not prove that the switch will activate (at lesser pressures) at the proper water level setting selected.

6. **Remove the water level control.** To remove the water level control, remove the wires and then remove the screws that hold the component to the console panel (Figure 15-29). Next, remove the plastic hose.

7. **Reinstall the water level control.** To reinstall the water level control, just reverse the disassembly procedure, and reassemble. Remember, you will have to identify the wires according to the wiring diagram in order to properly reinstall them on the water level control.

FIGURE 15-29 Disconnecting the wires and removing the screws that hold the control to the console.

FIGURE 15-30   Tape the lid closed. Then pry the top open and lift the top off.

Lid Switch

The typical complaints associated with the lid switch are:

- Washer will not spin.
- Washer will not agitate.

To handle these problems, perform the following steps:

1. **Verify the complaint.** Verify the complaint by closing the washer lid and turning the timer dial to start the spin cycle.

FIGURE 15-31   Removing the plastic shield to gain access to the lid switch.

2. **Check for external factors.** You must check for external factors not associated with the appliance. Is the appliance installed properly? Does the appliance have the correct voltage? Is there any physical damage to the component?

3. **Disconnect the electricity.** Before working on the washer, disconnect the electricity. This can be done by pulling the plug from the electrical outlet. Be sure that you only remove the washer plug. Or disconnect the electricity at the fuse panel or at the circuit breaker panel. Turn off the electricity.

4. **Gain access to the lid switch.** If the washer's top snaps in place, tape the lid shut. Use a putty knife to release the spring clips in each corner, and lift off the top. If the spring clips won't release, open the lid, pull the top toward you, and lift (Figure 15-30). Raise the washer top to gain access to the lid switch (Figure 15-31).

PART V

Pump

Pump retaining clips

Transmission

Motor

FIGURE 15-32 Remove the screws that hold the console and then remove the two clips that hold the cabinet in place and remove the cabinet.

On some models, the top is held down with two screws that are secured from underneath the top. The front of the cabinet is secured in place with two screws; take them out and remove the front panel. Then remove the screws that hold the top in place. On other models, the top is part of the cabinet (Figure 15-32). To gain access to the switch, the cabinet will have to be removed as shown in Figure 15-18.

5. **Test the lid switch.** To test the lid switch, remove the two wires from the switch (Figure 15-33). With a continuity tester, test for continuity between the two terminals of the switch while moving the lid switch lever. If the switch fails these tests, replace the switch.

6. **Replace the lid switch.** Remove the screws that hold the switch in place (Figure 15-34). Install the new switch, and connect the wires on the switch terminals. Then reconnect the electricity and test the washer.

Figure 15-33 Removing the wires to check for continuity of the switch.

Figure 15-34 Removing the two screws that hold the lid switch in place.

Water Pump

The typical complaints associated with the water pump are:

- Washer will not drain the water out.
- It smells like something is burning.
- The water in the washer will not recirculate.

To handle these problems, perform the following steps:

1. **Verify the complaint.** Verify the complaint by operating the washer.

2. **Check for external factors.** You must check for external factors not associated with the appliance. Is the appliance installed properly? Does the appliance have the correct voltage?

3. **Disconnect the electricity.** Before working on the washer, disconnect the electricity to the washer. This can be done by pulling the plug from the electrical outlet. Be sure that you only remove the washer plug. Or disconnect the electricity at the fuse panel or at the circuit breaker panel. Turn off the electricity.

4. **Access and remove the water pump.** In order to gain access to the water pump in this type of washer, the cabinet must be removed (see Figure 15-18). Remove the screws that hold the control panel in place. Next, lift the control panel up. Then remove the clips from each side that hold the cabinet in place. Pull back on the cabinet, and move it out of the way (Figure 15-32). With no water in the tub, remove the hoses from the pump ports (some water will spill out of the pump). Wipe the water up immediately. Check the water pump ports for any obstructions. If no obstructions are found, then disconnect the two clamps that hold the pump in place. Remove the pump.

FIGURE 15-35
Loosen the screw on each clamp and remove the flexible coupling.

To gain access to the water pump in this type of washer, remove the back panel. With no water in the tub, remove the hoses from the pump ports (some water will spill out of the pump). Wipe the water up immediately. If obstructions are found, remove the obstructions, reconnect the hoses, and test for proper operation. If the unit is still not operating properly, or if the obstruction could not be removed, the pump must be removed. Next, remove the flexible pump coupling (Figure 15-35). Now remove the bolts that hold the pump in place (Figure 15-36). Remove the pump.

5. **Install a new water pump.** To install the new pump, just reverse the disassembly procedure, and reassemble. Test the washer for any water leaks. The direct-drive washer can be operated with the cabinet removed by installing a jumper wire in the lid switch harness connector (Figure 15-37).

Inner Basket

The typical complaints associated with the inner basket are:

- Basket will not spin.

- Water will not drain out of the tub.

- Washer damages the clothing.

FIGURE 15-36 Removing the pump hold-down bolts.

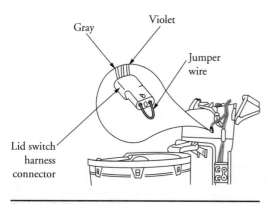

FIGURE 15-37   Installing a jumper wire in the lid switch harness connector to test the operation of a direct-drive washer.

To handle these problems, perform the following steps:

1. **Verify the complaint.** Verify the complaint by inspecting and operating the washer.

2. **Check for external factors.** You must check for external factors not associated with the appliance. Is the appliance installed properly? Does the appliance have the correct voltage?

3. **Disconnect the electricity.** Before working on the washer, disconnect the electricity. This can be done by pulling the plug from the electrical outlet. Be sure that you only remove the washer plug. Or disconnect the electricity at the fuse panel or at the circuit breaker panel. Turn off the electricity.

4. **Gain access to inner basket.** To gain access to the inner basket, you must first lift the top of the washer off (see Figure 15-30) and disconnect the fill hose from the water inlet, located on top of the splash guard (Figure 15-38). Remove the splash guard by removing the clips that hold it to the outer tub. Now that you have more room to work with, remove the agitator. On some models, the agitator is held down with a cap and stud assembly, or the agitator might just snap onto the transmission shaft (Figure 15-39). Next, use a spanner wrench to remove the locknut that holds the basket in place (Figure 15-40). On some models, the inner basket is held in place with four bolts. If so, remove the bolts. On some models, the agitator is part of the inner basket (Figure 15-41). Next, lift the inner basket out of the tub. Inspect the outer tub for debris and rust. Be sure that the tub drain opening is clear of debris.

5. **Reinstall the inner basket.** To reinstall the inner basket, just reverse the disassembly procedure, and reassemble. One important note: Do not overtighten the locknut. Just tighten the locknut enough to secure the basket in place. If you overtighten the locknut, the washer will not work properly.

6. **Test the washer.** After you have reassembled the inner basket, splash guard, and agitator, test the washer. Check for leaks around the top rim of the outer tub. Check for agitation and spin with a full load of clothes.

FIGURE 15-38   Removing the clamp from the inlet hose.

Figure 15-39
Removing the agitator.

Hoses

The typical complaints associated with the hoses are:

- Washer leaks water.
- Water will not pump out.
- Kinked or plugged hoses.

Figure 15-40
Always use a spanner wrench to remove the locknut.

FIGURE 15-41
Some models have eliminated the tall type agitator. This type of basket has more room in it for a larger wash load.

To handle these problems, perform the following steps:

1. **Verify the complaint.** Verify the complaint by inspecting and operating the washer.

2. **Check for external factors.** You must check for external factors not associated with the appliance. Is the appliance installed properly? Does the appliance have the correct voltage?

Pump

Snap clamp ring

FIGURE 15-42 Removing the snap ring clamp. If you are just replacing a part, always check the hose for any cracks where the clamp was resting on the hose.

3. **Disconnect the electricity.** Before working on the washer, disconnect the electricity. This can be done by pulling the plug from the electrical outlet. Be sure that you only remove the washer plug. Or disconnect the electricity at the fuse panel or at the circuit breaker panel. Turn off the electricity.

4. **Remove the defective hose.** To remove the defective hose, the tub must be empty of water. Loosen the clamps that hold the hose in place. On some models, the manufacturer uses a snap ring clamp. To remove this type of clamp, just squeeze the ends together (Figure 15-42).

Clutch

The typical complaints associated with the clutch are:

- Washer will not agitate.
- Washer will not spin.

To handle these problems, perform the following steps:

1. **Verify the complaint.** Verify the complaint by operating the washer.

2. **Check for external factors.** You must check for external factors not associated with the appliance. Is the appliance installed properly? Does the appliance have the correct voltage?

3. **Disconnect the electricity.** Before working on the washer, disconnect the electricity. This can be done by pulling the plug from the electrical outlet. Be sure that you only remove the washer plug. Or disconnect the electricity at the fuse panel or at the circuit breaker panel. Turn off the electricity.

4. **Gain access to the motor.** To gain access to the motor, the back panel must be removed. The back panel is held on with screws. Remove the screws, and remove the panel.

5. **Remove the motor and clutch.** In this type of washer, to remove the motor and clutch assembly, you must first disconnect the wires of the motor from the wiring harness. Then remove the flexible pump coupling (see Figure 15-35). Remove the belt from the clutch. Once the motor is isolated, remove the three bolts that hold the motor mounting plate to the suspension. Now remove the motor assembly from the washer. You are now ready to remove the clutch assembly. Remove the clutch drive plate by removing the roll pin. This is accomplished by using a drive pin tool on the roll pin (Figure 15-43) and hitting it with a hammer. On a stubborn clutch drive plate, the use of a wheel puller tool and a horseshoe collar tool will make it easier to remove (Figure 15-44). Disassemble and remove the clutch assembly, and replace with a new assembly.

Drive pin tool

Clutch drive plate

6. **Install the new clutch assembly.** To install the new clutch assembly, just reverse the disassembly procedure, and reassemble. Adjust the belt tension.

7. **Test the washer.** Test the washer operation by running the washer through a cycle. Be sure that you have some clothes in the washer when testing it.

Transmission

The typical complaints associated with the transmission are:

- Washer will not agitate.
- Washer will not spin.

To handle these problems, perform the following steps:

1. **Verify the complaint.** Verify the complaint by operating the washer.

FIGURE 15-44
Use a horseshoe
collar tool to remove a
stubborn clutch drum.

Wheel puller tool

Horseshoe
collar tool

Clutch drum

2. **Check for external factors.** You must check for external factors not associated with the appliance. Is the appliance installed properly? Does the appliance have the correct voltage?

3. **Disconnect the electricity.** Before working on the washer, disconnect the electricity. This can be done by pulling the plug from the electrical outlet. Be sure that you only remove the washer plug. Or disconnect the electricity at the fuse panel or at the circuit breaker panel. Turn off the electricity.

4. **Remove the inner basket.** To remove the inner basket, see the "Inner Basket" section earlier in this chapter.

Boot

Ring
clamp

FIGURE 15-45 Loosen the clamps, and lift the boot off. Inspect the boot for holes.

5. **Remove the transmission boot.** To remove the transmission boot, loosen the ring clamps and lift the boot off the transmission and the outer tub (Figure 15-45). Examine the boot: if it is damaged, replace it. If the ring clamps are rusted, replace them.

6. **Remove the transmission.** Remove the six bolts that hold the transmission to the washer's suspension. Remember the position that the transmission is in before you remove it. This will help you later for the reinstallation of the new transmission. Next, reach in and remove the drive belt from the transmission pulley. Lift the transmission out of the washer (Figure 15-46).

FIGURE 15-46 Removing bolts and lifting the transmission out of the tub.

7. **Install the transmission.** To reinstall the transmission, just reverse the disassembly procedure, and reassemble. Do not forget to reinstall the belt.

8. **Test the washer.** Test the washer operation by running the washer through a cycle. Be sure that you have some clothes in the washer when testing it.

Direct-Drive Washer Transmission and Brake/Drive Assembly

When handling problems related to the direct-drive washer transmission or brake/drive assembly, perform the following steps:

1. **Verify the complaint.** Verify the complaint by operating the washer.

2. **Check for external factors.** You must check for external factors not associated with the appliance. Is the appliance installed properly? Does the appliance have the correct voltage?

3. **Disconnect the electricity.** Before working on the washer, disconnect the electricity. This can be done by pulling the plug from the electrical outlet. Be sure that you only remove the washer plug. Or disconnect the electricity at the fuse panel or at the circuit breaker panel. Turn off the electricity.

4. **Remove the outer cabinet.** To remove the cabinet (see Figure 15-18), remove the screws that hold the control panel in place. Next, lift the control panel up. Then remove the clips from each side that hold the cabinet in place. Pull back on the cabinet, and move it out of the way (see Figure 15-32).

5. **Remove the agitator.** To remove the two-piece agitator (Figure 15-47), pull off the agitator cap and the inner cap. Remove the bolt that secures the agitator, pull up on the agitator, and remove it from the inner basket.

6. **Remove the tub ring, inner basket, and transmission.** Before you can remove the inner basket, you must first remove the tub ring. To do this, you must unsnap the tabs around the outer tub (see Figure 15-47). Out of the eight tabs, one of them is a locator tab (the smallest one). The locator tab will help align the tub ring properly on reassembly. Next, remove the spanner nut, expand the drive block, and lift the basket out of the tub. Then remove the motor and pump (Figures 15-19, 15-20, and 15-21) and lay the washer on its back. Locate and remove the three bolts that secure the transmission to the tub support. Pull the transmission out of the basket drive tube, being careful not to damage the bearing.

FIGURE 15-47   The agitator, tub ring, and inner basket.

Figure 15-48
Removing the brake/
drive assembly.

Brake and drive tube assembly

7. **Remove the brake/drive assembly.** To release the brake, turn the brake cam driver counterclockwise. This will allow you to pull the brake and drive tube out of the base assembly (Figure 15-48).

8. **Install the transmission and the brake/drive assembly.** To reinstall the transmission and the brake/drive assembly, just reverse the disassembly procedure, and reassemble.

9. **Test the washer.** Test the washer operation by running the washer through a cycle. Be sure that you have some clothes in the washer when testing it.

Diagnostic Charts

The following diagnostic flow charts will help you pinpoint the likely causes of the problem (Figures 15-49, 15-50, and 15-51).

Figure 15-49
Diagnostic flow chart:
Washer fills but will
not agitate.

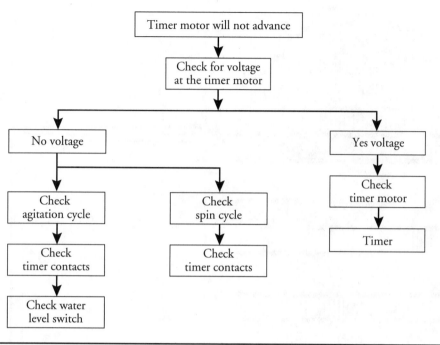

FIGURE 15-50   Diagnostic flow chart: Timer motor will not advance.

The wiring diagrams in this chapter are used as examples only. You must refer to the actual wiring diagram on the washer that you are servicing. Figure 15-52 depicts an actual wiring schematic. Figure 15-53 depicts an actual ladder diagram of Figure 15-52. A ladder diagram is generally easier to read.

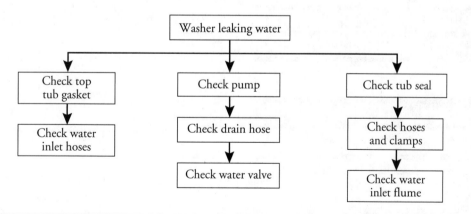

FIGURE 15-51   Diagnostic flow chart: Washer leaking water.

FIGURE 15-52   Automatic washer electrical schematic.

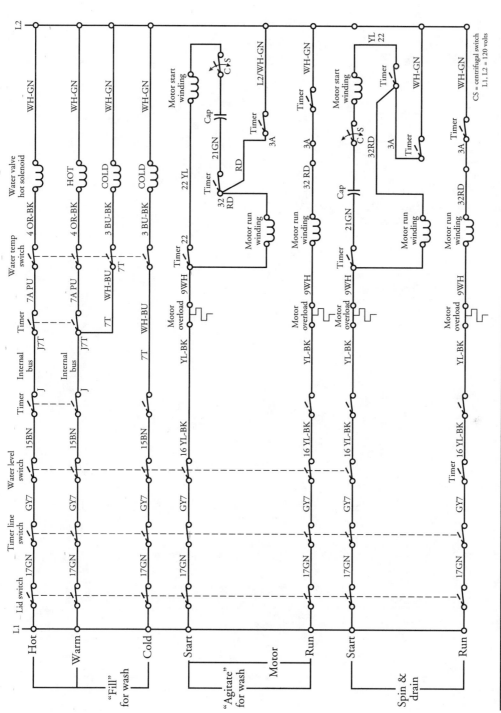

Figure 15-53 Automatic washer ladder diagram.

Automatic Electric Dryers

Automatic electric dryers are not complicated to repair. The more you know about the electrical, mechanical, and operational basics of the dryer, the easier it is to solve the problem. Some models are designed to operate on 240 volts; other models operate on 120 volts. To dry clothing, an automatic dryer must have drum rotation, electric heat, and air circulation. This chapter will provide the basic skills needed to diagnose and repair automatic electric dryers. Figure 16-1 identifies where some of the components are located within the automatic electric dryer. However, this illustration is used as an example only. The actual construction and features might vary, depending on what brand and model you are servicing.

Principles of Operation

The clothes are placed into the dryer according to the manufacturer's recommendation for the proper loading of the dryer. Next, the proper cycle is selected, and the dryer start button is pressed. The combination of the timer, the switches, and the thermostats regulates the air temperature within the drum and the duration of the drying cycle. During the drying cycle, room air is pulled into the dryer drum from the lower rear (or sometimes the front) of the cabinet, depending on which model the consumer owns. The air is pulled through the heater, drum, lint screen, down through the lint chute, and through the fan housing. It is then pushed out of the exhaust duct. The drive motor, blower wheel, belt, and pulleys cause the drum to turn and the air to move through the dryer. The belt wraps around the drum, motor pulley, and idler pulley. The blower wheel is secured to one end of the drive motor shaft. As the drive motor turns, the drum rotates, moving the clothes. At the same time, the blower wheel turns, moving the air. The heating element will cycle on and off, according to the temperature selected.

Functions and Cycles

Electric automatic dryers use three basic functions to operate:

- Heat is supplied by a resistance-type heating element.
- Air is drawn into the dryer. It is heated and circulated through the tumbling clothes. Then the warm, moisture-laden air is drawn through the lint screen and is vented through the duct system to the outside.

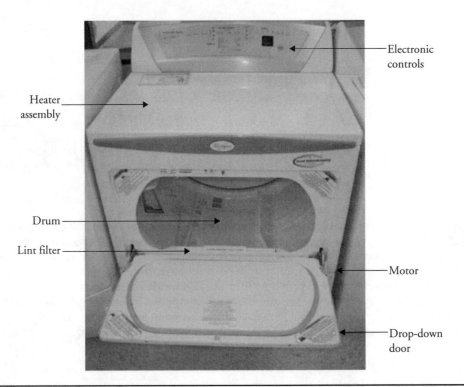

FIGURE 16-1   The location of components in a typical automatic electric dryer. The consumer has a choice between a drop-down or side-open door.

- Tumbling of the clothes is accomplished with a motor that drives a belt, which rotates the drum.

The cycles of an electric automatic dryer are as follows:

- Timed dry cycle: The timed dry cycle is controlled by the amount of time selected on the timer. The temperature in the dryer is controlled by a thermostat, which turns the heating element on and off throughout the timed cycle.

- Automatic dry cycle: The automatic dry cycle is not controlled by the timer. This cycle is controlled by the cycling thermostats. Heat is supplied to dry the clothes, and it will continue until the temperature in the drum reaches the selected cutout setting of the thermostat. When the thermostat is satisfied, the heater shuts off and the timer motor is activated. However, certain variables can control the cycle, which will cause the cycling of the thermostat before the end of the cycle.

- Permanent press and knit cycles: The permanent press and knit cycles are controlled by the amount of time selected on the timer. The temperatures of these cycles are controlled by the temperature rating of the cycling thermostats that are located within the cabinet, on the exhaust duct, and in the air supply. On some models, the user can select the desired type of heat setting with the temperature selector switch, located on the control panel.

- Air dry: The air dry cycle is controlled by the amount of time selected on the timer. This cycle uses the air to dry the clothes. The heating element is not used at all during this cycle.

Safety First

Any person who cannot use basic tools or follow written instructions should *not* attempt to install, maintain, or repair any automatic dryers. Any improper installation, preventive maintenance, or repairs could create a risk of personal injury or property damage.

If you do not fully understand the installation, preventive maintenance, or repair procedures in this chapter, or if you doubt your ability to complete the task on the automatic dryer, please call your service manager.

The following precautions should also be followed:

- Never bypass (or interfere with) the operation of any switch, component, or feature.
- The dryer exhaust should be vented properly. Never exhaust the dryer into a chimney, a common duct, an attic, or a crawl space.
- Be careful of sharp edges when working on the dryer.
- The dryer produces combustible lint, and the area should be kept clean.
- Never remove any ground wires from the dryer or the third (grounding) prong from the service cord.
- Never use an extension cord to operate a dryer.
- The wiring used in dryers is made with a special heat-resistant insulation. Never substitute it with ordinary wire.

Before continuing, take a moment to refresh your memory of the safety procedures in Chapter 2.

Automatic Electric Dryers in General

Much of the troubleshooting information in this chapter covers automatic electric dryers in general, rather than specific models, in order to present a broad overview of service techniques. The illustrations that are used in this chapter are for demonstration purposes only to clarify the description of how to service dryers. They in no way reflect upon a particular brand's reliability.

Electrical Requirements

Electrical dryers can be connected to an electrical power source in three ways:

- Electrical cord and plug, which plugs into a 240-volt wall receptacle.
- Directly wired to a fused electrical disconnect box with a built-in shut-off switch.

- Directly wired to a circuit breaker or to a fuse panel. All dryers must be properly wired, grounded, and polarized, according to the manufacturer's installation instructions and per all local codes and ordinances.

Location and Installation of Dryer

The following are some general principles that should be followed when installing a dryer:

- Locate the dryer where there is easy access to existing electrical lines.
- Be sure that you observe all local codes and ordinances for the electrical, plumbing, and venting connections.
- The dryer should be installed and leveled on a firm floor to minimize vibration during operation.
- To reduce the risk of fire, never install a dryer on any type of carpeting.
- For proper operation of the dryer, be sure that there is adequate make-up (new, replacement) air in the room where it is installed.
- Do not install a dryer in an area where the make-up air is below 45 degrees Fahrenheit. This will greatly reduce the drying efficiency and increase the cost of operating the dryer.
- Always follow the installation instructions that are provided with every automatic dryer purchased. If the installation instructions are not available, order a copy from the manufacturer.
- When installing a dryer in a mobile home, the dryer must have an outside exhaust. If the dryer exhaust goes through the floor, and the area under the mobile home is enclosed, the exhaust must terminate outside of the enclosed area.
- If the dryer is to be installed in a recessed area or closet area, follow the manufacturer's recommendation for its installation.
- When relocating the dryer to a new location, test the voltage at the new location, and be sure that it matches the dryer voltage specifications as listed on the nameplate (serial plate). Also inspect the entire vent system for any obstructions.

Proper Exhausting of the Dryer

Proper exhausting instructions for the model being installed are available through the manufacturer. Each manufacturer has its own specifications for the size and the length of the ductwork needed to run its dryer properly. The maximum length of the exhaust system depends upon the type of duct, the number of elbows, and the type of exhaust hood used. Figure 16-2 illustrates a typical dryer exhaust installation.

The following guide is recommended for the exhausting of a dryer:

- Keep the duct length as short as possible.
- Keep the number of elbows to a minimum to minimize the air resistance.
- Never reduce the diameter of round ductwork below four inches.
- Install all exhaust hoods at least 12 inches above ground level.

FIGURE 16-2
Typical dryer exhaust
installation.

Exhaust hood

Metal duct

2 Turns

Dryer

- The exhaust duct and exhaust hood should be inspected periodically and cleaned, if necessary.
- All duct joints should be taped. Never use screws to join the duct joints together. Screws protruding into the duct will cause lint buildup, and this will eventually clog the duct.
- Never exhaust the dryer into any wall, ceiling, attic, or under a building.
- Accumulated lint could become a fire hazard, and moisture could cause damage.
- If the exhaust duct is adjacent to an air conditioning duct, the exhaust duct must be insulated to prevent moisture buildup.

Step-by-Step Troubleshooting by Symptom Diagnosis

In the course of servicing an appliance, don't overlook the simple things that might be causing the problem. Step-by-step troubleshooting by symptom diagnosis is based on diagnosing malfunctions, with possible causes arranged into categories relating to the operation of the dryer. This section is intended only to serve as a checklist to aid you in diagnosing a problem. Look at the symptom that best describes the problem that you are experiencing with the dryer, and then correct the problem.

Dryer Will Not Run

- Do you have the correct voltage at the dryer?
- Test the door switch for continuity of the switch contacts.
- Test for continuity of the motor windings. Also check for a grounded motor.
- Test the timer for continuity of the switch contacts.
- Test the "push to start" switch for continuity of the switch contacts.

- Check for broken or loose wiring. Also check the wire terminal connections that connect to the different components.
- For electronic control models, check for an error code. Check the electronic control board.

No Heat

- Do you have the correct voltage? The motor in a dryer runs on 120 volts. The heating element works on 240 volts; compact models run on 120/240 volts. Check voltage at the wall receptacle. Also check the nameplate voltage rating for the model that you are working on.
- Check for any in-line fuses that might have blown. Refer to the wiring diagram, located on the rear of the dryer or behind the console panel.
- Are the thermostats functioning properly?
- Test the timer for continuity of the switch contacts.
- Test for continuity of the motor windings. Also test for a grounded motor.
- Test the heating element for continuity.
- Test the temperature selector switch for continuity.
- Check for broken or loose wiring.

Drum Will Not Rotate

- Check the drive belt. Is it broken or worn?
- Check the idler pulley. Is the pulley seized up?
- Check the drum support bearings. Be sure that they rotate freely.
- Is the dryer overloaded with clothing?
- Check for foreign objects that might be lodged between the drum and the bulkheads.

Dryer Is Noisy

- Check for loose components. Is everything secure and in its proper place?
- Check the idler pulley. Lint buildup can cause the idler pulley shaft to squeak.
- Check the drive belt. Is the belt partially torn?
- Check the drive motor. Is the pulley secured to the shaft? Also check the fan assembly. Is it loose?
- Check for lint or foreign objects lodged between the drum and the bulkheads.
- Check the front and rear bearings.
- Check the blower assembly. Be sure that the blower wheel is tight on the motor shaft end.
- Is the dryer level?

Dryer Runs and Heats, but the Clothes Won't Dry

- Check for a defective thermostat. Use a thermometer to check the duct temperatures during the cycling of the dryer.
- Check for a loose pulley or blower wheel. Inspect and tighten the blower wheel.
- Check for restricted air flow in the dryer.
- Check the vent system.

Clothes Too Hot or Fabric Damage

- Check the exhaust system for restrictions.
- Check for air leaks in the ductwork and the front and rear seals on the drum.
- Check the thermostats.
- Check the temperature setting with the customer.
- Check the door latch and strike.
- Inspect the interior of the drum for any foreign objects protruding into it.

Common Drying Problems

The drying of clothes can be affected by several factors:

- Clogged lint screen.
- Exhaust duct too long, collapsed, or crimped.
- Dryer air is exhausted into the area where the dryer is located. For example, the dryer is located inside a closet and operating with the closet door closed.
- Overloading of clothing in the dryer.
- Dryer is located in an area where the temperature is below 45 degrees Fahrenheit.
- Laundry washed in cold water will take slightly longer to dry.
- The moisture content of certain clothes. Towels and denim retain more moisture, for example.
- Wrong spin speed selected for the type of laundry being washed. For example, towels, denim, blankets, small rugs, etc. use the normal wash setting to extract most of the water from the laundry. If there is an extra option for high-speed spin it should also be selected to extract as much water as possible from the laundry. The drying cycle will be increased greatly if the permanent press cycle or the gentle cycle is selected for these items.
- Electric supply is less than what is needed to operate the dryer efficiently.
- Load type (towels, bedspreads, jeans, etc.)

All of these must be taken into account when diagnosing a complaint of clothes not drying properly.

Lint

Lint consists of fibers that have broken away from the fabric. It can collect inside the dryer cabinet and base and create a fire hazard. This lint should be removed every one to two years by cleaning it out. Some lint can also collect in the door opening, the drum, the heater assembly, the blower assembly, and the duct assembly. This accumulation of lint should be removed periodically when performing maintenance on the dryer; otherwise, it could create problems with the future use of the dryer.

The lint screen should be cleaned every time the dryer is used. The duct system that exhausts the air to the outside of the dwelling should also be cleaned out every one to two years.

Shrinkage

Shrinkage is caused by overdrying and by the type of fabric being dried. To reduce the shrinking of cotton and rayon knits, it is recommended that the user remove them from the dryer while the clothing is still damp and lay them on a flat surface to air-dry.

Also, you might suggest that the user set the heat setting on a lower setting. Before drying any synthetic clothing in a dryer, the user should read the label on the garment for the proper drying instructions and for the proper heat selection. The user should also read the use and care manual for the proper instructions when operating the dryer.

Stains

Greasy-looking stains are often caused by:

- Fabric softener that is designed for use in the dryer.
- Fabric softener that was undiluted when the clothes were washed.
- Not enough detergent being used in the wash. This will cause the soil in the water to stick to the outer tub and to return to the next load being washed.
- Drying of clothing that already has a stain on it.

Brown stains on the clothing are often caused by:

- Fabric softener that is designed for use in the dryer.
- Leaving the wash in the washer tub after the cycle is completed.
- Leaving the wet clothing in the dryer for an extended amount of time.
- A leaking transmission seal.

Static Electricity

Static electricity in the clothes is caused by overdrying them. To reduce this condition, you must instruct the user to reduce the amount of time the clothes are to be dried and to use a lower temperature setting. Static electricity in synthetics is normal. To reduce this condition, use a liquid fabric softener in the wash cycle or use a fabric softener that is designed for use in the dryer.

Wrinkling

Wrinkling of the clothes is often caused by:

- Not using the permanent press cycle, which has a cooling-down period during which the heating element is not on.
- Clothing not removed from the dryer after it has completed the cycle.
- Overloading the dryer. The clothes need room to tumble freely without getting tangled.
- The quality of the permanent press material in the garment might be poor.

Electric Dryer Maintenance

Automatic dryers must be cleaned periodically. The excess lint must be removed to prevent the possibility of a fire and the possibility of the dryer not functioning properly. Most models have a service panel on the front of the dryer for accessibility. The outside of the cabinet should be wiped with a damp cloth. It is recommended that the components be inspected for wear and tear; if repairs are needed, they should be made as soon as possible.

Repair Procedures

Each repair procedure is a complete inspection and repair process for a single dryer component.

Dryer Timer

The typical complaints associated with the dryer timer are:

- The dryer will not run at all.
- The clothes are not drying.
- The dryer will not stop at the end of the cycle.
- The timer will not advance through the cycle.

To handle these problems, perform the following steps:

1. **Verify the complaint.** Verify the complaint by operating the dryer through its cycles. Before you change the timer, check the other components controlled by the timer.
2. **Check for external factors.** You must check for external factors not associated with the appliance. Is the appliance installed properly? Does the appliance have the correct voltage?
3. **Disconnect the electricity.** Before working on the dryer, disconnect the electricity. This can be done by pulling the plug from the electrical outlet. Or disconnect the electricity at the fuse panel or at the circuit breaker panel. Turn off the electricity.
4. **Remove the console panel to gain access.** To gain access to the timer, remove the screws that secure the console to the top of the dryer (Figure 16-3). On other models, to gain access, remove the back panel behind the console.

(a)   (b)

Figure 16-3   To gain access to the component, remove the screws. Rest the console panel on its face to gain access to the component.

5. **Test the timer.** There are two ways to check the timer:
   - Connect a 120-volt fused service cord to the timer motor to see if the timer motor is operating (Figure 16-4). Connect the ground wire of the 120-volt fused service cord to the console ground. Be cautious when working with live wires. Avoid getting shocked. The timer motor operates on 120 volts. If the motor does not operate, replace the timer. If the timer motor runs but does not advance the cams, the timer has internal defects and should be replaced.
   - Set the ohmmeter range to $R \times 100$, disconnect the timer motor leads, and check for resistance (Figure 16-5). The meter should read between 2000 and 3000 ohms. Next, test the timer switch contacts with the wiring diagram's configuration for the affected cycle. Place the meter probe on each terminal being tested, and turn the timer knob. If the switch contact is good, your meter will show continuity.

6. **Remove the timer.** To remove the timer, pull the timer knob from the timer stem; then remove the timer mounting screws. Remove the wire lead terminals from the timer. Mark the wires as to their location on the timer. Some timers have a quick disconnect, instead of individual wires, which makes it easier to remove the timer wires (Figure 16-6).

7. **Install a new timer.** To install a new timer, just reverse the disassembly procedure, and reassemble. Replace the wires on the timer. Reinstall the console panel, and restore the electricity to the dryer. Test the dryer for proper operation.

Motor wire

Motor terminal

Motor wire

Grounded to
console chassis

Motor terminal

Grounded to
electrical box

24-inch
test leads

Electrical box

On/Off
switch

120-volt fused
service cord

20-amp
resettable fuse

6-foot
service cord

Male plug
with ground

120-volt
receptacle

FIGURE 16-4 The test cord attached to the timer motor leads.

FIGURE 16-5
Checking for continuity
between the timer
motor and the switch
contacts.

Multitester

Timer motor

Timer

Timer plug

Timer

Timer mounting screws

FIGURE 16-6   Remove the timer dial. Then remove the timer mounting screws, and lift the timer off the mounting bracket.

Electronic Control Board and User Interface Controls

The typical complaints associated with the electronic control board or the user interface controls are:

- The dryer won't run or power up.
- Unable to program the dryer.
- The display board will not display anything.
- One or more key pads will not accept commands.
- Unusual display readouts.

To prevent electrostatic discharge (ESD) from damaging expensive electronic components, follow the steps in Chapter 9.

To handle these problems, perform the following steps:

1. **Verify the complaint.** Verify the complaint by operating the dryer. Turn off the electricity to the appliance and wait for two minutes before turning it back on. If a fault code appears, look up the code. If the dryer will not power up, locate the technical data sheet behind the control panel for diagnostics information.

2. **Check for external factors.** You must check for external factors not associated with the appliance. Are there 240 volts of electricity to the dryer?

3. **Disconnect the electricity.** Before working on the dryer, disconnect the electricity. This can be done by pulling the plug out of the wall receptacle. Or disconnect the electricity at the fuse panel or circuit breaker panel. Turn off the electricity.

WARNING *Some diagnostic tests will require you to test the components with the power turned on. When you disassemble the control panel, you can position the panel in such a way so that the wiring will not make contact with metal. This will allow you to test the components without electrical mishaps.*

4. **Remove the console panel to gain access.** Begin by removing the screws from the dryer top (for front-loading models only) to gain access to the electronic control board. Remove the console panel screws to gain access to the user interface controls (Figure 16-7). On top-loading models, the console will roll upwards or toward you after you remove the console screws.

5. **Test the electronic control board and user interface controls.** If you are able to run the dryer diagnostic test mode, check the different functions of the dryer. Use the technical data sheet for the model you are servicing to locate the test points from the wiring schematic. Check all wiring connections and wiring. Using the technical data sheet, test the electronic control or user interface controls, input voltages, and output voltages. On some models, fuses are soldered to the printed circuit board (PCB). These fuses must be tested first before condemning the component.

FIGURE **16-7**
An exploded view of
the electronic controls
in a front-loading
dryer.

1. Electronic control board assembly
3. Knob
4. Program light pipe
5. Auxiliary interface board
6. Pin
8. Push button
9. Push button
11. Push button
12. Push button spring
13. Start light pipe
15. Light pipe

16. Light pipe
19. Screw
22. Interface board wiring harness
23. Wiring harness
29. Control panel
39. Screw
53. Top panel
55. Digital display lens
56. Digital display light pipe
57. Digital display (LED)

6. **Remove the electronic control board and user interface controls.** To remove the
defective component, remove the screws that secure the board to the control panel
or dryer frame. Disconnect the connectors from the electronic control board or user
interface control.

7. **Install the new component.** To install a new electronic control board or user
interface control, just reverse the disassembly procedure, and reassemble. Reinstall
the console panel, and restore the electricity to the dryer. Test the dryer operation.

Dryer Motor

The typical complaints associated with motor failure are:

- Fuse is blown or the circuit breaker trips.
- Motor will not start; it only hums.
- The dryer will not run.

To handle these problems, perform the following steps:

1. **Verify the complaint.** Verify the complaint by operating the dryer through its cycles. Listen carefully, and you will hear if there are any unusual noises or if the circuit breaker trips.

2. **Check for external factors.** You must check for external factors not associated with the appliance. Is the appliance installed properly? Does it have the correct voltage?

3. **Disconnect the electricity.** Before working on the dryer, disconnect the electricity. This can be done by pulling the plug from the electrical outlet. Or disconnect the electricity at the fuse panel or at the circuit breaker panel. Turn off the electricity.

4. **Gain access to the motor.** In order to gain access to the motor, the top must be raised (Figure 16-8) by removing the screws from the lint screen slot. Insert a putty knife about two inches from each corner; then disengage the retaining clips and lift the top. On some models, the top is held down with screws (see inset in Figure 16-8). Remove the screws and lift the top. Now the front panel of the dryer must be removed (Figure 16-9). To remove the top panel, insert a putty knife and disengage the retaining clip (Figure 16-9a).

Screw holes

Lint screen slot

FIGURE 16-8 Insert a putty knife to disengage the retaining clips that hold the top down. Don't forget to remove the two screws under the lint screen cover. On some models, remove the screws from underneath the top (see insert).

On some models, the top panel is held in place with two screws. As Figure 16-9b shows, remove the screws that hold the lower part of the front panel in place. Next, remove the screws that hold the upper part of the front panel in place (Figure 16-9c), and then disconnect the door switch wires (label them). With the front panel out of the way, you can now disconnect the drive belt (Figure 16-10). Push on the idler pulley to release the tension from the drive belt, and remove the belt from the motor pulley. Grab the drum and remove it from the cabinet (Figure 16-11). On some models, you will have to remove the back panel because the drum comes out through the rear of the cabinet.

5. **Disconnect the motor wire leads.** Disconnect the motor wire leads from the wiring harness. Set the ohmmeter on R × 1. Figure 16-12 illustrates testing the motor

FIGURE 16-9
Remove the top panel.
Remove the screws
from the lower front
panel. Remove the
screws from the upper
front panel.
Remember to remove
the door switch wires.

(a)

(b)

(c)

FIGURE 16-10
Push the idler pulley
assembly toward the
drive motor pulley to
release the belt
tension. Then
disconnect the drive
belt from the idler
pulley and drive motor
pulley.

FIGURE 16-11 Remove the drum from the dryer cavity.

FIGURE 16-12 Test the centrifugal switch and drive motor windings.

windings and the centrifugal switch for continuity. When testing for resistance on the motor, test from the common wire lead to the run winding. Then test for resistance from the common wire lead to the start winding. Next, test for resistance from the start winding to the run winding. To test for a grounded winding in the motor, take the ohmmeter probes and test from each motor wire lead terminal to the motor housing. The ohmmeter will indicate continuity if the windings are grounded. If the ohmmeter reading shows no resistance between the motor windings, then replace the motor. If the motor checks out okay, then check the timer.

6. **Remove the motor.** To remove this type of motor, you must first disconnect the blower assembly by holding the motor shaft stationary and then turning the blower wheel to remove it from the rear of the motor shaft (Figure 16-13). Then remove the spring clamps that hold the motor in the motor bracket (Figure 16-14). On some models, the motor pulley must be removed (new motors come without the pulley attached) by loosening the allen set screw.

FIGURE 16-13 Disconnect the blower assembly from the drive motor shaft.

FIGURE 16-14
Hold the motor and remove the spring clamps.

Rubber cushion

Spring clamp

Motor pulley

Motor bracket

7. **Install the new motor.** To install the new motor (Figure 16-15), just reverse the disassembly procedure, and reassemble. Then reassemble the dryer in the reverse order of its disassembly. Restore the electricity to the dryer and test the motor.

Drive Belt

The typical complaints associated with belt failure are:

- The drum will not turn.
- Dryer motor spins freely.
- It smells like something is burning.

FIGURE 16-15
Place the motor in the cradle and position it to fit into the slots properly.

To handle these problems, perform the following steps:

1. **Verify the complaint.** Verify the complaint by operating the dryer through its cycles. Listen carefully, and you will hear if there are any unusual noises. Then, with the door open, press the door switch and start the dryer. The drum should rotate.

2. **Check for external factors.** You must check for external factors not associated with the appliance. Is the appliance installed properly? Does the appliance have the correct voltage? Check for foreign objects lodged between the drum and bulkhead, etc.

3. **Disconnect the electricity.** Before working on the dryer, disconnect the electricity. This can be done by pulling the plug from the electrical outlet. Or disconnect the electricity at the fuse panel or at the circuit breaker panel. Turn off the electricity.

4. **Gain access to the drive belt.** To gain access to the drive belt, the top must be raised (see Figure 16-8) by removing the screws from the lint screen slot. Then insert a putty knife about two inches from each corner, disengage the retaining clips, and lift the top. On some models, the top is held down with screws (see the inset in Figure 16-8). Remove the screws and lift the top. Now the front panel of the dryer must be removed (see Figure 16-9). To remove the top panel, insert a putty knife and disengage the retaining clip (see Figure 16-9a). As shown in Figure 16-9b, remove the screws that hold the lower part of the front panel in place. Next, remove the screws that hold the upper part of the front panel in place (see Figure 16-9c), and then disconnect the door switch wires.

5. **Remove the drive belt.** To remove the drive belt on this type of dryer, you can disconnect the belt (see Figure 16-10). Push on the idler pulley to release the tension from the drive belt. Now remove the belt from the motor pulley and from around the drum (Figure 16-16). If the drive belt is broken, just remove the belt.

6. **Install a new drive belt.** To install the new drive belt, just reverse the disassembly procedure, and reassemble. Then reassemble the dryer in the reverse order of its disassembly. Restore the electricity to the dryer and test.

FIGURE 16-16   Removing the drive belt from around the drum.

Heater Element

The typical complaints associated with heater failure are:

- The dryer will not dry the clothes properly.

- There is no heat at all when a heat cycle is selected.

FIGURE 16-17   Remove the screws that hold the back panel in place.

To handle these problems, perform the following steps:

1. **Verify the complaint.** Verify the complaint by operating the dryer through its cycles. Then open the dryer door, and place your hand inside to see if it is warm in the drum.

2. **Check for external factors.** You must check for external factors not associated with the appliance. Is the appliance installed properly? Is the exhaust vent clogged? Check the voltage to the dryer. *Note:* The dryer motor runs on 120 volts, but the heater works on 240 volts.

3. **Disconnect the electricity.** Before working on the dryer, disconnect the electricity. This can be done by pulling the plug from the electrical outlet. Or disconnect the electricity at the fuse panel or at the circuit breaker panel. Turn off the electricity.

4. **Gain access to the heater assembly.** You must gain access to the heater assembly through the back. Remove the exhaust duct from the dryer. Then remove the screws from the back panel (Figure 16-17).

5. **Test the heater element.** To test the heater element, set the ohmmeter on R × 1. Remove the wires from the heater (Figure 16-18), and test for continuity. If it checks out okay, check the thermostats.

FIGURE 16-18
Test the heater element for continuity.

PART V

FIGURE 16-19   Remove the heater box.

FIGURE 16-20   Remove the heater coil from the heater box.

6. **Remove the heater element.** To remove the heater in this type of dryer, you must first remove the screw and the bracket that holds the heater box in place. It is located on top of the heater box. You can gain access through the back of the dryer or lift the top and reach in near the back bulkhead. Then slide the heater box up, and pull it away at the bottom (Figure 16-19). With the heater box out of the dryer, remove the screw that holds the heater element in place (Figure 16-20). Pull the heater element out of the heater box.

7. **Install new heater element.** To install the new heater element, just reverse the disassembly procedure, and reassemble. Then reinstall the back panel. Restore the electricity to the dryer and test it.

Door Switch

The typical complaints associated with switch failure are:

- Dryer will not operate at all.
- Dryer light does not work when the door is open.

To handle these problems, perform the following steps:

1. **Verify the complaint.** Verify the complaint by operating the dryer through its cycles. Open the door to see if the light is working.

2. **Check for external factors.** You must check for external factors not associated with the appliance. Is the appliance installed properly? Does the dryer have the correct voltage supply?

3. **Disconnect the electricity.** Before working on the dryer, disconnect the electricity. This can be done by pulling the plug from the electrical outlet. Or disconnect the electricity at the fuse panel or at the circuit breaker panel. Turn off the electricity.

4. **Gain access to the door switch.** You must gain access to the door switch. On this model, the top must be raised (see Figure 16-8) by removing the screws from the lint screen slot. Then insert a putty knife about two inches from each corner, disengage the retaining clips, and lift the top. On some models, the top is held down with screws (see the inset in Figure 16-8). Remove the screws and lift the top.

5. **Test the door switch.** The door switch is located in one of the upper corners of the inside of the front panel (Figure 16-21a). Once you have located the door switch, disconnect the wires from the terminals. Set the ohmmeter on R × 1, attach the probes of the ohmmeter to the terminals of the door switch, and then close the door. With the door closed, there should be continuity. With the door open, the ohmmeter should not show continuity. Some dryer models have a light inside the drum. This light circuit is also part of the door switch circuitry. Three terminals are located on the door switch. Take your ohmmeter probe and place it on the common terminal of the door switch. To locate the common terminal, read the wiring diagram. It will indicate which terminal is the common, which is the light, and which is for the motor circuit. Take your probe and place it on the switch terminal (the light circuit); then close the dryer door. The reading should indicate no continuity. Now open the dryer door—you should have a continuity reading on the meter.

(a) (b)

(c) (d)

Figure 16-21 (a) Test the door switch for continuity. (b) Remove the screws that secure the door switch. (c) Removing a cylindrical door switch. (d) Removing a hinge-mounted door switch.

6. **Remove the door switch.** To remove the lever-type door switch (Figure 16-21b), remove the screws that hold the switch in place. Now lift the switch out from behind the front panel. To remove the cylindrical type of door switch (Figure 16-21c), squeeze the retaining clips on the back side of the front panel, and pull out the switch. To remove the hinge type of door switch (Figure 16-21d), the front panel must be removed. Next, remove the screws that hold the switch assembly in place.

7. **Install the new door switch.** To install the new door switch, reverse the disassembly procedure, and reassemble. Reassemble the dryer, and install the wires on the new switch terminals. Plug in the dryer and test it.

Thermostat

The typical complaints associated with thermostat failure are:

- The clothes are not drying.
- The dryer will not shut off.
- The dryer will not heat at all.

- The drying temperature is too high.

- Moisture retention (of fabrics) is unsatisfactory.

To handle these problems, perform the following steps:

1. **Verify the complaint.** Verify the complaint by operating the dryer through its cycles.

2. **Check for external factors.** You must check for external factors not associated with the appliance. Is the appliance installed properly? Does the dryer have the correct voltage supply? Is the exhaust vent blocked?

3. **Disconnect the electricity.** Before working on the dryer, disconnect the electricity. This can be done by pulling the plug from the electrical outlet. Or disconnect the electricity at the fuse panel or at the circuit breaker panel. Turn off the electricity.

4. **Gain access to the thermostat.** You must gain access to the thermostat through the back on this model. Remove the screws from the back panel (see Figure 16-17).

5. **Test the thermostat.** To test the thermostat for continuity, remove the wires from the thermostat terminals. With a continuity tester or an ohmmeter, test the thermostat for continuity (Figure 16-22). Do this to all of the thermostats in the dryer. On some models, the thermostats might be located on the heater housing, behind the drum, or in the lint screen opening (Figure 16-23). If they all check out okay, then reassemble the dryer and test for temperature operation. *Note:* Do not reinstall the back panel at this point. Take a piece of paper and write down the temperature ratings of the thermostats. These ratings are printed on the thermostats (L140, L290, etc.). To test for temperature operation, you will need a voltmeter and a temperature tester.

FIGURE 16-22
Use a continuity tester to check the thermostats.

FIGURE 16-23   The location of thermostats on some models.

First, set up the instruments for testing the thermostatic operation. Take the temperature tester thermocouple lead and insert it between the thermostat mounting ear (Figure 16-24) and the plate against which it mounts. Then connect the voltmeter probes across the thermostat terminals. Use alligator clips attached to the probe tips. This will allow you freedom of movement. Do not disconnect the wires from the thermostat. Set the voltmeter range to AC voltage and the selector switch on 300 volts. Remember: If there is more than one thermostat (in parallel with other thermostats), the thermostats not under test must be electrically isolated. Review the wiring schematic to determine which wires to remove and which ones to isolate. With the test meters in place, you are now ready to remove the exhaust vent duct from the dryer. Seal off 75 percent of the exhaust opening from the dryer (Figure 16-25). This will simulate a load of clothing in the dryer.

Restore the electricity to the dryer. This test requires that the electricity be turned on for its duration. Always be cautious when working with live wires. Avoid getting shocked. Set the controls on the dryer to operate at a high heat. Turn on the dryer and let it cycle. When the voltmeter is reading voltage, the thermostat has opened. When there is no voltage reading on the voltmeter, the thermostat is closed. As the dryer is cycling via the thermostat, record the temperature.

The thermostat should open at the preset temperature listed on the thermostat. Table 16-1 illustrates the types of thermostats, along with their opening and closing temperature settings. The temperature range of the thermostat should be within ±10 percent of the printed setting. If not, replace the thermostat.

CAUTION   *Turn off the electricity before replacing any parts of the dryer.*

FIGURE 16-24
Insert the thermocouple lead under the thermostat ear.

PART V

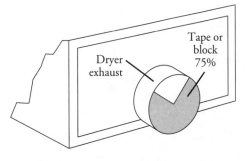

6. **Replace the thermostat.** Remove the screws that hold the thermostat in place. Replace the thermostat with an exact replacement with the same temperature rating. Reconnect the wires to their correct terminal positions. Then reverse the disassembly procedure to reassemble the dryer, and test the thermostat.

7. **Test the new thermostats.** To test the new thermostats, repeat step 5.

FIGURE 16-25   Block (or tape) 75 percent of the dryer exhaust to simulate a load.

Start Switch

The typical complaint associated with the start switch is that the dryer will not start. To handle this problem, perform the following steps:

1. **Verify the complaint.** Verify the complaint by operating the dryer. Before you change the start switch, check the other components that are in the start circuit.

2. **Check for external factors.** You must check for external factors not associated with the appliance. Is the appliance installed properly? Does it have the correct voltage?

Open	Close	Type	Open	Close	Type
260°	210°	L260	120°	110°	L120
270°	230°	L270	125°	115°	L125
290°	250°	L290	130°	115°	L130
300°	250°	L300	135°	120°	L135
310°	270°	L320	140°	130°	L140
340°	300°	L340	145°	125°	L145
120°	105°	LD120	150°	130°	L150
130°	115°	LD130	155°	135°	L155
140°	120°	LD140	160°	120°	L160
145°	125°	LD145	170°	130°	L170
155°	135°	LD155	180°	170°	L180
170°	150°	LD170	190°	150°	L190
200°	160°	LD200	200°	160°	L200
225°	185°	LD225	205°	165°	L205
240°	195°	LD240	225°	185°	L225
270°	225°	LD270	240°	200°	L240
290°	250°	LD290	250°	210°	L250

TABLE 16-1   Thermostat Opening and Closing Temperatures.

3. **Disconnect the electricity.** Before working on the dryer, disconnect the electricity. This can be done by pulling the plug from the electrical outlet. Or disconnect the electricity at the fuse panel or at the circuit breaker panel. Turn off the electricity.

4. **Remove the console panel to gain access.** Begin by removing the console panel to gain access to the start switch. With this type of dryer, remove the screws from the console (see Figure 16-3a). On some models, the console will be able to lie flat (see Figure 16-3b).

5. **Test the start switch.** Remove the wires from the start switch terminals. Set the ohmmeter on R × 1. Begin testing the start switch for continuity by placing one probe of the ohmmeter on the common terminal (C) of the switch; then connect the other probe to the normally open (NO) terminal. There should be no continuity. With the ohmmeter probes attached to the switch, press the start switch button. You should have continuity (Figure 16-26). If the switch fails the test, replace it.

6. **Remove the start switch.** To remove the start switch, pull the knob off the switch stem, remove the start switch mounting screws, and remove the switch.

7. **Install a new start switch.** To install a new start switch, just reverse the disassembly procedure, and reassemble. Replace the wires on the switch. Reinstall the console panel, and restore the electricity to the dryer. Test the dryer operation.

Drum Roller

The typical complaints associated with the drum roller are:

- Dryer is very noisy when operating.
- Dryer has a burning smell.

To handle these problems, perform the following steps:

1. **Verify the complaint.** Verify the complaint by operating the dryer through its cycles.

2. **Check for external factors.** You must check for external factors not associated with the appliance. Is the appliance installed properly? Does the dryer have the correct voltage supply?

FIGURE 16-26
Test the continuity of the switch contacts on a start switch.

Start switch

3. **Disconnect the electricity.** Before working on the dryer, disconnect the electricity. This can be done by pulling the plug from the electrical outlet. Or disconnect the electricity at the fuse panel or at the circuit breaker panel. Turn off the electricity.

4. **Gain access to the drum roller.** To gain access to the drum roller, the top must be raised (see Figure 16-8) by removing the screws from the lint screen slot. Then insert a putty knife about two inches from each corner to disengage the retaining clips, and lift the top. On some models, the top is held down with screws (see the inset in Figure 16-8). Remove the screws and lift the top. Now the front panel of the dryer must be removed (see Figure 16-9). To remove the top panel, insert a putty knife and disengage the retaining clip (see Figure 16-9a). As in Figure 16-9b, remove the screws that hold the lower part of the front panel in place; then remove the screws that hold the upper part of the front panel in place (see Figure 16-9c). Disconnect the door switch wires.

5. **Remove the drive belt and drum.** To remove the drive belt and the drum on this type of dryer, you must first disconnect the belt (see Figure 16-10). Push on the idler pulley to release the tension from the drive belt. Remove the belt from the motor pulley and from around the drum (see Figure 16-16). Grab hold of the drum and remove it from the cabinet (see Figure 16-11).

6. **Remove the drum roller.** Once the drum has been removed, the drum rollers, which are located on the rear bulkhead of this model (Figure 16-27), can then be removed. Take a pair of needle nose pliers and remove the tri-ring from the shaft. Then slide the roller off the shaft.

7. **Install a new drum roller.** To install the new roller, clean the shaft and lubricate it with a small amount of oil. Slide the new drum roller onto the shaft, and reinstall the tri-ring. To reassemble the dryer, just reverse the disassembly procedure, and reassemble. Test the dryer for proper operation.

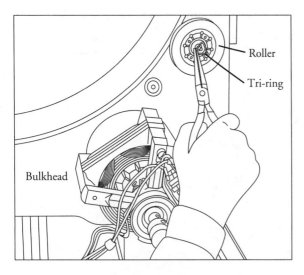

Figure 16-27   You must remove the tri-ring first in order to remove the drum roller.

Idler Pulley

The typical complaints associated with the idler pulley are:

- When dryer is running, you will hear a squealing noise.

- The dryer belt burns and might snap.

- The dryer drum will not turn.

To handle these problems, perform the following steps:

1. **Verify the complaint.** Verify the complaint by operating the dryer through its cycles.

2. **Check for external factors.** You must check for external factors not associated with the appliance. Is the appliance installed properly? Does the dryer have the correct voltage supply?

3. **Disconnect the electricity.** Before working on the dryer, disconnect the electricity. This can be done by pulling the plug from the electrical outlet. Or disconnect the electricity at the fuse panel or at the circuit breaker panel. Turn off the electricity.

4. **Gain access to the idler pulley.** To gain access to the idler pulley, in this type of dryer, remove the top panel, insert a putty knife, and disengage the retaining clip (see Figure 16-9a).

5. **Remove the idler pulley.** To remove the idler pulley, you must first remove the drive belt. To disconnect the belt on this model (see Figure 16-10), push on the idler pulley to release the tension from the drive belt. Now remove the belt from the motor pulley and from around the idler pulley. If the drive belt is broken, just remove the belt. Next, lift the idler pulley up and out of the dryer (Figure 16-28). This type of idler pulley is replaced as a complete assembly. Another type of idler pulley (Figure 16-29a) has a tension spring. To remove this type of idler, you must first remove the tension spring (Figure 16-29b), and then lift the idler assembly out of the dryer. Inspect the pulley for wear. If it is worn, remove the screw from the axle, and pull the axle out (Figure 16-29c). On this type of idler pulley, only replace the worn-out part.

6. **Install the new idler pulley.** To install the repaired idler pulley, just reverse the disassembly procedure, and reassemble it. Reassemble the dryer in the reverse order of its disassembly, and test it for proper operation.

Temperature Selector Switch

The typical complaints associated with the temperature selector switch are:

- Inability to select a certain temperature setting.
- No heat.

To handle these problems, perform the following steps:

1. **Verify the complaint.** Verify the complaint by operating the dryer through its cycles. Before you change the temperature selector switch, check the other components controlled by this switch.

Figure 16-28
After removing the drive belt, lift the idler pulley off the base.

FIGURE 16-29
(a) This type of idler pulley uses a spring to hold tension against the drive belt. (b) Remove the spring and lift the idler assembly out of the dryer for inspection. (c) Remove the axle to replace the pulley.

(a)

(b)

(c)

2. **Check for external factors.** You must check for external factors not associated with the appliance. Is the appliance installed properly? Does it have the correct voltage?

3. **Disconnect the electricity.** Before working on the dryer, disconnect the electricity to it. This can be done by pulling the plug from the electrical outlet. Or disconnect the electricity at the fuse panel or at the circuit breaker panel. Turn off the electricity.

4. **Gain access to the temperature selector switch.** To access the temperature selector switch, remove the console panel. On this type of dryer, remove the screws in the console (see Figure 16-3a). On some models, the console will be able to lie flat (see Figure 16-3b). On other models, your access is through the rear panel on the console.

FIGURE 16-30
Test the temperature selector switch for continuity.

Temperature selector switch

5. **Test the temperature selector switch.** To test the temperature selector switch, locate the selector switch circuit on the wiring diagram. Identify the terminals that are regulating the temperature setting to be tested. Set the ohmmeter on the R × 1 scale. Next, place the ohmmeter probes on those terminals. Then select that temperature setting by either rotating the dial or by depressing the proper button on the switch (Figure 16-30). If the switch contacts are good, your meter will show continuity. Test all of the remaining temperature settings on the temperature selector switch. Remember to check the wiring diagram for the correct switch contact terminals (those that correspond to the setting that you are testing).

6. **Remove the temperature selector switch.** To remove the temperature selector switch, remove the screws that hold the switch to the console base, and remove the switch.

7. **Install the new temperature selector switch.** To install the new temperature selector switch, just reverse the disassembly procedure, and reassemble it. Then reattach the wires to the switch terminals according to the wiring diagram. Reassemble the console panel. Be sure when reassembling the console panel that the wires do not become pinched between the console panel and the top of the dryer.

Diagnostic Charts

The following diagnostic flow charts will help you to pinpoint the likely causes of dryer problems (Figures 16-31, 16-32, and 16-33).

The wiring diagram in this chapter is only an example. You must refer to the wiring diagrams on the dryer that you are servicing. Figure 16-34 depicts an actual wiring schematic diagram and an actual pictorial electrical wiring diagram. Figure 16-35 depicts an actual ladder diagram. A ladder diagram is simpler and is usually easier to read.

PART V

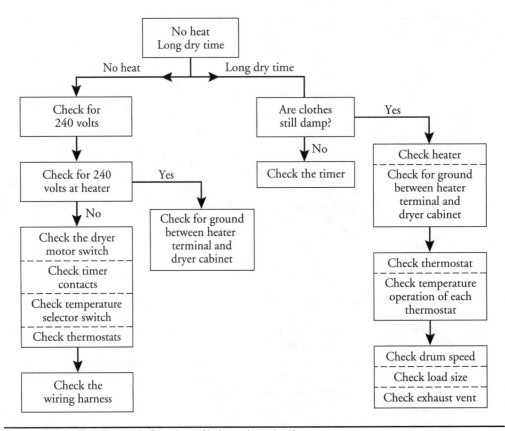

Figure 16-31   The diagnostic flow chart: No heat, long dry time.

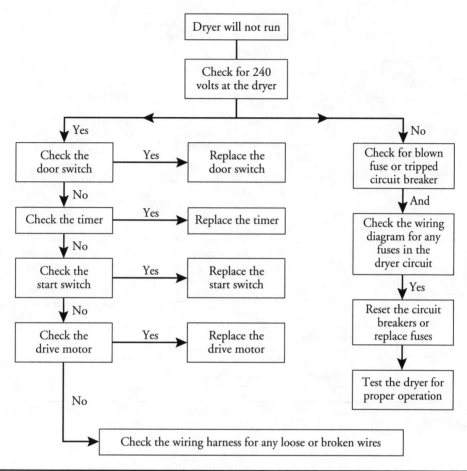

FIGURE 16-32   The diagnostic flow chart: Dryer will not run.

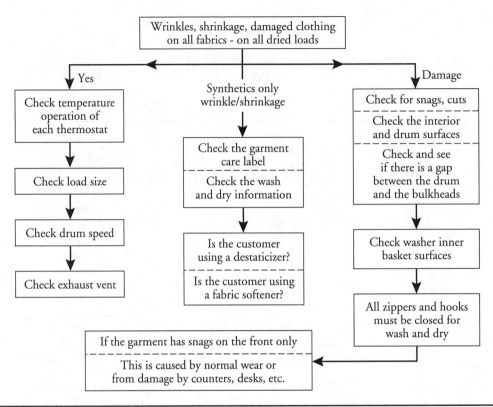

FIGURE 16-33 The diagnostic flow chart: Wrinkles, shrinkage, and damaged clothing with all fabrics on all dried loads.

FIGURE **16-34**
The automatic dryer
wiring diagram.

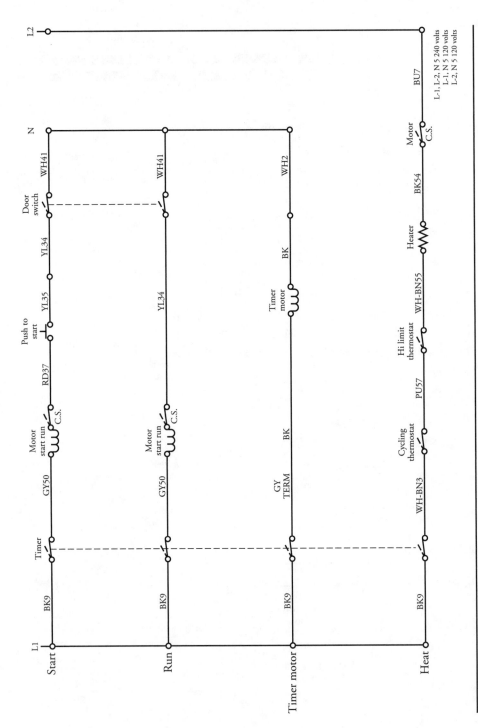

Figure 16-35   The automatic dryer ladder diagram.

Automatic Gas Dryers

G as and electric dryers are similar in operation, with the exception of how the air is heated for drying the clothes. The electric components in a gas dryer will operate on 120 volts only. They use liquefied petroleum gas (LP or LPG) or natural gas as a fuel, air (oxygen), and an ignition source to produce a flame. The gas burner can produce between 18,000 and 30,000 BTU/hr of heat, whereas the electric dryer produces between 17,000 and 19,000 BTU/hr of heat. All of the mechanics of the electric dryer, with the exception of the electric heater, are incorporated into the gas dryer.

The dryer will circulate the air, apply heat, and tumble the clothes to increase the surface exposure to remove water from the clothes by evaporation, and then vent the combustion by-products to the outside of the residence. The gas burner assembly includes a gas shutoff valve, burner, automatic solenoid gas control valve, pressure regulator, an ignition system, temperature controls, and safety limit controls.

The operating controls will prevent burner from coming on if certain conditions are not met. Before the dryer can start, the door must be closed (closing the door switch), and then the consumer will set the timer and press the start button, starting the cycle. If, for any reason, the motor does not start, the burner operation will not commence. On some models, a belt switch is incorporated into the circuitry to prevent the burner from operating if the dryer belt breaks.

The safety switches will prevent burner operation if the temperature exceeds their pre-set limit. These safety switches are located on the burner chamber, bulkhead, and exhaust passages.

Top-of-the-line models offer electronic controls, moisture sensors to sense how much water is remaining in the clothes, and automatic cycles designed for custom drying a variety of fabrics and load sizes.

This chapter will provide the basic skills needed to diagnose and repair automatic gas dryers. Figure 17-1 identifies where the components are located within the automatic gas dryer. However, this illustration is used as an example only. The actual construction and features might vary, depending on what brand and model you are servicing.

Electronic controls

Belt

Drum and drying rack

Side-open door

Lint filter

Gas burner assembly

Figure 17-1   The location of components in a typical automatic gas dryer. The consumer has a choice between a drop-down or side-open door.

Safety First

Any person who cannot use basic tools or follow written instructions should *not* attempt to install, maintain, or repair any automatic dryers. Any improper installation, preventive maintenance, or repairs could create a risk of personal injury or property damage.

If you do not fully understand the installation, preventive maintenance, or repair procedures in this chapter, or if you doubt your ability to complete the task on the automatic dryer, please call your service manager.

Do not store any flammable materials near a gas dryer, and avoid ignition sources in the event you smell gas leaking. Make sure you have proper ventilation when servicing the gas dryer to prevent carbon monoxide (CO) poisoning.

The following additional precautions should also be followed:

- Never bypass (or interfere with) the operation of any switch, component, or feature.
- Be careful of sharp edges when working on the dryer.
- The dryer produces combustible lint, and the area should be kept clean.
- Never remove any ground wires from the dryer or the third (grounding) prong from the service cord.

- Never use an extension cord to operate a dryer.

- The wiring used in dryers is made with a special heat-resistant insulation. Never substitute it with ordinary wire.

- Always follow the use and care guide instructions from the dryer manufacturer.

- Always keep combustible products away from the gas dryer.

- Keep the dryer clean of soot, grease, and food spillages.

- Teach your children not to play near or with the dryer.

- Always have a fire extinguisher nearby just in case of mishaps that might lead to a fire.

- Have a smoke detector and a carbon monoxide detector installed in the home, and check the batteries yearly.

- Make sure the gas dryer is properly vented according to the manufacturer's recommendations. Never exhaust the dryer into a chimney, a common duct, an attic, or a crawl space.

Before continuing, take a moment to refresh your memory of the safety procedures in Chapter 2.

Automatic Gas Dryers in General

Much of the troubleshooting information in this chapter covers automatic gas dryers in general, rather than specific models, in order to present a broad overview of service techniques. The illustrations that are used in this chapter are for demonstration purposes only to clarify the description of how to service dryers. They in no way reflect upon a particular brand's reliability.

Gas Supply Requirements

The gas dryer is manufactured and shipped for natural gas installation. If the gas supply to the home is liquefied petroleum gas (LP or LPG), the dryer will then have to be converted. When installing the gas supply line to the dryer, the shutoff valve must be within six feet of the product. This will make it easier to locate and shut off the supply of gas to the product for servicing. The recommended gas supply line size should be ½-inch IPS pipe; it is acceptable to use 3/8-inch tubing for less than 20 feet to connect to the shutoff valve, only if it is approved by the local codes in your area. The gas supply line must include a 1/8-inch NPT-plugged connection between the dryer and shutoff valve for testing the gas pressure (Figure 17-2).

For more information regarding gas appliance installations and gas supply line installations, refer to the following sources:

- National Fuel Gas Code
- International Fuel Gas Code
- The local gas codes in your area
- The manufacturer's recommendations
- The AGA (American Gas Association)

PART V

3/8" tubing

1/8" NPT plugged test
gauge connection

Gas shutoff valve

3/8" pipe to flare adapter fitting

½" NPT-approved gas
supply line

Figure 17-2 A typical gas supply line to the dryer. The plugged test gauge connection will be used for testing the gas pressure.

Gas Dryer Maintenance

Automatic dryers must be cleaned periodically. The excess lint must be removed to prevent the possibility of a fire and the possibility of the dryer not functioning properly. Most models have a service panel on the front of the dryer for accessibility. The outside of the cabinet should be wiped with a damp cloth. It is recommended that the components be inspected for wear and tear; if repairs are needed, they should be made as soon as possible. The gas burner assembly must be inspected and cleaned yearly. The dryer gas line must also be inspected and checked for gas leaks yearly—sooner if you suspect a gas leak.

Vent System

A clogged or restricted vent system can cause the dryer not to function properly. The following symptoms can occur:

- The clothes take too long to dry.
- The clothes are too hot at the end of the cycle.
- Unusual noises coming from the dryer.
- At the end of the cycle, the clothes are damp and hot.
- The sensors and thermostats fail to operate properly.
- Complications in the burner assembly.

Step-by-Step Troubleshooting by Symptom Diagnosis

In the course of servicing an appliance, don't overlook the simple things that might be causing the problem. Step-by-step troubleshooting by symptom diagnosis is based on diagnosing malfunctions, with possible causes arranged into categories relating to the operation of the dryer. This section is intended only to serve as a checklist to aid you in diagnosing a problem. Look at the symptom that best describes the problem that you are experiencing with the dryer, and then correct the problem.

Gas Odors

- Check the pilot light and gas valve.
- Check for a gas leak in the supply line.
- Check the ventilation system.

Dryer Will Not Run

- Do you have 120 volts at the dryer?
- Test the door switch for continuity of the switch contacts. In addition, make sure the dryer door closes properly.
- Test for continuity of the motor windings. Also check for a grounded motor.
- Test the timer for continuity of the switch contacts.
- Test the "push to start" switch for continuity of the switch contacts.
- Check for broken or loose wiring. Also check the wire terminal connections that connect to the different components.
- Check the thermal fuse.

No Heat

- Check the controls setting.
- Do you have 120 volts at the dryer?
- Check the gas valve. Is the gas turned on?
- Check the thermal fuse.
- Check the thermostats.
- Check the ignitor. Does it light up?
- Check the flame sensor.
- Check for proper ventilation.

Dryer Will Not Turn Off

- Check the control settings. Does the timer advance?
- Check the electronics. Run the self-diagnostic test mode.
- Check the thermistor. Is it open or shorted?
- Check the thermostats. Check continuity.

Dryer Runs and Heats, but the Clothes Won't Dry

- Check for a defective thermostat. Use a thermometer to check the duct temperatures during the cycling of the dryer.
- Check for a loose pulley or blower wheel. Inspect and tighten the blower wheel.

- Check for restricted air flow in the dryer.
- Check the vent system. Is the vent hose behind the dryer kinked?
- Check the thermostats.
- Check the lint filter. Does the lint filter need cleaning?

The Burner Will Not Ignite

- Make sure that the gas supply to the dryer is turned on.
- Check the gas line for debris.
- Check the line to the burner for a blockage. If blocked, locate the source and correct.
- Test the value of the thermocouple; replace if necessary.
- Test the thermostat; replace if necessary.
- Check the installation and fresh air supply.

The Pilot Will Not Light or Stay Lit (Older Models)

- Make sure that the gas supply to the dryer is turned on all the way.
- Check the gas pressure to the dryer.
- Check for air or debris in the lines. Purge the gas line.
- Check the pilot line and orifice for debris.
- Test the thermocouple, thermostat, and ignitor. Replace if necessary.
- Check for drafts that can extinguish the pilot.
- Check the intake air for obstruction. This is the air that is needed for combustion.
- Check for flammable vapors.

The Burner Flame Is Too High

- Check the dryer installation. Is the orifice too large? You can change the orifice and retest the burner.
- Check for sufficient secondary air supply.
- Check the venting system.
- Check the burner air shutter. Is it adjusted correctly?

The Pilot Flame Is Too Small

- Check the pilot line and orifice for debris.
- Check the gas pressure to the dryer.

A Yellow Burner Flame

- Check for sufficient secondary air supply.
- Check the gas pressure.
- Check the venting system.

- Check the burner line for debris.
- Check the dryer installation.
- Check the burner orifice for obstruction.

The Burner Flame Floats and Lifts Off the Ports

- Check the dryer installation. Is the orifice too large? You can change the orifice and retest the burner.
- Check for high gas pressure.
- Check the vent system and burner. Locate the source and correct the problem.
- Check for drafts that can blow the flame off the burner.

The Formation of Soot (Smoke and Carbon)

- Check for sufficient secondary air supply.
- Check the ventilation system.
- Check for a yellow burner flame.
- Check for low gas pressure to the dryer.
- Is the dryer installed in a confined space?

Combustion Odors

- Check for sufficient secondary air.
- Check the ventilation system and the burner.
- Is the dryer installed in a confined space?

For additional symptoms, see Chapter 16.

Repair Procedures

Each repair procedure is a complete inspection and repair process for a single dryer component, containing the information you need to test a component that might be faulty and to replace it, if necessary. In this chapter, I will discuss the gas heating section only. For additional repair procedures, see Chapter 16.

Electronic Control Board and User Interface Controls

The typical complaints associated with the electronic control board or the user interface controls are:

- The dryer won't run or power up.
- Unable to program the dryer.
- The display board will not display anything.
- One or more key pads will not accept commands.
- Unusual display readouts.

PART V

To prevent electrostatic discharge (ESD) from damaging expensive electronic components, follow the steps in Chapter 9.

To handle these problems, perform the following steps:

1. **Verify the complaint.** Verify the complaint by operating the dryer. Turn off the electricity to the appliance and wait for two minutes before turning it back on. If a fault code appears, look up the code. If the dryer will not power up, locate the technical data sheet behind the control panel for diagnostics information.

2. **Check for external factors.** You must check for external factors not associated with the appliance. Are there 120 volts of electricity to the dryer?

3. **Disconnect the electricity.** Before working on the dryer, disconnect the electricity. This can be done by pulling the plug out of the wall receptacle. Or disconnect the electricity at the fuse panel or circuit breaker panel. Turn off the electricity.

WARNING *Some diagnostic tests will require you to test the components with the power turned on. When you disassemble the control panel, you can position the panel in such a way so that the wiring will not make contact with metal. This act will allow you to test the components without electrical mishaps.*

4. **Shut off the gas supply.** Before you begin servicing any gas components, shut off the supply of gas to the appliance.

5. **Remove the console panel to gain access.** Begin by removing the screws from the dryer top (for front-loading models only) to gain access to the electronic control board. Remove the console panel screws to gain access to the user interface controls (see Figure 16-7). On top-loading models, the console will roll upwards or toward you after you remove the console screws.

6. **Test the electronic control board and user interface controls.** If you are able to run the dryer diagnostic test mode, check the different functions of the dryer. Use the technical data sheet for the model you are servicing to locate the test points from the wiring schematic. Check all wiring connections and wiring. Using the technical data sheet, test the electronic control or user interface controls, input voltages, and output voltages. On some models, fuses are soldered to the printed circuit board (PCB). These fuses must be tested first before condemning the component.

7. **Remove the electronic control board and user interface control.** To remove the defective component, remove the screws that secure the board to the control panel or dryer frame. Disconnect the connectors from the electronic control board or user interface control.

8. **Install the new component.** To install a new electronic control board or user interface control, just reverse the disassembly procedure, and reassemble. Reinstall the console panel, and restore the electricity to the dryer. Test the dryer operation.

Burner Assembly

The typical complaints associated with the burner assembly are:

- Burner will not light.
- The burner goes on and off.

- Soot buildup.
- The ignitor will not glow.
- The burner trips on the high-limit thermostat.

To handle these problems, perform the following steps:

1. **Verify the complaint.** Verify the complaint by operating the dryer through its cycles. Then open the dryer door, and place your hand inside to see if it is warm in the drum.

2. **Check for external factors.** You must check for external factors not associated with the appliance. Is the appliance installed properly? Is the exhaust vent clogged? Check the voltage and gas supply to the dryer.

3. **Disconnect the electricity.** Before working on the dryer, disconnect the electricity. This can be done by pulling the plug from the electrical outlet. Or disconnect the electricity at the fuse panel or at the circuit breaker panel. Turn off the electricity.

4. **Shut off the gas supply.** Before you begin servicing any gas components, shut off the supply of gas to the dryer.

5. **Gain access to the burner assembly.** To gain access to the burner assembly, you must remove the top panel (or lower panel or access panel). The panel is held in place with two retaining clips or screws, depending on the model. Using a flat-blade screwdriver, insert it about 4 inches from each side (Figure 17-3). Depress the clips, pull the panel toward you, and remove the panel.

6. **Remove the burner assembly.** The burner assembly is located on the lower-front side of the cabinet on the right. Some models place the burner assembly on the lower-front side on the left. Remove the wires from the flame sensor (Figure 17-4a), and then disconnect the wiring harness connector from the burner assembly (Figure 17-4b). Disconnect the gas line from the gas valve. Locate the two screws that secure the burner assembly to the dryer frame, and remove the screws. Remove the burner assembly by pulling it toward you carefully (Figure 17-4c). With the burner assembly removed from the cabinet, you can now inspect and access the burner, high-limit thermostat, ignitor, gas valve, pilot assembly (older models), and flame sensor. Before you reinstall the burner assembly, do the maintenance on the burner assembly and the lower cabinet, removing lint and debris.

FIGURE 17-3
To open the top panel, insert a flat-blade screwdriver, depress the retaining clips, and remove the panel.

PART V

Gas burner tube

High-limit thermostat

Mounting screw

Flame sensor

(a)

Ignitor

Ignitor mounting screw

Ignitor wiring connector

(b)

Gas valve mounting screws

(c)

FIGURE 17-4   Removing a typical gas burner assembly.

7. **Test the gas valve solenoid coils.** In order to check the solenoid coils on the gas valve, remove the wire leads (and label them) that connect to the coils from the wiring harness. These are slide-on terminal connectors attached to the ends of the wire. Set your ohmmeter on R × 100, and attach the meter probes as follows (Figure 17-5):

- Terminals 1 and 2: The resistance will be between 1340 and 1390 ohms.

- Terminals 1 and 3: The resistance will be between 535 and 585 ohms.

- Terminals 4 and 5: The resistance will be between 1170 and 1270 ohms.

If the meter readings differ, replace the solenoid coils.

8. **Test the ignitor.** Disconnect the ignitor wiring connection, and measure the resistance between the terminals on the ignitor. Set your ohmmeter on R × 100—the resistance will be between 50 and 250 ohms. If the meter readings differ, replace the ignitor (Figure 17-6).

9. **Test the flame sensor.** A flame sensor (Figure 17-7) is a thermostatically controlled single-pole, single-throw, normally closed switch. This device is mounted over a window cut out in the burner tube. The switch will open within 15 to 90 seconds, when the switch senses the burner flame, and the ignitor temperature will be at 1800 to 2500 degrees Fahrenheit. To test the flame sensor, set the ohmmeter on R × 1, remove the wire terminals from the sensor, and measure the resistance between the terminals. The flame sensor should measure zero ohms, indicating that the sensor is good. If the meter readings differ, replace the flame sensor.

10. **Test the high-limit thermostat.** The high-limit thermostat is located on the burner tube (see Figure 17-4a) or in the

Figure 17-5

The test points on the solenoid coils, which are mounted on the gas valve.

rear of the dryer (Figure 17-8); or there may be two high-limit thermostats mounted in both places, depending on the model. If the temperature exceeds the pre-set limit, the high-limit thermostat will open to shut down the gas burner. To test the thermostat, remove the wire terminals from the high-limit thermostat. Set the ohmmeter on $R \times 1$, and measure the resistance between the thermostat terminals. The ohmmeter will read continuity, indicating that the high-limit thermostat is good. If the meter readings differ, replace the thermostat.

Figure 17-6

The gas flame ignitor.

PART V

Flame sensor mounting screw

Flame sensor

Burner tube

Burner assembly
mounting bracket

FIGURE 17-7   The flame sensor is mounted over a slot opening in the side of the burner tube.

11. **Install the burner assembly.** To install the burner assembly (see Figure 17-4), just reverse the disassembly procedure, and reassemble. If you are replacing the gas valve, make sure that you use approved pipe joint compound or Teflon tape on the pipe connections (according to local codes) and on the threads of the gas valve before you install the valve. Turn on the gas supply. Check for gas leaks by using a chloride-free soap solution. *Do not use a flame to check for gas leaks.* The soap solution will begin to bubble if a leak is present. Restore the electricity to the dryer, and test the burner assembly. Check the burner flame for proper operation. Then reinstall the top panel (see Figure 17-3) in the reverse order of its disassembly.

High-limit
thermostat

Lint
chute

Heater
box

High-limit
thermostat

Operating
thermostat and
bias heater

Blower
housing

Exhaust

Gas line hook-up

FIGURE 17-8 The rear view of a dryer, illustrating the location of the high-limit thermostat and the location of the other components.

Diagnostic Charts

The following diagnostic flow charts will help you to pinpoint the likely causes of various dryer problems (Figures 17-9 and 17-10; see also Figure 16-33).

The wiring diagram in this chapter is only an example. You must refer to the actual wiring diagrams on the dryer that you are servicing. Figures 17-11, 17-12, and 17-13 depict actual strip circuit diagrams. The strip circuit diagram is simpler, and it is usually easier to read. Figure 17-14 depicts an actual wiring schematic diagram.

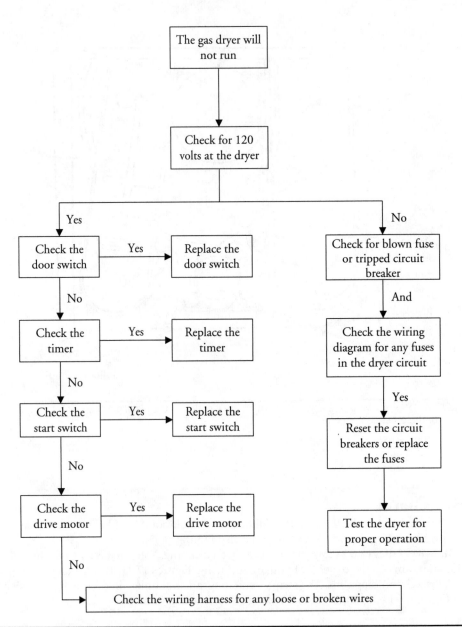

FIGURE 17-9   Diagnostic flow chart: The gas dryer will not run.

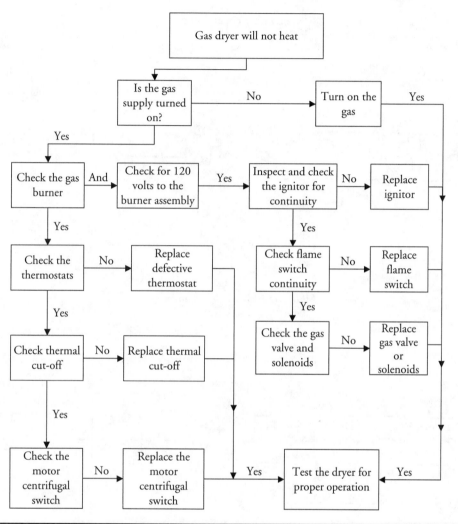

FIGURE 17-10 Diagnostic flow chart: The gas dryer will not heat.

FIGURE 17-12 Typical strip circuits illustrating the temperature control circuit, heating circuit, and the push-to-start circuit.

Figure 17-13 Typical strip circuits illustrating the buzzer circuit, drum lamp circuit, and the timer circuit.

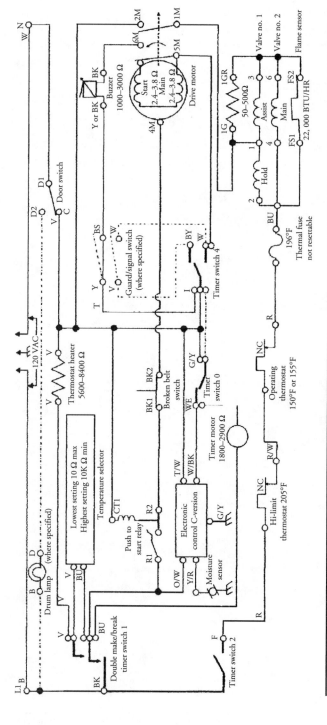

Figure 17-14   A typical gas dryer wiring schematic with electronic components.

Electric Ranges and Ovens

lectric ranges (Figures 1-2 and 18-1), cooktops (Figure 1-4), and ovens (Figure 1-6) are available in a variety of styles. Electric ranges and ovens might seem to be complicated, but they are not. The more you know about the electrical and mechanical operation of the product, the easier it will be to solve its problems. An electric range, cooktop, or oven operates on 240 volts for the heating elements and 120 volts for the accessories (clock, lights, etc.). Most repairs are electrical in nature. This chapter will provide the basics needed to diagnose and repair these appliances. Figure 18-2 identifies where components are located within the range—this illustration is used as an example only. The actual construction and features might vary, depending on what brand and model you are servicing.

Principles of Operation

The surface units are simple to operate. After placing a pot on the surface element, you turn the surface unit switch to the desired setting. When the surface unit switch is on, current flows from the wall receptacle, through the wiring and the surface unit switch, and then through the heating element. To properly cook on the surface unit, the pot or pan must lie flat across the heating element (Figure 18-3), making contact with the entire cooking surface. If not, the food will not cook evenly and there will be greater heat losses (resulting in wasted electricity).

To bake in the oven, place the food on the oven rack, and close the oven door. Set the oven selector switch to bake and set the oven temperature control on the desired temperature. Current flows from the wall receptacle, through the wiring and the selector switch, through the thermostat, and to the bake element. When the selected temperature in the oven is reached, the bake element will go off. When the temperature in the oven decreases, the thermostat then reactivates the bake element. On some models, both the bake and broil elements operate at the same time to first preheat the oven cavity.

The broiling operation in the range is accomplished by placing the food on the top oven rack and partially closing the oven door. When closing the door, there is usually a stop in the hinge that allows the door to remain open a certain amount; this is done so that the broiler element will not cycle on temperature. The broiler element stays on for continuous operation until the user turns off the controls. When broiling, current will flow from the wall receptacle, through the wiring and the selector switch, through the thermostat, and then to the broil element.

FIGURE 18-1 An example of a freestanding electric range.

Self-cleaning ranges and ovens differ—both from manufacturer to manufacturer and from model to model—in how the heating element operates to clean the oven. Pyrolytic cleaning is the true self-cleaning system. It uses high heat during a special one- to three-hour cycle to decompose food soil and grease. During the cycle, which is clock-controlled, the oven door is latched and locked. It cannot be opened until the oven cools down. All of the oven walls, racks, and the door (except for a small area outside the door gasket) are completely cleaned. After cleaning, you might find a small bit of white ash, which can be easily wiped out.

Catalytic, or continuous, cleaning uses a special porous coating on the oven walls that partially absorbs and disperses the soil. This process occurs during normal baking and keeps the oven presentably clean; however, the racks and door parts must be cleaned by hand. Some manufacturers recommend occasionally operating an empty oven at 500 degrees Fahrenheit to remove any buildup of soil. This special oven coating cannot be cleaned with soap, detergent, or commercial oven cleaners without causing permanent damage.

Surface element

Surface element receptacle

Drip pan

Surface element

Main top

Latch cover

Terminal blocks

Latch arm

Latch

Latch support

Ballast

Latch link

Vent/Flu

Broil support

Latch switch

Hinge

Hinge spring

Broil element

Side panel

Broil shield

Inner side liner

Bake element

Example only: Construction and features may vary depending on brand and model

FIGURE 18-2 An example of a range chassis.

On pyrolytic clean models, you simply close and latch the oven door and then set the controls to clean. Before using the self-cleaning cycle, the consumer must read the use and care manual in order to set the controls properly. If the controls are not set according to the manufacturer's recommendations, the oven will not be properly cleaned.

FIGURE 18-3
The cookware must make full contact with the surface unit.

Next, you set the timer clock for the proper time when you want the clean cycle to begin. The clock in a self-cleaning oven has two functions: it controls when the oven will operate in the timed-bake cycle and it controls the clean cycle. The bake element will come on, and the temperature will begin to rise. This will take approximately 60 minutes. The reason for the slow temperature rise is to prevent damage to the oven cavity and the door.

When the temperature is above 550 degrees Fahrenheit, the oven lock light will come on, indicating that the latch on the oven door cannot be opened until the oven cavity cools below that temperature. When the oven cavity temperature stabilizes (between 840 and 920 degrees Fahrenheit), the cleaning process begins. This process requires approximately two to three hours. Some models have a cooling fan in the circuit to aid in keeping the exterior temperature low. Self-cleaning ovens have a nonelectric, "catalytic" smoke eliminator in the vent to consume the smoke from the soil load. The catalytic smoke eliminator will begin to operate when the oven cavity temperature is between 300 and 400 degrees Fahrenheit. The types of oven door lock mechanisms used on self-cleaning models are:

FIGURE 18-4   The manual latch system for the self-cleaning range is used on some models.

- Manual door lock: With this type of system, the user manually moves the oven door latch assembly to lock the door (see Figure 18-4).

- Electromechanical door lock: With this type of system, the user will either press a lock switch button (located on the control panel) in order to move the latch handle or, alternatively, move the latch handle to lock or unlock the oven door. This action will activate a solenoid coil, allowing the door latch mechanism and linkage to operate (Figure 18-5).

- Electric door lock: With this type of system, the user sets the cleaning controls, and the door lock assembly will operate through the control of an electric motor to activate the locking mechanism. This system is similar to the electromechanical door lock; however, instead of a latch solenoid, an electric motor is used.

Solenoid
voice
rivet

Solenoid
lock cam

Lock cam
pin

Limit switch
(solenoid
actuating)

Switch
mounting
screw

Limit switch
(door lock
innerlock)

Insulation board

Solenoid
switch

Solenoid
mounting
bracket

Linkage arm

Door latch
plate

Oven door
latch assembly

FIGURE 18-5   The latch solenoid system for the self-cleaning range is used on some models.

Safety First

Any person who cannot use basic tools or follow written instructions should *not* attempt to install, maintain, or repair any electric range, oven, or cooktop. Any improper installation, preventive maintenance, or repairs could create a risk of personal injury or property damage.

If you do not fully understand the installation, preventive maintenance, or repair procedures in this chapter, or if you doubt your ability to complete the task on your appliance, please call your service manager.

The following precautions should also be followed:

- Never use a range to heat the home; it simply wasn't designed for that purpose.
- Keep the cooking area clean of spills and grease.
- Do not use flammable liquids near a cooking appliance.
- When repairing the range, always use the proper tools.
- Always reconnect the ground wire to the range after repairs have been made.
- Never use aluminum foil to line drip bowls; it could cause an electrical shock or become a fire hazard.

Before continuing, take a moment to refresh your memory of the safety procedures in Chapter 2.

PART V

Electric Ranges, Ovens, and Cooktops in General

Much of the troubleshooting information in this chapter covers electric ranges, ovens, and cooktops in general, rather than specific models, in order to present a broad overview of service techniques. The illustrations that are used in this chapter are for demonstration purposes only to clarify the description of how to service these appliances. They in no way reflect on a particular brand's reliability.

Please note that *eye-level ranges* are also referred to as *high-low ranges* or *tri-level ranges*.

Location and Installation of Electric Range, Oven, or Cooktop

Locate the range, oven, or cooktop where it will be well lighted and have access to proper ventilation. The range must be level for proper baking and cooking results. The range might be installed adjacent to the left and/or right base cabinets and against a rear vertical wall (for the "antitip" cleat). A wall oven must be installed on a supporting surface that is strong enough to support the weight of the oven and its contents while remaining level from side to side and from front to rear. A cooktop must be installed on a flat surface, supported by the countertop, and it should be level.

Contact your flooring company to check if the flooring can withstand a minimum of 200 degrees Fahrenheit. It is also recommended that you contact your builder to determine if the cabinets and walls can withstand the heat produced by the appliance. Some kitchen cabinets and building materials were not designed to withstand the heat that is given off by the product.

The proper installation instructions for your model are included in the use and care manual. These instructions will assist you with the installation requirements (dimensions, electrical requirements, cutout dimensions, venting, etc.) needed to complete the installation according to the manufacturer's specifications.

Step-by-Step Troubleshooting by Symptom Diagnosis

When servicing an appliance, don't overlook the simple things that might be causing the problem. Step-by-step troubleshooting by symptom diagnosis is based on diagnosing malfunctions, with their possible causes arranged into categories relating to the operation of the range. This section is intended only to serve as a checklist to aid you in diagnosing a problem. Look at the symptom that best describes the problem that you are experiencing with the range, oven, or cooktop, and then proceed to correct the problem.

Oven Will Not Heat

- Are the oven controls set properly?
- If the entire range is inoperative, check for voltage and check the circuit breakers or fuses.
- Test oven thermostat switch contacts for continuity.
- Test the heating elements for continuity and for a short.
- Check for loose or broken wiring connections.
- Test the other components in the circuits that operate the heating elements.
- Test the oven selector switch.

Oven Temperature Is Not Accurate

- Test the oven thermostat for accuracy by using an oven temperature tester.
- Check to be sure that the oven door is closed for baking.
- Is the oven vent blocked with aluminum foil?

Oven Will Not Cycle Off

- Test the oven thermostat switch contacts for continuity.
- Test the oven selector switch contacts for continuity.
- Test the time clock switch contacts for continuity.
- Test the oven cycling relay.

Broil Element Is Not Heating

- Are the broil controls set properly?
- If the entire range is inoperative, check for voltage and check circuit breakers or fuses.
- Test the oven thermostat switch contacts for continuity.
- Test the heating element for continuity and for a short.
- Check for loose or broken connections.
- Test other components in the circuits that operate the heating elements.
- Does the user have the door closed completely?
- Check the oven selector switch.
- Check to be sure that the oven door is open for broiling.

Surface Unit Will Not Operate

- Is the right surface unit switch turned on?
- Check for loose or broken wiring connections.
- Test the surface element for continuity and for a short.
- If the entire range is inoperative, check for voltage and check circuit breakers or fuses.
- Test the surface unit switch for continuity.

Heat Is Escaping from Oven Door

- Is the oven door aligned properly with the range body?
- Check the oven door gasket.
- Check the oven door hinge for damage.
- Check the oven door spring. Is it broken?
- Is the consumer closing the door completely for baking?

Cooking Performance

If there are no mechanical problems with the oven's operation but it will not bake the food properly or if the food is partially cooked, etc., your next step will be to look at the symptom that best describes the problem that you are experiencing with the oven. Then correct the problem. If necessary, instruct the user on how to get better results from the oven. This information is covered in the use and care manual.

Satisfactory Baking Results

To have satisfactory baking results, the following conditions must exist:

- Proper oven door seal.
- The oven vent not blocked off.
- The oven thermostat must be calibrated properly.
- Food preparation must be done correctly.
- The proper type of cookware must be used.
- Follow the manufacturer's recommended cooking instructions.

Satisfactory Broiling Results

To have satisfactory broiling results, the following conditions must exist:

- On most models, the oven door must be open partially to ensure that the broiling element will not cycle on temperature.
- Food preparation must be done correctly.
- The proper type of cookware must be used.
- Follow the manufacturer's recommended broiling instructions.

Satisfactory Surface Cooking Results

To have satisfactory surface cooking results, the following conditions must exist:

- The reflector pans (bowls) must be used under all of the surface units.
- Use cookware with flat bottoms that make correct contact with the heating element.
- Use cookware large enough to entirely cover the surface units.
- Food preparation must be done correctly.
- Follow the manufacturer's recommended cooking instructions.

Surface Cookware

One of the keys to successful cooking is the use of proper cookware. For efficiency and best results, use pans that have tight-fitting lids and lightweight handles that do not tilt the pans.

The following is a brief review of the various types of cookware that are available. The different characteristics of the cookware will aid you when diagnosing cooking complaints.

- **Aluminum** is a metal that spreads heat quickly and evenly, and responds to temperature changes. This cookware is best for frying, braising, and pot roasts. The inside of an aluminum pan might be a natural finish or a nonstick coating.

- **Cast iron** is slow to change temperature and holds heat. This makes good cookware for browning, frying, stewing, and other cooktop cooking. Cast iron cookware is also available with an enamel finish.

- **Copper** is excellent for gourmet cooking, wine sauces, and egg cookery, and it is quick to change temperature.

- **Glass cookware** is slow to change temperature. This type of cookware works best for a long time on a low heat when cooking with a liquid.

- **Porcelain enamel** is long-lasting and is used for cooking soups and other liquids.

- **Stainless steel** is exceptionally strong. It is used for frying, sauces, soups, vegetables, and egg cooking.

Oven Cookware

Sometimes, a technician must educate the consumer on how to be a successful cook. The use of the correct type of cookware is very important. The following are some guidelines for choosing cookware:

- Always use the correct size pan as given in the recipe.

- When baking foods, use flat-bottomed pans to keep the food level.

- Aluminum oven cookware that is shiny (not cast) produces delicate browning, tender crusts, and reduces the spattering of roasts. This type of cookware is best for cakes, muffins, some quick breads, cookies, and roasting.

- Pottery, ceramic, cast metal, and ovenproof glass cookware give food a deep, crusty brown surface. When this type of cookware is used, the oven temperature should be reduced by 25 degrees Fahrenheit.

- The use of dull or darkened cookware is suitable for pies and other foods baked in pastry shells.

- Shallow-sided pans and flat baking sheets are best for cookies and biscuits, where top and side browning is wanted.

Food Preparation

Proper food preparation is the other key to successful cooking. If you suspect a baking problem, ask the cook to see the cooked food. If needed, ask the cook to prepare some food for baking, and watch how the food is prepared. The following descriptions will assist you in determining the problem.

Flat Cake

The cake comes out of the oven, and you notice that the cake is flat or that it has no volume to it. The reasons are:

- There might be too much liquid or not enough liquid.

- During preparation, the mix was overbeaten or underbeaten.

- The pan used to make the cake in was too large.

- The cook did not adjust the oven racks properly or placed the pan on the wrong rack.

- The oven temperature is too high or too low.
- The mix was stored improperly and/or was exposed to high humidity.
- The cook might have forgotten to use eggs.

Cake Has Fallen

The cake comes out of the oven, and you notice that the cake has fallen or the center has a dip in it. The reasons are:

- The cake is underbaked.
- There might be too much liquid or not enough liquid.
- The cook tested the cake too soon.
- The cake pan was too small.
- The cook moved the cake before it was completely baked.

Sticky Crust

The cake comes out of the oven, and you notice that it has a sticky top crust. The reasons are:

- The cake was underbaked.
- The cook stored the cake in a sealed container when the cake was still warm.
- The humidity in the air is high.
- Too much liquid in the mix.
- The cook might have substituted sweet fruit juices for other liquids.

Holes in Cake

The cake comes out of the oven, and you notice that it has holes or tunnels in it. The reasons are:

- Baking temperature selected was too high.
- The cook did not adjust the oven racks properly or placed the pan on the wrong rack.
- During preparation, the mix was overbeaten or underbeaten.
- Very large air bubbles became trapped in the batter.
- The cook might have used the wrong type of pan, which might have caused uneven temperature conductance.

Cake Shrinkage

The cake comes out of the oven, and you notice that it shrank or pulled away from the sides of the pan. The reasons are:

- The cake was overbaked.
- The cake pan was too close to the oven wall or too close to other pans.
- Extreme overbeating of the mixture.

Cake Peaked in the Center

The cake comes out of the oven, and you notice that the center of the cake has a peak in it. The reasons are:

- The oven temperature selected was too high.
- The pan was too small.
- The cook might have used the wrong type of pan, which might have caused uneven temperature conductance.
- There might not have been enough liquid used in the mixture.
- The cake pan was too close to the oven wall or too close to other pans.
- Extreme overbeating of the mixture.

Crust Not Brown

The cake comes out of the oven, and you notice that the crust on top of the cake is not brown. The reasons are:

- Oven door opened too many times during baking.
- Too much liquid in the mix.
- Oven temperature too low when baking.
- The cake pan was too deep.
- Extreme overbeating of the mixture.

Cake Too Dry

The cake comes out of the oven, and you notice that it is too dry and that it falls apart. The reasons are:

- The cake was overbaked.
- Not enough liquid in the mix.
- The cook might have forgotten to use eggs.

Cookies Brown Rapidly

The cookies come out of the oven, and you notice that they are browned more at the sides and/or the end of the cookie sheet. The reasons are:

- The cook might have used the wrong size cookie sheet, which might have caused uneven temperature conductance.
- Cookie sheets with sides will cause rapid browning at the edges. Are the cookies placed at least one inch from the sides?

Cookies Brown Slowly

The cookies come out of the oven, and you notice that they browned slowly. The reasons are:

- The heat might be leaking out of the oven. Check the oven door for heat leakage.
- The racks are uneven. Check the oven racks. Are they level when cold and when hot?

Cookies Are Dark on the Bottom

The cookies come out of the oven, and you notice that they are dark on the bottom. The reasons are:

- The oven racks were not adjusted properly, or the cookie sheet was placed on the wrong rack.
- More than one cookie sheet was placed in the oven.

Cooking Meats

The cook complains that the meats are not cooking properly; ask to see some of the food. This section is intended only to serve as a checklist to aid you in diagnosing a problem. Look at the symptom that best describes the problem that the cook is experiencing with the meats, and then correct the problem.

The Meat Burns on the Bottom

The possible reasons are:

- Check the size of the pan used to cook the meat. There should be at least one or two inches of oven rack visible around the pan when it is placed on the oven rack so that the oven cavity is evenly heated.
- Was the oven preheated?
- Was the door opened and closed frequently?
- Was the meat elevated off the bottom of the pan?

Meats Are Undercooked

The possible reasons are:

- Check the oven temperature. Is it calibrated correctly?
- Check the size of the pan used to cook the meat. There should be at least one or two inches of oven rack visible around the pan when it is placed on the oven rack so that the oven cavity is evenly heated.
- Was the type of pan used too deep? Was the pan covered?

Roasting of Meats Takes Too Long

The possible reasons are:

- Check the oven temperature. Is it calibrated correctly?
- Check for inadequate ventilation.
- Check for the improper use of aluminum foil.

Range, Oven, and Cooktop Maintenance

The range, oven, or cooktop can be cleaned with warm water, mild detergent, and a soft cloth on all cleanable parts, as recommended in the use and care manual. Never use scouring pads on these surfaces, except where recommended in the use and care manual. Also, never use abrasive cleaners that are not specifically recommended by the manufacturer. Do not allow grease spillovers to accumulate after cooking on top of the range; this will become a fire hazard.

On glass surfaces, you can use a glass cleaner to clean any soil stains. Stubborn soil stains on glass surfaces can be removed with a paste of baking soda and water. Never use abrasive cleaners on glass surfaces. Never clean the heating elements. When you turn on the elements, the soil will burn off.

On self-cleaning models, never use oven cleaners to clean the oven cavity. This will cause hazardous fumes when the oven is in the cleaning cycle. After the oven cavity has cooled, use only soap and water to clean small spills on the inside of the cavity.

Repair Procedures

Each repair procedure is a complete inspection and repair process for a single range, oven, or cooktop component. It contains the information you need to test and replace components.

Infinite (Burner) Switch

The typical complaints associated with failure of the infinite (burner) switch are:

- The surface element will not heat at all.
- The surface element will stay on high in any position.
- Intermittent switch operation.

To handle these problems, perform the following steps:

1. **Verify the complaint.** Verify the complaint by turning the infinite switch on. Is the surface element heating?
2. **Check for external factors.** You must check for external factors not associated with the appliance. Is the appliance installed properly? Does it have the correct voltage?
3. **Disconnect the electricity.** Before working on the range or cooktop, disconnect the electricity to the appliance. This can be done by pulling the plug from the receptacle. Or disconnect the electricity at the fuse panel or at the circuit breaker panel. Turn off the electricity.
4. **Gain access to the infinite switch.** Depending on the model you are repairing, you can access the component by removing the back panel (Figure 18-6). On models with front-mounted controls (Figure 18-7), the panel is attached with screws on both ends. Remove the screws, and tilt the control panel. Be very careful not to let the wires disengage from the components.

   Some built-in models have a removable backsplash (Figure 18-8); just lift the backsplash, and rest it on the cooktop. It would be a good idea to place something on the cooktop to protect it from getting damaged. Next, remove the screws from the backsplash, which hold the rear panel, to access the components. If you are repairing a wall oven or an eye-level range, the control panel can be removed (Figure 18-9) by opening the door and removing the screws that secure the panel. These screws might be underneath the front of the exhaust hood or just below the control panel. Some control panels are hinged—just tilt the control panel toward you for servicing. To access the components on other models, both the rear panel and the front control panel (usually glass) will have to be removed. To pull out the glass, remove the screws that secure the trim piece that holds the glass in place.

FIGURE **18-6**
Remove the screws
from the back panel
to gain access to the
components.

5. **Test the infinite switch.** To test the infinite switch for continuity between the switch contacts, remove all of the wires from the switch terminals (label them). Set the infinite switch in the "high" position. Using your ohmmeter, place the probes on the L1 and H1 (Figure 18-10a) terminals; there should be continuity. Next, place the probes on the L2 and H2 (Figure 18-10b) terminals; there should be continuity. Now place the probes on the P and H1 (Figure 18-10c) terminals; there

FIGURE **18-7**   Remove the screws from both ends to remove the control panel and gain access to the components.

FIGURE **18-8**   Remove the screws from the back panel to gain access to the components.

Figure 18-9   Remove the screws that secure the control panel to the oven to gain access to the components.

should be continuity. The infinite switch is defective if there is no continuity between L1 and H1 or between L2 and H2. If there is no continuity between P and H1, the indicator light circuit is defective.

6. **Remove the infinite switch.** To remove the infinite switch on most models, you must remove the two screws that secure the switch to the control panel. On some models, the infinite switch is secured with a nut to the control panel. Unscrew the nut to remove this type of infinite switch.

7. **Install a new infinite switch.** To install the new infinite switch, just reverse the disassembly procedure, and reassemble. Be sure that you install the wires onto the correct terminals according to the wiring diagram.

Surface (Burner) Element

The typical complaints associated with surface element (burner) failure are:

- The surface element will not heat at all.
- When the surface element is turned on, it trips the circuit breaker or blows the fuses.
- On dual-surface elements, only part of the element heats up.

(a)   (b)   (c)

Figure 18-10   Testing an infinite switch for continuity.

PART V

FIGURE 18-11 To remove a plug in a surface element, just lift and slide the element out of the receptacle.

To handle these problems, perform the following steps:

1. **Verify the complaint.** Verify the complaint by turning on the infinite switch. Is the surface element heating?

2. **Check for external factors.** You must check for external factors not associated with the appliance. Is the appliance installed properly? Does it have the correct voltage?

3. **Disconnect the electricity.** Before working on the range or cooktop, disconnect the electricity to the appliance. This can be done by pulling the plug from the receptacle. Or disconnect the electricity at the fuse panel or at the circuit breaker panel. Turn off the electricity.

4. **Remove the surface element.** On most models, the surface element can be removed by simply pulling the element out of its receptacle (Figure 18-11). Other models have the surface elements connected to the cooktop and directly wired to their infinite switch wires (Figure 18-12). To remove this type of element, remove the screw that secures the element to the cooktop. Then remove the clips that hold the insulators to the element terminals (see inset in Figure 18-12). Next, unscrew the wires from the terminals without bending the terminals.

FIGURE 18-12
To remove a wired-in surface element, remove the screw that secures the element to the cooktop. Then remove the insulator by prying off the clip.

FIGURE 18-13 Testing the surface element for resistance.

FIGURE 18-14 Testing the surface element for ground.

5. **Test the surface element.** Using the ohmmeter, set the range scale on R × 1. Place the probes on the element terminals (Figure 18-13). The meter reading should show continuity. The actual readings will vary from manufacturer to manufacturer, according to the size of the unit and the wattage used, but the readings should generally be between 19 and 115 ohms. To test for a grounded surface element, place one probe on the sheath and the other probe on each element terminal, in turn (Figure 18-14). If continuity exists at either terminal, the element is shorted and should be replaced.

6. **Install a new surface element.** To install the new surface element, just reverse the disassembly procedure, and reassemble. Be sure that you install the wires on their correct terminals, according to the wiring diagram.

Solid-Disc Elements

The typical complaints associated with solid-disc elements are the same as those for the traditional surface elements. To handle problems pertaining to them, perform the following steps:

1. **Verify the complaint.** Verify the complaint by turning the infinite switch on. Is the solid-disc element heating?

2. **Check for external factors.** You must check for external factors not associated with the appliance. Is the appliance installed properly? Does it have the correct voltage?

3. **Disconnect the electricity.** Before working on the range or cooktop, disconnect the electricity from the appliance. This can be done by pulling the plug from the receptacle. Or disconnect the electricity at the fuse panel or the circuit breaker panel. Turn off the electricity.

FIGURE 18-15   Remove the screws from under the front edge of the cooktop to gain access to the solid-disc elements.

4. **Gain access to the solid-disc element.** To access the element on this model, you will have to raise the cooktop. To do this, remove the screws from under the front edge of the cooktop (Figure 18-15); then lift the cooktop and prop it up. On some models, the trim might have to be removed first.

5. **Test the solid-disc element.** Remove the wires from the element. Using the ohmmeter, set the range scale on R × 1. Place the probes on the element terminals; there should be continuity. The actual readings will vary from manufacturer to manufacturer and with the size of the unit and the wattage used, but the readings should generally be between 19 and 115 ohms. To test for a grounded element, place one probe on the sheath, and attach the other probe to each element terminal. If continuity exists at either terminal, the element is shorted and should be replaced.

6. **Remove the solid-disc element.** To remove the solid-disc element, remove the wires from the element (Figure 18-16). Remove any brackets that secure the element to the cooktop. The element is removed from the top of the cooktop (see inset in Figure 18-16).

Element

Receptacle

FIGURE 18-16   Remove the wires from the solid-disc element before you test for continuity. Remove the screws that secure the element; then remove the element from the top of the cooktop.

7. **Install a new solid-disc element.** To install the new solid-disc element, just reverse the disassembly procedure, and reassemble. Be sure that you install the wires on their correct terminals according to the wiring diagram.

Radiant Heating Element

The typical complaints associated with the radiant heating element are the same as those for surface elements. To handle problems pertaining to it, perform the following steps:

1. **Verify the complaint.** Verify the complaint by turning the infinite switch on. Is the radiant element heating?

2. **Check for external factors.** You must check for external factors not associated with the appliance. Is the appliance installed properly? Does it have the correct voltage?

3. **Disconnect the electricity.** Before working on the range or cooktop, disconnect the electricity to the appliance. This can be done by pulling the plug from the receptacle. Or disconnect the electricity at the fuse panel or at the circuit breaker panel. Turn off the electricity.

4. **Gain access to the radiant heating element.** To access the element, you will have to raise the cooktop (see Figure 18-15). On some models, you will have to remove the trim first. Remove the screws from under the front edge of the cooktop; then lift the cooktop and disconnect the element's wiring harness plug. Remove the cooktop, and place it upside-down on a table. Don't forget to place a blanket under the cooktop to prevent damage.

5. **Test the radiant heating element.** In order to test the element, remove the element mounting bracket from the cooktop. This is held on with two screws on either end of the bracket. Then carefully turn over the heating element assembly. This will give you access to the limiter and the element (Figure 18-17). Be careful—the element is embedded in an insulated casting, which is used to prevent the cooktop glass from overheating. Remove the wires from the element. On the ohmmeter, set the range scale on R × 1. Place the probes on the element terminals; there should be continuity. The actual readings will vary from manufacturer to manufacturer and according to the size of the unit and the wattage used, but the readings should generally be between 19 and 115 ohms. If the element fails the test, replace it along with the limiter.

6. **Remove the radiant heating element.** To remove the element, remove the screws that secure the element to the bracket. Lift the element out. The new element comes as a complete assembly.

7. **Install a new radiant heating element.** To install the new radiant heating element, just reverse the disassembly procedure, and reassemble. The element must be flat against the glass and located under the heater panel (see Figure 18-17). Be sure that you install the wires on their correct terminals according to the wiring diagram.

FIGURE 18-17
The radiant heating element must be flat against the glass.

Glass cooktop

Temperature limiter

240-volt element

Bake Element

The typical complaints associated with bake element failure are:

- The bake element will not heat at all.

- When the bake element is turned on, it trips the circuit breaker or it blows the fuses.

To handle these problems, perform the following steps:

1. **Verify the complaint.** Verify the complaint by turning the selector switch to "bake" and setting the thermostat. Is the bake element heating? On some models, the clock must be set to "manual." Check the use and care manual for the model you are servicing.

2. **Check for external factors.** You must check for external factors not associated with the appliance. Is the appliance installed properly? Does it have the correct voltage?

3. **Disconnect the electricity.** Before working on the range, disconnect the electricity to it. This can be done by pulling the plug from the receptacle. Or disconnect the electricity at the fuse panel or at the circuit breaker panel. Turn off the electricity.

4. **Gain access to the bake element.** To access the bake element, open the oven door and remove the oven racks. Begin by removing the two screws that secure the element to the oven cavity (Figure 18-18).

5. **Remove and test the bake element.** Slide the element forward, and remove the wires from the bake element terminals (label them), either by removing the screws from the terminals or by pulling the wires off the bake element terminals (Figure 18-19). Using the ohmmeter, set the range scale on R × 1. Place the probes on the element terminals (Figure 18-20); there should be continuity. The actual readings will vary from manufacturer to manufacturer and according to the size of the unit and the wattage used, but the readings should generally be between 19 and 115 ohms. To test

FIGURE 18-18 Remove the two screws and slide the element forward to get at the element terminals.

FIGURE 18-19 Remove the screws and wires from the element before testing.

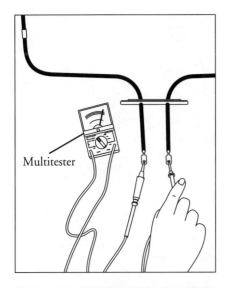

Multititester

FIGURE 18-20   Testing the bake element for resistance.

FIGURE 18-21   Testing the element for ground.

for a grounded bake element, place one probe on the sheath and the other probe on an element terminal (Figure 18-21). If continuity exists, the element is shorted and should be replaced. Test both terminals.

6. **Install a new bake element.** To install the new bake element, just reverse the disassembly procedure, and reassemble. Be sure that you install the wires on their correct terminals according to the wiring diagram.

Broil Element

The typical complaints associated with failure of the broil element are:

- The broil element will not heat at all.
- When the broil element is turned on, it trips the circuit breaker or it blows the fuses.
- On dual broil elements, only half of the element heats.

To handle these problems, perform the following steps:

1. **Verify the complaint.** Verify the complaint by turning the selector switch to "broil" and/or by setting the thermostat to "broil." Is the broil element heating? On some models, the clock must be set to "manual." Check the use and care manual for the model you are servicing.

2. **Check for external factors.** You must check for external factors not associated with the appliance. Is the appliance installed properly? Does it have the correct voltage?

3. **Disconnect the electricity.** Before working on the range, disconnect the electricity. This can be done by pulling the plug from the receptacle. Or disconnect the electricity at the fuse panel or at the circuit breaker panel. Turn off the electricity.

FIGURE 18-22   Testing the broil element for resistance.

FIGURE 18-23   Testing a dual broil element for resistance between the A and C terminals.

4. **Gain access to the broil element.** To access the broil element, open the oven door and remove the oven racks. The broil element is located at the top of the oven cavity. Begin by removing the two screws that secure the element to the oven cavity. Then remove the holding brackets from the element and slide the element forward.

5. **Remove and test the broil element.** Remove the wires from the broil element terminals, either by removing the screws from the terminals or by pulling the wires off the broil element terminals (label them). Using the ohmmeter, set the range scale on R × 1. Place the probes on the element terminals (Figure 18-22); there should be continuity. The actual readings will vary from manufacturer to manufacturer and according to the size of the unit and the wattage used. The readings should generally be between 19 and 115 ohms. To test for a grounded broil element, place one probe on the sheath and the other probe on an element terminal (see Figure 18-21). If continuity exists, the element is shorted and should be replaced. Test both terminals.

To test a dual broil element, remove the wires from their terminals. Using the ohmmeter, set the range scale on R × 1. Place the probes on terminals A and C (Figure 18-23); there should be continuity. Then place the probes on terminals A and B (Figure 18-24); there should be continuity. Finally, place the probes on terminals B and C; there should be continuity. To test for a grounded element, place one probe on the sheath. With the other probe, touch terminal A, then B, then C (Figure 18-25). If continuity exists, the element is shorted and should be replaced.

FIGURE 18-24   Testing a dual broil element for resistance between the A and B terminals.

FIGURE 18-25   Testing the dual broil element for ground.

6. **Install a new broil element.** To install the new broil element, just reverse the disassembly procedure, and reassemble. Be sure that you install the wires on their correct terminals according to the wiring diagram.

Oven Selector Switch

The typical complaints associated with failure of the oven selector switch are:

- Bake element will not heat.
- Broil element will not heat.
- Self-cleaning feature will not work.
- Timed baking will not work.

To handle these problems, perform the following steps:

1. **Verify the complaint.** Verify the complaint by turning the oven selector switch to the desired setting. Set the thermostat to the desired temperature setting. Does that part of the range/oven work? Test all of the cycles on the selector switch.

2. **Check for external factors.** You must check for external factors not associated with the appliance. Is the appliance installed properly? Does it have the correct voltage?

3. **Disconnect the electricity.** Before working on the range or oven, disconnect the electricity to the appliance. This can be done by pulling the plug from the receptacle. Or disconnect the electricity at the fuse panel or at the circuit breaker panel. Turn off the electricity.

4. **Gain access to the oven selector switch.** Depending on which model you are repairing, you can access the component by removing the back panel (see Figure 18-6) on a freestanding range. On models with front-mounted controls (see Figure 18-7), the panel is attached with screws on both ends. Remove the screws, and tilt the control panel. Be careful not to let the wires come off their components. Some built-in models have a removable backsplash (see Figure 18-8). Lift the backsplash and rest it on the cooktop. It would be a good idea to place something on the cooktop first to protect it from damage. Next, remove the screws from the backsplash that hold the rear panel to gain access to the components. If you are repairing a wall oven or an eye-level range, the control panel can be removed (see Figure 18-9) by opening the door and removing the screws that secure the panel. These might be underneath the front of the exhaust hood or just below the control panel. Some control panels are hinged; just tilt the control panel toward you for servicing. On other models, both the rear panel and the front control panel (usually glass) will have to be removed to gain access to the components. To remove the glass, remove the screws that secure the trim piece that holds the glass in place, and then remove the trim ring. On some models, the trim might have to be removed first.

Figure 18-26   When testing the oven selector switch, only remove one wire from each pair of terminals being tested.

5. **Test the oven selector switch.** To test the oven selector switch for continuity between certain switch contacts, remove only those wires that are being tested and label them (Figure 18-26). To test each switch contact, refer to the wiring diagram for the proper switch contact terminals. Only remove one pair of wires at a time to test for continuity, as the oven selector switch is checked in each position.

6. **Remove the oven selector switch.** To remove the oven selector switch on most models, you must remove the two screws that secure the switch to the control panel. On some models, the oven selector switch is secured to the control panel with a nut. Unscrew this nut to remove the oven selector switch (Figure 18-27). Do not remove the wires until the new switch is ready for installation. Then transfer the wires to the correct terminals.

7. **Install a new oven selector switch.** To install the new oven selector switch, just reverse the disassembly procedure, and reassemble. Be sure that you install the wires on the correct terminals according to the wiring diagram.

Figure 18-27   Removing the two screws that secure the oven selector switch from the control panel.

Oven Thermostat

The typical complaints associated with failure of the oven thermostat are:

- Bake element will not heat.

- Broil element will not heat.

- Timed baking will not work.

- Oven temperature is not accurate.

To handle these problems, perform the following steps:

1. **Verify the complaint.** Verify the complaint by turning the oven selector switch to the desired setting and setting the oven thermostat to the desired temperature. Do those functions of the range/oven work?

On some models, the clock must be set to "manual." Check the use and care manual for the model you are servicing.

2. **Check for external factors.** You must check for external factors not associated with the appliance. Is the appliance installed properly? Does it have the correct voltage?

3. **Disconnect the electricity.** Before working on the range or oven, disconnect the electricity to the appliance. This can be done by pulling the plug from the receptacle. Or disconnect the electricity at the fuse panel or at the circuit breaker panel. Turn off the electricity.

4. **Gain access to the oven thermostat.** You can gain access to the component by removing the back panel (see Figure 18-6) on a freestanding range. On models with front-mounted controls (see Figure 18-7), the panel is attached with screws on both ends. Remove the screws, and tilt the control panel. Be careful not to let the wires come off their components. Some built-in models have a removable backsplash (see Figure 18-8); lift the backsplash and rest it on the cooktop. It would be a good idea to place something on the cooktop first to protect it from damage. Next, remove the screws from the backsplash, which holds the rear panel, to access the components. If you are repairing a wall oven or an eye-level range, the control panel can be removed (see Figure 18-9) by opening the door and removing the screws that secure the panel. These might be underneath the front of the exhaust hood or just below the control panel. Some control panels are hinged—just tilt the control panel toward you for servicing. On other models, both the rear panel and the front control panel (usually glass) will have to be removed to gain access to the components. To remove the glass, remove the screws that secure the trim that holds the glass in place. Then remove the trim piece. On some models, the trim might have to be removed first.

5. **Test the oven thermostat switch contacts.** To test the oven thermostat switch for continuity between certain switch contacts, remove only those wires that are being tested from their terminals (Figure 18-28). To test each switch contact, refer to the wiring diagram for the proper switch contact terminals. Only remove one pair of wires at a time (label them) to test for continuity.

FIGURE 18-28
Testing the oven thermostat switch contacts for continuity.

Multitester

Temperature control

6. **Calibrate the oven thermostat.** Before making any adjustments to the thermostat, test the oven temperature. With an oven temperature tester, place the thermocouple tip in the center of the oven cavity. Be sure that it does not touch any metal. Close the oven door, set the oven to bake, and adjust the thermostat setting to the 350-degree mark. Let the oven cycle for 20 to 30 minutes. Then record the minimum and maximum temperatures of three cycles. Next, add these temperatures and divide by 6. This will give you the average temperature of the oven:

$$\frac{370+335+350+340+360+335}{6}=348.3 \text{ degrees Fahrenheit}$$

The average temperature calculated should be within ±25 degrees of the temperature setting selected (rotary dial type). If not, try calibrating the thermostat. To calibrate the thermostat, pull the oven thermostat knob off. Turn it over (Figure 18-29). The calibration ring is on the back of the knob. Loosen the two screws, and move the pointer in the direction needed. Each line that the pointer is moved indicates 10 degrees of change in the calibration. Tighten the screws and place the knob back on the thermostat stem. Retest the oven temperature. On models that do not have the calibration ring on the back of the dial, replace the thermostat if the temperature is more than 25 degrees out of calibration.

7. **Remove the oven thermostat.** To remove the oven thermostat on most models, you must disconnect the thermostat capillary tube from its supports (Figure 18-30) and push it through the back wall of the oven cavity. Be careful. Do not break the capillary tube wire because the contents inside are flammable. Next, remove the two screws that secure the thermostat to the control panel. Remove the thermostat (Figure 18-31). Leave the wires on the thermostat for now.

Figure 18-29 Turn the oven thermostat knob over. On the back of the knob are the words "increase" and "decrease." By moving the pointer, you can increase or decrease the temperature setting of the thermostat. Each line indicates 10-degree increments.

FIGURE 18-30 Lift the thermostat capillary tube off the supports, being careful not to break it.

8. **Install a new oven thermostat.** To install the new oven thermostat, just reverse the disassembly procedure, and reassemble. Transfer the wires, one at a time, from the old thermostat to the new one. Be sure that you install the wires on their correct terminals according to the wiring diagram. Reassemble the control panel in the reverse order of disassembly, and test.

Oven Cycling Relay

The typical complaints associated with oven cycling relay failure are:

- Bake element will not heat.
- Broil element will not heat.
- Bake element stays on all the time.
- Oven temperature is not accurate.

To handle these problems, perform the following steps:

1. **Verify the complaint.** Verify the complaint by turning the oven selector switch to the desired setting and setting the oven thermostat to the desired temperature. Do these functions of the oven work? Test all of the cycles.

FIGURE 18-31 Remove the oven thermostat from the range. Leave the wires on the thermostat until the new thermostat is ready for installation. Then transfer the wires to the correct terminals.

2. **Check for external factors.** You must check for external factors not associated with the appliance. Is the appliance installed properly? Does it have the correct voltage?

3. **Disconnect the electricity.** Before working on the range or oven, disconnect the electricity to the appliance. This can be done by pulling the plug from the receptacle. Or disconnect the electricity at the fuse panel or at the circuit breaker panel. Turn off the electricity.

4. **Gain access to the oven cycling relay.** To access the oven cycling relay (Figure 18-32), remove the back panel (see Figure 18-6). On some models, this relay is located under the oven cavity, where the utility drawer is usually found.

FIGURE 18-32 The oven cycling relay.

5. **Test the oven cycling relay.** To test the oven cycling relay, remove all of the wires from the relay (label them). Using the ohmmeter, set the range scale on R × 1. Place the probes on the L and H1 terminals; there should not be continuity. Next, place the probes on the L and H2 terminals; there should not be continuity. Then place the probes on the H1 and H2 terminals. There should not be continuity. Now place the probes on the S and R terminals. There should be continuity. If your meter indicates no reading at all between the S and R terminals, this indicates that the heater wire inside the relay is defective and that the relay should be replaced.

6. **Remove the oven cycling relay.** To remove the oven cycling relay, remove the two screws that secure the relay to the cabinet.

7. **Install a new oven cycling relay.** To install the new oven cycling relay, just reverse the order of step 6. Transfer the wires from the old relay to the new relay. Be sure that you install the wires on their correct terminals according to the wiring diagram. Reassemble the remainder of the range/oven in the reverse order of its disassembly, and test.

Range/Oven Time Clock

The typical complaints associated with failure of the range/oven time clock are:

- Timed bake cycle will not operate.
- Unable to use the self-cleaning cycle.
- Clock loses time.
- Clock is not functioning.
- The minute reminder is not functioning.

To handle these problems, perform the following steps:

1. **Verify the complaint.** Verify the complaint by turning the clock controls to the correct time. Does the clock time advance? Some models have the minute reminder located on the same clock. Set this timer to test it. Does it advance?

2. **Check for external factors.** You must check for external factors not associated with the appliance. Is the fuse blown? Does the appliance have the correct voltage?

3. **Disconnect the electricity.** Before working on the range or oven, disconnect the electricity to the appliance. This can be done by pulling the plug from the receptacle. Or disconnect the electricity at the fuse panel or at the circuit breaker panel. Turn off the electricity.

4. **Gain access to the range/oven clock.** You can access the clock by removing the back panel (see Figure 18-6) on a freestanding range. On models with front-mounted controls (see Figure 18-7), the panel is attached with screws on both ends. Remove the screws, and tilt the control panel. Be careful not to let the wires come off their components. Some built-in models have a removable backsplash (see Figure 18-8); just lift the backsplash and rest it on the cooktop. It would be a good idea to place

something on the cooktop first to protect it from damage. Next, remove the screws from the backsplash that hold the rear panel to gain access to the clock. If you are repairing a wall oven or an eye-level range, the control panel can be removed (see Figure 18-9) by opening the door and removing the screws that secure the panel. These might be underneath the front of the exhaust hood or just below the control panel. Some control panels are hinged; just tilt the control panel toward you for servicing. On other models, to gain access to the clock, both the rear panel and the front control panel (usually glass) will have to be removed. Remove the screws that secure the trim piece that holds the glass in place. Then remove the trim piece. On some models, the trim might have to be removed first. Figure 18-33 illustrates the different types of clock faces available on ranges/ovens.

FIGURE 18-33
(a) Standard time clock. (b) Analog clock with minute reminder, timed baking, and self-cleaning. (c) and (d) Digital clock with timed baking, self-cleaning, and minute reminder. (e) Electronic clock, with or without additional features.

PART V

FIGURE 18-34 Clocks are secured to the control panel by clips, screws, or nuts.

5. **Test the range/oven clock.** On some models, if the clock does not run, check for a fuse in the circuit. Locate the clock motor wire leads, and isolate them from the circuit (label them). Using the ohmmeter, set the range scale on R × 1. Place the probes on the clock motor leads; there should be continuity. If not, replace the clock. On some models, the clock is used to control the timed baking and the self-cleaning cycles (see Figures 18-33b, 18-33c, and 18-33d). To check the switch mechanism of the clock, remove the wires from the switch terminals (label them). Using the ohmmeter, set the range scale on R × 1. Place the probes on the terminals. Look at the wiring diagram for the correct terminals to test. Some models have one switch; other models have two sets of switches. Test for continuity of the switch contacts when you push in and turn the start and stop knobs on the clock and when the knobs pop out. On electronic control models, refer to the technical data sheet located in the range for the testing procedure.

6. **Remove the range/oven time clock.** First remove the clock knobs from the clock stems. To remove the range/oven time clock in this model (Figure 18-34), use a screwdriver and depress the clips that hold the clock to the control panel. On other models, the clock is secured to the control panel by screws or nuts. Pull the clock toward the front of the appliance (Figure 18-35).

7. **Install a new range/oven clock.** Transfer the wires from the old clock to the new clock. Be sure that you install the wires on their correct terminals according to the wiring diagram. To install the new range/oven clock, just reverse the disassembly procedure, and reassemble.

Electronic Control Board and User Interface Controls

The typical complaints associated with the electronic control board or the user interface controls are:

- The range or oven won't run or power up.

- Unable to program the range or oven.

FIGURE 18-35 Removing the clock from the front of the control panel.

- The display board will not display anything.
- One or more key pads will not accept commands.
- Unusual display readouts.

To prevent electrostatic discharge (ESD) from damaging expensive electronic components, follow the steps in Chapter 9.

To handle these problems, perform the following steps:

1. **Verify the complaint.** Verify the complaint by operating the range or oven. Turn off the electricity to the appliance and wait for two minutes before turning it back on. If a fault code appears, look up the code. If the range or oven will not power up, locate the technical data sheet behind the control panel for diagnostics information.

2. **Check for external factors.** You must check for external factors not associated with the appliance. Are there 240 volts of electricity to the range or oven?

3. **Disconnect the electricity.** Before working on the range or oven, disconnect the electricity. This can be done by pulling the plug out of the wall receptacle. Or disconnect the electricity at the fuse panel or circuit breaker panel. Turn off the electricity.

WARNING *Some diagnostic tests will require you to test the components with the power turned on. When you disassemble the control panel, you can position the panel in such a way that the wiring will not make contact with metal. This will allow you to test the components without electrical mishaps.*

4. **Remove the console panel to gain access.** You can access the electronic control (some models have multiple circuit boards) by removing the back panel (see Figure 18-6) on a freestanding range. On models with front-mounted controls (see Figure 18-7), the panel is attached with screws on both ends. Remove the screws, and tilt the control panel. Be careful not to let the wires come off their components. Some built-in models have a removable backsplash (see Figure 18-8); just lift the backsplash and rest it on the cooktop. It would be a good idea to place something on the cooktop first to protect it from damage. Next, remove the screws from the backsplash that hold the rear panel to gain access to the electronic control. If you are repairing a wall oven or an eye-level range, the control panel can be removed (see Figure 18-9) by opening the door and removing the screws that secure the panel. These might be underneath the front of the exhaust hood or just below the control panel. Some control panels are hinged; just tilt the control panel toward you for servicing. On other models, to gain access to the electronic control, both the rear panel and the front control panel (usually glass) will have to be removed. Remove the screws that secure the trim piece that holds the glass in place. Then remove the trim piece. On some models, the trim might have to be removed first.

5. **Test the electronic control board and user interface controls.** If you are able to run the range/oven diagnostic test mode, check the different functions of the range/oven. Use the technical data sheet for the model you are servicing to locate the test points from the wiring schematic. Check all wiring connections and wiring. Using the technical data sheet, test the electronic control or user interface controls,

input voltages, and output voltages. On some models, fuses are soldered to the printed circuit board (PCB). These fuses must be tested first before condemning the component.

6. **Remove the electronic control board or user interface control.** To remove the defective component, remove the screws that secure the board to the control panel or range/oven frame. Disconnect the connectors from the electronic control board or user interface control.

7. **Install the new component.** To install a new electronic control board or user interface control, just reverse the disassembly procedure, and reassemble. Reinstall the console panel, and restore the electricity to the range/oven. Test the range operation.

Diagnostic Charts and Wiring Diagrams

The following diagnostic charts will help you to pinpoint the likely causes of the various problems associated with electric ranges and ovens (Figures 18-36, 18-37, 18-38, and 18-39).

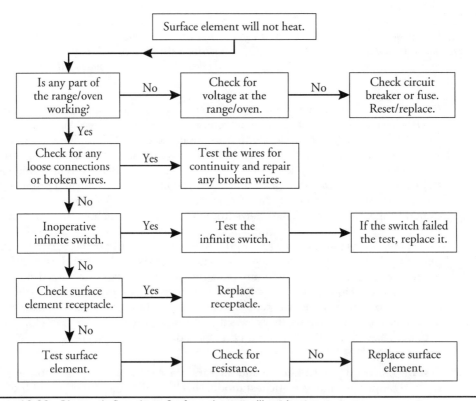

FIGURE 18-36   Diagnostic flow chart: Surface element will not heat.

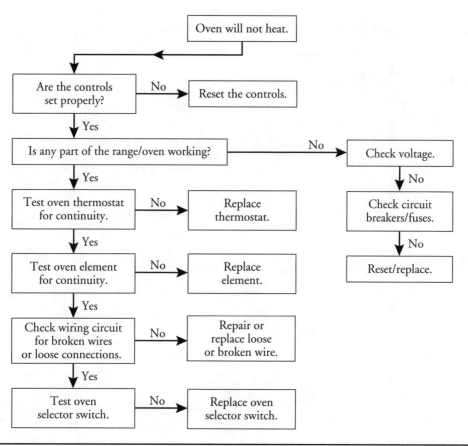

FIGURE 18-37   Diagnostic flow chart: Oven will not heat.

The wiring diagrams in this chapter are examples only. You must refer to the actual wiring diagram on the range, oven, or cooktop that you are servicing. Figures 18-40 and 18-41 are for an identical appliance. Figure 18-40 depicts an actual wiring schematic. Figure 18-41 depicts a cross-section of a ladder wiring diagram (strip circuit). This type of diagram is easier to read and understand.

Table 18-1 illustrates a self-cleaning cycle chart indicating a light soil cycle. On some models, the cleaning cycle will take two hours to complete. The heavy soil cycle will take two to three hours to complete. The average stabilized temperature in a self-cleaning cycle should be between 840 and 920 degrees Fahrenheit. The door can only be opened when the oven temperature has dropped below 520 degrees Fahrenheit.

PART V

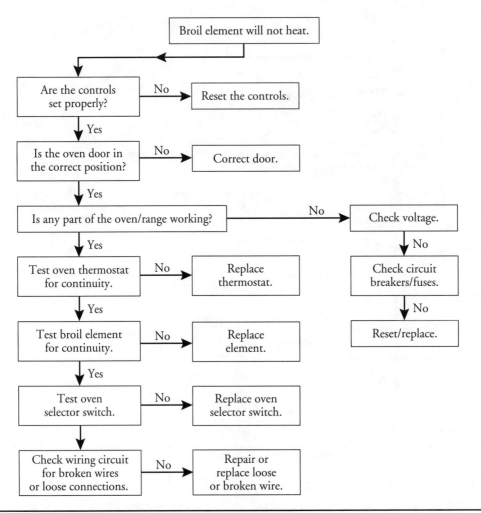

FIGURE 18-38 Diagnostic flow chart: Broil element will not heat.

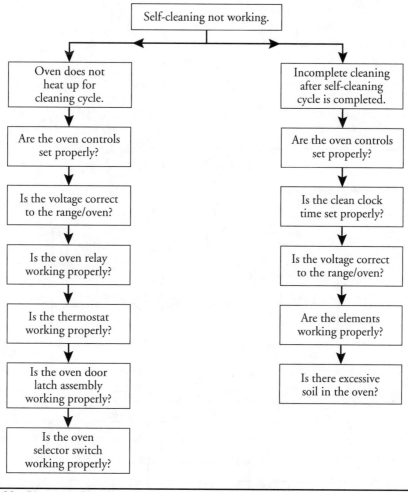

FIGURE 18-39 Diagnostic flow chart: Self-cleaning function not working.

FIGURE 18-40   Wiring schematic of a typical electric range.

Example only: refer to the diagram on the unit when servicing
L-1, L-2 = 240 V
N, L-1 = 120 V
N, L-2 = 120 V

FIGURE 18-41   The oven light, bake, broil, and surface element circuits.

TABLE 18-1 Self-cleaning Cycle Chart.

19
CHAPTER

Gas Ranges and Ovens

G as ranges, ovens, and cooktops are designed on the basis of using the heat generated by combustion to cook the food. The types of fuels that feed gas appliances are:

- Natural gas: Natural gas is lighter than air and has a heating value between 900 and 1200 BTUs.

- Liquefied petroleum (LP or LPG): LP gas is heavier than air and has a heating value between 2500 and 3200 BTUs.

- Mixed gas: Mixed gas is lighter than air and has a heating value between 700 and 900 BTUs.

- Manufactured gas: Manufactured gas is lighter than air and has a heating value between 500 and 700 BTUs.

To sustain combustion in a range, oven, or cooktop, an ignition source, such as a flame or by electrical means, is used to ignite the gas vapors. To ignite these gases, the temperature will have to be between 900 and 1200 degrees Fahrenheit.

The pressure of natural gas that is supplied to a residence will vary between a five- and nine-inch water column. LP or LPG gas pressure for residential appliances, as established by the gas industry, will be between a nine- and eleven-inch water column. To determine the correct pressure rating, the technician must refer to the manufacturer's specifications or the installation instructions for that product. The two most common types of gas used in homes are LP and natural gas.

On some models, electricity (120 volts) is supplied to the gas appliance for ignition, temperature control, safety valve, electronic controls, and accessories (clock, lights, etc.). Gas cooking appliances are available in a wide variety of styles also. Figures 1-1, 1-4, and 19-1 are just a small sampling.

This chapter will provide the basics needed to diagnose and repair gas cooking appliances. Figure 19-2 identifies where components are located within the gas range. This illustration is used as an example only. The actual construction and features might vary, depending on the brand and model you are servicing.

FIGURE 19-1 A five-burner gas cooktop with sealed burners.

FIGURE 19-2 The locations of components in a gas freestanding range.

Principles of Operation

When the gas is turned on (standing pilot ignition system), whether by a knob or automatically by a control device, gas will begin to flow from the supply line through a gas pressure regulator into the gas manifold for distribution to the burners or to the oven safety valve (Figure 19-3). The gas travels from the manifold through the controls, passing through the orifice into the venturi throat and mixing tube. The air-gas mixture then enters the burner head, exits through the burner ports, and travels through the flash tube to the pilot flame (Figure 19-4). Once ignited, the flame begins to burn evenly around the burner head.

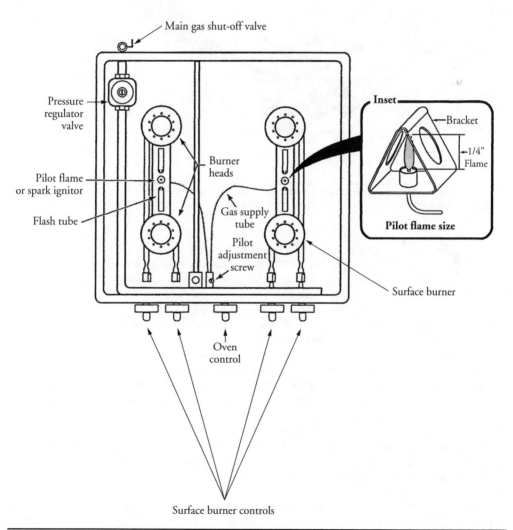

FIGURE 19-3   A top view of a standing pilot ignition system, illustrating the component locations in a freestanding gas range. See inset for pilot flame height.

FIGURE 19-4 The air-gas mixture is traveling through the burner ports and flash tube, encountering the pilot flame and igniting the burner.

When the oven thermostat is turned on, gas flows from the manifold (Figure 19-5) through the thermostat and to the safety valve. The thermostat also feeds gas through the pilot gas line to the pilot assembly. The pilot flame will increase in size, heating the safety valve-sensing bulb (Figure 19-6), opening up the safety valve and allowing gas to flow to the burner. When the air-gas mixture flows out of the oven burner, it encounters the pilot flame and the burner ignites. When the oven cavity temperature is satisfied, the gas flow will stop to the oven burner and the gas flow to the pilot assembly will decrease to the standing pilot light.

On models with electronic ignition, the standing pilot is replaced with ignitor switches, surface burner electrodes, and a spark module (Figures 19-7, 19-8, 19-9, and 19-10). When the consumer turns the burner control knob to the "lite" position, a switch within the ignitor switch will close, supplying 120 volts through the ignitor switch to the spark module. Some spark modules will produce a high-voltage pulse to the ignitor every two seconds at very low amperage (4 milliamps). Some models use a spark module that will produce two to three sparks per second. This pulse will produce a spark between the ignitor electrode and the ignitor-grounding strap (see Figure 19-8). At the same time, the gas will begin to flow from the supply line through a gas pressure regulator into the gas manifold for distribution to the burners. When the burner is lit, the control knob will then be turned to the "on" position, where the height of the flame can be adjusted. When this occurs, the switch within the ignitor switch will open, turning off the voltage to the spark module (see Figure 19-7).

Pilot gas line

Pilot flame
sensing bulb and
capillary tube

Oven burner

Pilot
assembly

Main gas line

Safety
valve

Manifold

Gas in

Sensing bulb and
capillary tube

Oven
thermostat

FIGURE 19-5   The components that make up the oven burner system with a standing pilot ignition system.

When the oven thermostat is turned on and set to the desired temperature, a switch contact from within the thermostat will close. This action will complete the circuit, supplying 120 volts to the glow-bar ignition system (Figure 19-11). The glow-bar ignitor (Figure 19-12) must heat up to 2000 degrees Fahrenheit, allowing 2.5 to 3.0 amps to flow to the safety valve. As the bi-metal warps, the safety valve will open, allowing the gas to flow through the orifice into the venturi throat and mixing tube. The air-gas mixture then enters the oven burner head, exits through the burner ports, and travels to the glow-bar ignitor, igniting the oven burner. Once ignited, the flame begins to burn evenly around the burner head. The entire process takes about 60 seconds to complete.

The self-cleaning operation in a gas range is similar to that of the electric range (see Chapter 18). The only difference is that some models use two burners (Figure 19-13). The broil burner allows the consumer to broil food in the same oven cavity. In the self-cleaning mode, both burners do not come on at the same time. As Figure 19-13 shows, the burners are controlled by the electronic control to operate the dual safety valve and the glow-bar ignitors. When the oven cavity temperature reaches 600 degrees Fahrenheit, the door latch

FIGURE 19-6   When the thermostat control knob is in the "off" position, the pilot flame stands up. With the thermostat control knob turned on, the pilot flame will increase in size, heating the safety valve-sensing bulb.

lock light will come on, locking the oven door and preventing the consumer from opening it. When the oven cavity temperature stabilizes (between 840 and 920 degrees Fahrenheit), the cleaning process begins. This cleaning process requires approximately two to three hours. When the self-cleaning cycle is complete, the burners turn off and the oven begins to cool down. When the oven cavity temperature drops below 600 degrees Fahrenheit, the oven latch lock light will go off, allowing the consumer to open the oven door.

Safety First

Any person who cannot use basic tools or follow written instructions should *not* attempt to install, maintain, or repair any gas range, oven, or cooktop. Any improper installation, preventive maintenance, or repairs could create a risk of personal injury or property damage.

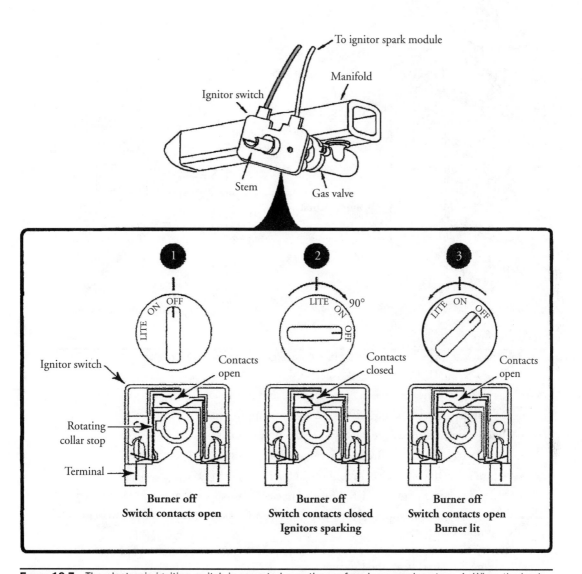

FIGURE 19-7   The electronic ignition switch is mounted over the surface burner valve stem. 1. When the knob is in the "off" position, the contact switch is open and there is no gas flow. 2. The knob is set in the "lite" position, the contact switch is closed, and gas flows to the burner. 3. When the knob is turned to the "on" position, the ignitor switch will open, breaking the circuit, and gas flows to the burner.

If you do not fully understand the installation, preventive maintenance, or repair procedures in this chapter, or if you doubt your ability to complete the task on your appliance, please call your service manager.

The following precautions should also be followed:

- Always follow the use and care guide instructions from the manufacturer.
- Never use a range to heat the home; it simply was not designed for that purpose.

To ignitor switches

Spark module

To ignitor electrode

(a)

Ground strap (bracket)

Ceramic insulated body

Metal rod

Surface burner ignitor (electrode)

Electrode wire

To spark module

(b)

FIGURE 19-8 (a) A typical spark module. (b) The surface burner ignitor and grounding bracket.

- Keep the cooking area clean of soot, spills, and grease.
- Do not use flammable liquids near a cooking appliance.
- When repairing the range, always use the proper tools.
- Always reconnect the wires to the range after repairs have been made.
- Before servicing any gas components in the range, turn off the gas supply.
- Always have a fire extinguisher nearby in case of mishaps that might lead to a fire.
- Make sure that gas appliances have proper venting according to the manufacturer's recommendations.
- To eliminate personal injury or fire when operating the appliance, avoid using the storage cabinets above the product.

Before continuing, take a moment to refresh your memory of the safety procedures in Chapter 2.

Do not store any flammable materials near a gas appliance, and avoid ignition sources in the event you smell gas leaking. Make sure you have proper ventilation when servicing the gas appliance to prevent carbon monoxide (CO) poisoning.

Gas Ranges, Ovens, and Cooktops in General

Much of the troubleshooting information in this chapter covers gas ranges, ovens, and cooktops in general, rather than specific models, in order to present a broad overview of service techniques. The illustrations that are used in this chapter are for demonstration

FIGURE 19-9   The components that make up the standard burner system using an electronic ignition system to light the burner.

purposes only to clarify the description of how to service these appliances. They in no way reflect on a particular brand's reliability.

Please note that *eye-level ranges* are also referred to as *high-low ranges* or *tri-level ranges*.

Location and Installation of Gas Range, Oven, and Cooktop

Locate the range, oven, or cooktop where it will be well lighted and have access to proper ventilation. The range must be level for proper baking and cooking results. The range might be installed adjacent to the left and/or right base cabinets and against a rear vertical wall (for the "anti-tip" cleat). A wall oven must be installed on a supporting surface that is strong

PART V

Figure 19-10   The components that make up the sealed burner system using an electronic ignition system to light the burner.

enough to support the weight of the oven and its contents while remaining level from side to side and from front to rear. A cooktop must be installed on a flat surface, supported by the countertop, and it should be level. Contact your flooring company to check if the flooring can withstand a minimum of 200 degrees Fahrenheit. It is also recommended that you contact your builder to determine if the cabinets and walls can withstand the heat produced by the gas appliance. Some kitchen cabinets and building materials were not designed to withstand the heat that is given off by the product.

The proper installation instructions for your model are included with the appliance. These instructions will assist you with the installation requirements (dimensions, electrical requirements, cutout dimensions, venting, etc.) needed to complete the installation according to the manufacturer's specifications.

Some gas appliance models are equipped with electronic controls and electronic ignition. The electrical plug that is considered for use must be properly polarized and grounded in order for the electronic controls to function properly. The appliance chassis must also be grounded so that the electronics in the product will function properly.

FIGURE 19-11   The components that make up the oven burner system using a glow-bar ignition system to light the burner.

Step-by-Step Troubleshooting by Symptom Diagnosis

When servicing an appliance, don't overlook the simple things that might be causing the problem. Step-by-step troubleshooting by symptom diagnosis is based on diagnosing malfunctions, with their possible causes arranged into categories relating to the operation

FIGURE 19-12
A typical glow-bar
ignitor.

FIGURE 19-13 The components that make up the self-cleaning oven system using two glow-bar ignitors to light the burners. The consumer can bake or broil in the same oven cavity.

of the gas cooking appliance. This section is intended only to serve as a checklist to aid you in diagnosing a problem. Look at the symptom that best describes the problem that you are experiencing with the range, oven, or cooktop, and then proceed to correct the problem.

Gas Odors

- Check the pilot light.
- Check for a gas leak in the supply line.
- Check the ventilation system.
- Check the burner controls. When not in use, the controls must be completely turned off.

Surface Burner Will Not Light

- Check the pilot light.
- Check the pilot burner port holes for a blockage.
- Check for 120 volts to the appliance. Check fuses or circuit breakers.
- On electronic models, check the ignition system.
- Check the burner openings for a blockage.
- Check the air-gas mixture.
- Check for a blockage in the gas supply to the appliance and burner.

The Pilot for the Surface Burner Will Not Stay Lit

- Check the burner for blockages.
- Check the setting and adjust the pilot light.
- Check the air-gas mixture.
- Check for a blockage in the gas supply to the appliance and burner.

The Surface Burner Flame Is Too Low

- Check the burner assembly for blockages.
- Check the air-gas mixture.
- Check for a blockage in the gas supply to the appliance and burner.

The Surface Burner Flame Is Too High

- Check the air-gas mixture.
- Test the gas pressure to the appliance and make corrections, if necessary.

The Oven Burner Will Not Light or Stay Lit

- Check the pilot light. If the pilot is too small, adjust the height.
- Check the control settings on the clock or electronic controls.
- Check for 120 volts to the appliance.
- Check the ignition system.
- Check the fuses in the appliance.
- Check the glow-bar ignitor.
- Check the selector switch, thermostat, and the safety valve for proper operation.
- Check for a blockage in the pilot light.

Oven Temperature Will Not Hold

- Check the oven temperature and thermostat control.
- Check for a blockage in the oven burner.
- Check the oven door and gasket.

Self-Cleaning Function Is Not Working

- Check the controls for the proper settings.
- Check the door latch assembly. Make sure the latch engages.
- Check the thermostat and the selector switch for proper operation.
- Check for 120 volts to the appliance.
- Check the door alignment and gasket.

The Burner Orifice Squeals

- Check the orifice opening for debris.
- Make sure that the correct orifice size is in use.
- Test the gas pressure to the appliance and make corrections, if necessary.

Supply Air Is Noisy

- Check the venturi throat and mixing tube for debris and rough edges.
- Check the burner positioning over the orifice.

Erratic Burner Flame

- Check the gas pressure at the gas regulator.
- Check the burner controls for proper operation.

Yellow or Sooty Burner Flame

- Make sure that the burner is in the correct position.
- Check the air-gas mixture.
- Test the gas pressure to the appliance and make corrections, if necessary.

Gas Appliance Maintenance

To maintain gas appliances, always follow the manufacturer's recommendations for periodic maintenance as stated in the use and care manual. The range, oven, or cooktop can be cleaned with warm water, mild detergent, and a soft cloth on all cleanable parts, as recommended in the use and care manual. Also, never use abrasive cleaners that are not recommended by the manufacturer.

Do not allow grease spillovers to accumulate on top of the range after cooking; they will become a fire hazard. When cleaning the burners, always make sure that all of the portholes are free of debris. If, for any reason, the burner portholes are blocked, the flame appearance will be different. Blocked portholes will reduce gas flow and the heating valve of the burner will be reduced.

Repair Procedures

Each repair procedure is a complete inspection and repair process for a single range, oven, or cooktop component, containing the information you need to test a component that might be faulty and replace it, if necessary.

Gas Burner Valve

The typical complaints associated with the burner valve are:

- It is hard to turn on.
- When the gas burner valve is on, little or no gas comes out.
- There is a gas smell around the burner valve.

To handle these problems, perform the following steps:

1. **Verify the complaint.** Verify the complaint by testing for gas leaks. Next, turn on the burners. Is the pilot light on? Are the electronic switches working?

2. **Check for external factors.** You must check for external factors not associated with the appliance. Is the appliance installed properly? Are there 120 volts to the appliance? Is the gas turned on?

3. **Disconnect the electricity.** Before working on the appliance, disconnect the electricity. This can be done by pulling the plug from the electrical outlet. Or disconnect the electricity at the fuse panel or at the circuit breaker panel. Turn off the electricity.

4. **Shut off the gas supply.** Before you begin servicing any gas components, shut off the supply of gas to the appliance.

5. **Gain access to the gas burner valve.** To gain access to the gas burner valve, lift the cooktop, pull off the control knobs, and then remove the screws from the front panel (Figure 19-14). Remove the control panel. This will expose the gas burner valve and manifold (Figure 19-15).

6. **Remove the gas burner valve.** To remove the gas burner valve, remove the screw that secures the burner to the burner bar, and then lift the burner head up and out of the range (Figure 19-16). With the gas burner valve still attached to the gas manifold, remove the orifice with a wrench, turning it counterclockwise. Now remove the gas burner valve by removing the mounting screw that is secured to the manifold right above the valve (see Figure 19-15). Remove the valve.

7. **Install a new gas burner valve.** To install a new gas burner valve, just reverse the disassembly procedure, and reassemble. Be careful not to damage the washers when reinstalling the gas burner valve (see Figure 19-15). Most gas ranges, ovens, and cooktops have universal orifices. When you remove the orifice, you must reinstall that orifice for the type of gas supplied to the range. For natural gas, the orifice must be tightened down so that the gas LP insert protrudes through the orifice cap (Figure 19-17a). For LP gas, the orifice will tighten down only two and a half times, resting against the gas LP insert (Figure 19-17b). If the orifice is tightened down too much the gas flow will be restricted. Reinstall the front panel, and turn on the gas supply to the appliance. Test for any gas leaks. Then turn on the electricity, test the operation, and adjust the burner flame if needed. For additional information, see Chapter 6.

FIGURE 19-14   A top view of the components in an electronic ignition system.

Ignitor Switch

The typical complaints associated with the ignitor switch are:

- The ignitor switch is hard to turn on.
- The surface burner will not light.

To handle these problems, perform the following steps:

1. **Verify the complaint.** Verify the complaint by turning on the surface burners. Do any of the burners come on? Do you hear or see the ignitors sparking? Are the ignitor electrodes clear of debris? Is there good contact between the ignitor bracket and the burner bar?

2. **Check for external factors.** You must check for external factors not associated with the appliance. Is the appliance installed properly? Are there 120 volts to the appliance?

FIGURE 19-15
An exploded view
of the surface
burner valve.

- Mounting screw
- Rubber washer
- Orifice
- Gas manifold pipe
- Rubber washer
- Gas valve housing
- Gas "on" position
- Gas control seat
- Gas openings
- Spring
- Detent collar
- Stem housing
- Screw
- Stem

Burner head

Venturi

Air shutter

FIGURE 19-16 To remove the surface burner, remove the screw that secures the burner to the burner bar, and then lift the burner head up and out of the range.

FIGURE 19-17   (a) The universal orifice in the natural gas position. (b) The universal orifice in the LP position.

3. **Disconnect the electricity.** Before working on the range or cooktop, disconnect the electricity to the appliance. This can be done by pulling the plug from the receptacle. Or disconnect the electricity at the fuse panel or at the circuit breaker panel. Turn off the electricity.

4. **Shut off the gas supply.** Before you begin servicing any gas components, shut off the supply of gas to the appliance.

5. **Gain access to the ignitor switch.** To gain access to the ignitor switch, lift the cooktop, pull off the control knobs, and then remove the screws from the front panel (see Figures 19-7 and 19-14). Remove the control panel to expose the ignitor switches.

6. **Test the ignitor switch.** Remove the wires from the ignitor switch terminals. Set your ohmmeter to the R × 1 scale; reinstall the surface burner knob on the gas valve. Place the ohmmeter probes on the ignitor switch terminals. With the burner knob in the "off" position, the ohmmeter should read no continuity. Turn the burner knob to the "lite" position; the ohmmeter should read continuity. If your readings differ, replace the ignitor switch.

7. **Install a new ignitor switch.** To install a new ignitor switch, remove the screws that secure the ignitor switch, reverse the disassembly procedure, and reassemble. Turn on the electricity and test the surface burner. On some models, the ignitor will produce between two and three sparks per second.

Surface Burner Ignitor

The typical complaints associated with the surface burner ignitor are:

- The surface burner will not light.
- Only one surface burner ignitor is working.
- Intermittent operation of the surface burner ignitor.

To handle these problems, perform the following steps:

1. **Verify the complaint.** Verify the complaint by turning on the surface burners. Do the burners come on? Do you hear or see the ignitors sparking? Are the ignitor electrodes clear of debris? Is there a good ground contact between the ignitor bracket and the burner bar? If all of the ignitors are sparking, check for a shorted ignitor switch.

2. **Check for external factors.** You must check for external factors not associated with the appliance. Is the appliance installed properly? Are there 120 volts to the appliance?

3. **Disconnect the electricity.** Before working on the range or cooktop, disconnect the electricity to the appliance. This can be done by pulling the plug from the receptacle. Or disconnect the electricity at the fuse panel or the circuit breaker panel. Turn off the electricity.

4. **Shut off the gas supply.** Before you begin servicing any gas components, shut off the supply of gas to the appliance.

5. **Gain access to the ignitor.** To gain access to the ignitor (see Figure 19-8), lift up the cooktop (see Figure 19-14). The ignitor is located on the burner bar, between the two burners' flash tubes.

6. **Test the ignitor.** Set your ohmmeter on R × 100. Test the ignitor bracket to ground; the meter should indicate continuity. Disconnect the ignitor electrode wire and test for continuity.

7. **Install a new ignitor.** Remove the screws that secure the ignitor bracket to the burner bar, and disconnect the ignitor electrode wire. To reinstall the ignitor, just reverse the disassembly procedure, and reassemble. Test for a good ground between the ignitor bracket and the burner bar. Turn on the electricity and test the surface burners. The ignitor should produce two sparks per second or, on some models, two to three sparks per second.

Spark Module

The typical complaints associated with the spark module are:

- You hear a "clicking" noise but the surface burner will not light.
- You hear an erratic "clicking" noise from the surface burner ignitor.
- You do not hear the "clicking" noise and the surface burner will not light.

To handle these problems, perform the following steps:

1. **Verify the complaint.** Verify the complaint by turning on the surface burners. Do the burners come on? Do you hear or see the ignitors sparking? Are the ignitor electrodes clear of debris? Is there good contact between the ignitor bracket and the burner bar? On models with a direct-spark ignition electronic circuit board, refer to the technical data sheet, located in the range, for the proper diagnostics test points.

2. **Check for external factors.** You must check for external factors not associated with the appliance. Is the appliance installed properly? Are there 120 volts to the appliance? Does the electrical outlet have the correct polarity? Is the gas to the appliance turned on?

3. **Disconnect the electricity.** Before working on the range or cooktop, disconnect the electricity to the appliance. This can be done by pulling the plug from the receptacle. Or disconnect the electricity at the fuse panel or at the circuit breaker panel. Turn off the electricity.

4. **Shut off the gas supply.** Before you begin servicing any gas components, shut off the supply of gas to the appliance.

5. **Gain access to the spark module.** You can access the component by removing the back panel (see Figure 18-6). Locate the spark module (see Figure 19-8).

6. **Test the spark module.** Before you test the spark module, test the surface burner igniter electrodes and the surface burner igniter switches first. Next, check the ignition system wiring for a short or loose wiring. If they test okay, proceed to test the spark module (Figures 19-8 and 19-18). Set your voltmeter on the 120-volt scale. Remove the L1 wire from the spark module, and attach one of the meter probes to that wire. Attach the other meter probe to the neutral terminal on the spark module. Next, turn on the electricity to the appliance, turn on the surface burner control knob to the "lite" position, and test for 120 volts. The voltmeter should read 120 volts. If it does not, check the wiring connections and wiring.

7. **Remove the spark module.** With the electricity turned off, disconnect the wiring terminals from the spark module one at a time and label them. Remove the screws that secure the spark module to the appliance.

8. **Install a new spark module.** To install the new spark module, just reverse the disassembly procedure, and reassemble. Be sure that you install the wires onto the correct terminals according to the wiring diagram for the product you are servicing. Turn on the electricity, and test the surface burners. Make sure that the spark at the ignitor has a sharp blue appearance. This indicates a good working ignition system.

Glow-Bar Ignitor

The typical complaints associated with the glow-bar ignitor are:

- The oven/broiler will not light.
- The glow-bar ignitor is on but, the oven/broiler burner will not ignite.

To handle these problems, perform the following steps:

1. **Verify the complaint.** Verify the complaint by turning on the oven. Does the glow-bar ignitor turn an orange-red color?

2. **Check for external factors.** You must check for external factors not associated with the appliance. Is the appliance installed properly? Are there 120 volts to the appliance? Is the gas to the appliance turned on?

3. **Disconnect the electricity.** Before working on the range or cooktop, disconnect the electricity to the appliance. This can be done by pulling the plug from the receptacle. Or disconnect the electricity at the fuse panel or at the circuit breaker panel. Turn off the electricity.

4. **Shut off the gas supply.** Before you begin servicing any gas components, shut off the supply of gas to the appliance.

Figure 19-18   A typical wiring schematic for a range with electronic ignition.

5. **Gain access to the glow-bar ignitor.** To gain access to the glow-bar ignitor (see Figures 19-11 and 19-13), open the oven door, remove the oven racks, and remove the oven floor panel. Next, remove the nut (Figure 19-19) from the oven burner baffle, and remove the baffle. You now have access to all of the components shown in Figure 19-19.

6. **Test the glow-bar ignitor.** Disconnect the glow-bar ignitor wires from the wiring harness. Set your ohmmeter on the R × 100 scale, and check for continuity at the glow-bar ignitor wire terminals. Read the meter scale—if it is over 1100 ohms, replace the ignitor.

7. **Remove the glow-bar ignitor.** To remove the ignitor, remove the screws that secure the glow-bar ignitor to the oven burner (see Figures 19-11, 19-13, and 19-19).

Figure 19-19 An exploded view of the components that make up the oven burner system.

8. **Install the new glow-bar ignitor.** To install the new glow-bar ignitor, just reverse the disassembly procedure, and reassemble. Be careful—do not touch the silicon carbide bar within the metal enclosure. If you do, it will create a hot spot on the silicon carbide bar (from body oils), and the ignitor might burn out.

Pressure Regulator Valve

The typical complaints associated with the pressure regulator valve are:

- The oven / broiler burner will not ignite. No gas is supplied to the burner.
- Erratic oven / broiler burner flame. The flame will fluctuate between high and low.

To handle these problems, perform the following steps:

1. **Verify the complaint.** Verify the complaint by turning on the appliance. Do the surface burners come on? Are the burner flames small and blue in color? Did you check the gas pressure at the gas supply line, surface burners, and the oven burner? Do the burner flames fluctuate in size? Is the pilot light staying lit? Is there a gas smell near the pressure regulator valve?

2. **Check for external factors.** You must check for external factors not associated with the appliance. Is the appliance installed properly? Are there 120 volts to the appliance? Is the gas to the appliance turned on?

3. **Disconnect the electricity.** Before working on the range or cooktop, disconnect the electricity to the appliance. This can be done by pulling the plug from the receptacle. Or disconnect the electricity at the fuse panel or at the circuit breaker panel. Turn off the electricity.

4. **Shut off the gas supply.** Before you begin servicing any gas components, shut off the supply of gas to the appliance.

5. **Gain access to the pressure regulator valve.** Locate the pressure regulator valve. On some models, lift the cooktop or gain access through the oven cavity (see Figures 19-3, 19-11, 19-14, and 19-19).

6. **Test the pressure regulator.** The pressure regulator valve is a non-serviceable component. If you suspect a problem with the regulator, test the gas pressures with a manometer or a magnehelic gauge (see Figures 6-22 and 6-24). Test the gas pressure at the supply line before the pressure regulator valve and at the surface burner orifices. On some models using natural gas, the reading will be between a four- and six-inch water column. For LP models, the pressure will be between a nine- and eleven-inch water column. If the readings are not within the ranges mentioned, replace the pressure regulator.

7. **Remove the gas pressure regulator valve.** To remove the gas pressure valve (see Figure 19-14), pull out the range from the wall and lift up the cooktop. Disconnect the gas supply line from the range. Remove the gas line from the intake side of the regulator. Remove the surface burners, pull off the control knobs, and then remove the screws from the front panel (see Figure 19-14). This will expose the gas burner valves, thermostat, and manifold. Next, remove the thermostat from the manifold. If the model you are servicing has electronic ignition, remove the wires from the

ignitor switches and label them. Finally, remove the screws that secure the manifold to the range. Figure 6-20 shows an exploded view of the gas components. Remove the manifold with the pressure regulator. Be careful—do not damage the capillary tube in the thermostat. To remove the gas pressure regulator valve (see Figures 19-11, 19-13, and 19-19), disconnect the gas line to the regulator. Remove the oven burner and ignitor. Next, remove the screws that secure the safety valve bracket to the range, and remove the safety valve with the regulator attached.

8. **Install the new gas pressure regulator.** To install the gas pressure regulator, just reverse the disassembly procedure, and reassemble. Make sure that you use approved pipe joint compound or Teflon tape on the pipe connections (according to local codes) and on the threads of the pressure regulator valve before you install the valve in the range. Turn on the gas supply to the range. Check for gas leaks by using a chloride-free soap solution. *Do not use a flame to check for gas leaks.* The soap solution will begin to bubble if a leak is present.

Safety Valve

The typical complaints associated with the safety valve are:

- The oven /broiler burner will not ignite. No gas is supplied to the burner.
- Broken capillary tube.

To handle these problems, perform the following steps:

1. **Verify the complaint.** Verify the complaint by turning on the appliance. Does the oven or broil burner come on?

2. **Check for external factors.** You must check for external factors not associated with the appliance. Is the appliance installed properly? Are there 120 volts to the appliance? Is the gas to the appliance turned on?

3. **Disconnect the electricity.** Before working on the range or cooktop, disconnect the electricity to the appliance. This can be done by pulling the plug from the receptacle. Or disconnect the electricity at the fuse panel or at the circuit breaker panel. Turn off the electricity.

4. **Shut off the gas supply.** Before you begin servicing any gas components, shut off the supply of gas to the appliance.

5. **Gain access to the safety valve.** To gain access to the safety valve (see Figure 19-20), open the oven door, remove the oven racks, and remove the oven floor panel. Next, remove the nut (see Figure 19-19) from the oven burner baffle, and remove the baffle. You now have access to all of the components shown in Figure 19-19.

6. **Test the safety valve.** The standing pilot safety valve cannot be tested: it is a temperature-controlled device. If you notice that the capillary tube is broken, replace the safety valve. Other models use an electrically controlled safety valve (bi-metaloperated). To test this type of valve (see Figure 19-20), set your ohmmeter to the R × 1 scale. Disconnect the wires from the valve terminals, and touch the ohmmeter probes to the terminals. The reading should be between 1 and 5 ohms. If your meter reads an open circuit or high resistance, the safety valve will have to be replaced.

Figure 19-20 (a) A standing pilot safety valve. (b) A combination bake/broil safety valve. (c) A dual gas safety valve with the pressure regulator attached.

7. **Remove the safety valve.** To remove the gas safety valve (see Figures 19-11, 19-13, and 19-19), disconnect the gas line to the regulator. Remove the oven burner and ignitor. Next, remove the screws that secure the safety valve bracket to the range, and remove the safety valve with the regulator attached.

8. **Install the new safety valve.** To install the safety valve, just reverse the disassembly procedure, and reassemble. Make sure that you use approved pipe joint compound or Teflon tape on the pipe connections (according to local codes) and on the threads of the pressure regulator valve and the safety valve before you install the valve(s) in the range. Turn on the gas supply to the range. Check for gas leaks by using a chloride-free soap solution. *Do not use a flame to check for gas leaks.* The soap solution will begin to bubble if a leak is present.

Electronic Control Board and User Interface Controls

The typical complaints associated with the electronic control board or the user interface controls are:

- The range or oven won't run or power up.
- Unable to program the range or oven.
- The display board will not display anything.
- One or more key pads will not accept commands.
- Unusual display readouts.

To prevent electrostatic discharge (ESD) from damaging expensive electronic components, follow the steps in Chapter 9.

To handle these problems, perform the following steps:

1. **Verify the complaint.** Verify the complaint by operating the range or oven. Turn off the electricity to the appliance and wait for two minutes before turning it back on. If a fault code appears, look up the code. If the range or oven will not power up, locate the technical data sheet behind the control panel for diagnostics information.

2. **Check for external factors.** You must check for external factors not associated with the appliance. Are there 120 volts of electricity to the range or oven?

3. **Disconnect the electricity.** Before working on the range or oven, disconnect the electricity. This can be done by pulling the plug out of the wall receptacle. Or disconnect the electricity at the fuse panel or circuit breaker panel. Turn off the electricity.

WARNING *Some diagnostic tests will require you to test the components with the power turned on. When you disassemble the control panel, you can position the panel in such a way that the wiring will not make contact with metal. This will allow you to test the components without electrical mishaps.*

4. **Shut off the gas supply.** Before you begin servicing any gas components, shut off the supply of gas to the appliance.

5. **Remove the console panel to gain access.** You can access the electronic control board (some models have multiple circuit boards) by removing the back panel (see Figure 18-6) on a freestanding range. On models with front-mounted controls (see Figure 18-7), the panel is attached with screws on both ends. Remove the screws, and tilt the control panel. Be careful not to let the wires come off their components. Some built-in models have a removable backsplash (see Figure 18-8); just lift the backsplash and rest it on the cooktop. It would be a good idea to place something on the cooktop first to protect it from damage. Next, remove the screws from the backsplash that hold the rear panel to gain access to the electronic control board. If you are repairing a wall oven or an eye-level range, the control panel can be removed (see Figure 18-9) by opening the door and removing the screws that secure the panel. These might be underneath the front of the exhaust hood or just below the control panel. Some control panels are hinged; just tilt the control panel toward you for servicing. On other models, to gain access to the electronic control board, both the rear panel and the front control panel (usually glass) will have to be removed. Remove the screws that secure the trim piece that holds the glass in place.

6. **Test the electronic control board and user interface control.** If you are able to run the range/oven diagnostic test mode, check the different functions. Use the technical data sheet for the model you are servicing to locate the test points on the wiring schematic. Check all wiring connections and wiring. Using the technical data sheet, test the electronic control or user interface controls, input voltages, and output voltages. On some models, fuses are soldered to the printed circuit board (PCB). These fuses must be tested first before condemning the component.

7. **Remove the electronic control board or user interface control.** To remove the defective component, remove the screws that secure the boards to the control panel or range/oven frame. Disconnect the connectors from the electronic control board or user interface control.

8. **Install the new component.** To install a new electronic control board or user interface control, just reverse the disassembly procedure, and reassemble. Reinstall the console panel, and restore the electricity to the range/oven. Test the range/oven operation.

Oven Temperature Sensor

The typical complaints associated with the oven temperature sensor are:

- Erratic temperature in the oven cavity.
- The oven/broiler burner will not ignite.
- Error code flashing.

To handle these problems, perform the following steps:

1. **Verify the complaint.** Verify the complaint by turning on the appliance. Does the oven operate? Test the oven temperature.

2. **Check for external factors.** You must check for external factors not associated with the appliance. Is the appliance installed properly? Are there 120 volts to the appliance? Is the gas to the appliance turned on?

3. **Disconnect the electricity.** Before working on the range or cooktop, disconnect the electricity to the appliance. This can be done by pulling the plug from the receptacle. Or disconnect the electricity at the fuse panel or at the circuit breaker panel. Turn off the electricity.

4. **Shut off the gas supply.** Before you begin servicing any gas components, shut off the supply of gas to the appliance.

5. **Gain access to the oven temperature sensor.** To gain access, open the oven door. The temperature sensor is located in the upper corners of the oven cavity.

6. **Remove the oven temperature sensor.** Remove the oven racks. Remove the two screws that secure the sensor to the oven cavity wall. Pull the sensor toward you gently, disconnect the connector, and remove the temperature sensor from the oven. There may be times when the connector will not pass through the hole in the oven cavity. If this is the case, then you will have to access the connector from the rear of the range/oven.

7. **Test the oven temperature sensor.** Set your ohmmeter on the R × 10 scale, and measure the resistance between the two pin connectors on the sensor wire connector. For most sensors, the meter reading at room temperature should match the chart in Figure 19-21. For example: If the room temperature is approximately 75 degrees Fahrenheit, the resistance in the sensor will be 1100 ohms. If the reading is less than 1000 ohms, replace the sensor. If the reading is above 1100 ohms at 75 degrees, replace the sensor.

8. **Install the new oven temperature sensor.** To install the oven temperature sensor, just reverse the disassembly procedure, and reassemble. Test the oven temperature. With an oven temperature tester, place the thermocouple tip in the center of the oven cavity. Be sure that the thermocouple tip does not touch any metal. Close the

FIGURE 19-21
A typical oven temperature sensor. This type of sensor is only used in ranges and ovens with electronic controls.

Oven temperature sensor   Connector

Temperature (°F)	Resistance (Ω)
32	1000
75	1100
250	1450
350	1650
450	1850
550	2050
650	2230
900	2700

oven door, set the oven to bake, and adjust the thermostat setting to the 350-degree mark. Let the oven cycle for 20 to 30 minutes. Then record the minimum and maximum temperatures of three cycles. Next, add these temperatures, and divide by 6. This will give you the average temperature of the oven:

$$\frac{370 + 335 + 350 + 340 + 360 + 335}{6} = 348.3 \text{ degrees Fahrenheit}$$

The average temperature calculated should be within ±5 degrees of the temperature setting selected. Oven temperature sensors (RTDs) are used with electronically controlled ovens. To make any temperature adjustments in the electronic control, the technician needs to locate the technical data sheet. This sheet will show you how to make the temperature adjustments for the model you are servicing.

Thermostat (Standing Pilot)

The typical complaints associated with the thermostat are:

- The oven/broiler burner will not come on.
- Erratic oven cavity temperatures.

To handle these problems, perform the following steps:

1. **Verify the complaint.** Verify the complaint by turning on the appliance. Does the oven operate? Test the oven temperature.
2. **Check for external factors.** You must check for external factors not associated with the appliance. Is the appliance installed properly? Are there 120 volts to the appliance? Is the gas to the appliance turned on?
3. **Disconnect the electricity.** Before working on the range or cooktop, disconnect the electricity to the appliance. This can be done by pulling the plug from the receptacle. Or disconnect the electricity at the fuse panel or at the circuit breaker panel. Turn off the electricity.
4. **Shut off the gas supply.** Before you begin servicing any gas components, shut off the supply of gas to the appliance.
5. **Gain access to the oven thermostat.** To gain access to the oven thermostat, lift up the cooktop, pull off the control knobs, and then remove the screws from the front panel (see Figure 19-14). Remove the control panel. This will expose the gas burner valves, thermostat, and manifold (see Figures 6-20 and 19-5).
6. **Remove the thermostat.** Before removing the thermostat, remove the capillary tube and sensor from the oven cavity. The sensing bulb will be attached by two retaining clips. Squeeze the clips, pull away from the oven cavity, and remove the sensing bulb and capillary tubing. Feed the capillary tubing through the oven cavity to the top of the range. Disconnect the main gas line and the pilot gas line from the thermostat. Then remove the screws that secure the thermostat to the manifold (see Figures 6-20 and 19-5).
7. **Install the new thermostat.** To install the thermostat, just reverse the disassembly procedure, and reassemble. Make sure that you use approved pipe joint compound or Teflon tape on the pipe connections (according to local codes) and on the threads

of the gas lines (if needed) before you install the thermostat on the manifold. Turn on the gas supply to the range. Check for gas leaks by using a chloride-free soap solution. *Do not use a flame to check for gas leaks.* The soap solution will begin to bubble if a leak is present.

8. **Test the oven temperature.** With an oven temperature tester, place the thermocouple tip in the center of the oven cavity. Be sure that the thermocouple tip does not touch any metal. Close the oven door, set the oven to bake, and adjust the thermostat setting to the 350-degree mark. Let the oven cycle for 20 to 30 minutes. Then record the minimum and maximum temperatures of three cycles. Next, add these temperatures, and divide by 6. This will give you the average temperature of the oven:

$$\frac{370 + 335 + 350 + 340 + 360 + 335}{6} = 348.3 \text{ degrees Fahrenheit}$$

The average temperature calculated should be within ±25 degrees of the temperature setting selected. If not, try calibrating the thermostat. To calibrate the thermostat, pull the oven thermostat knob off, and look at the thermostat (see inset in Figure 19-22). The slotted screw on the thermostat is for converting the thermostat for the type of gas supplied to the range. Set it to the proper setting. If you look into the thermostat stem, you will see a small adjustment screw. This screw is used for calibrating the thermostat. Turning the screw slightly to the right will increase the temperature. To decrease the temperature setting, turn the screw to the left slightly. For every adjustment made, you must test the oven temperature as stated above.

FIGURE 19-22
When converting the range to the type of gas that is supplied to the home, you must convert the thermostat as well.

Range/Oven Time Clock

See Chapter 18 for testing and removal procedures.

Flame Adjustments

There may be times when the surface, oven, or broil burner flame needs to be adjusted. The proper surface burner flame should be about 5/8 inch high with a well-defined blue flame. If the gas pressure to the burner is correct, all you have to do is adjust the air shutter on the surface burner (see Figure 19-16) to obtain the correct size and color. The oven and broil burners' flame should be about 3/4 inch high with a well-defined blue flame. With the correct gas pressure to the burners, all that is needed is to adjust the air shutters on the burners (Figures 19-13 and 19-23) to obtain the correct size and color. For sealed burner cooktops, you cannot make any air adjustments.

Diagnostic Charts and Wiring Diagrams

The following diagnostic flow charts will help you to pinpoint the likely causes of the various problems associated with gas ranges and ovens (Figures 19-24 and 19-25). The wiring diagrams in this chapter are examples only. You must refer to the actual wiring diagram for the range, oven, or cooktop that you are servicing. Figure 19-26 depicts samples of actual strip circuit diagrams. The strip circuit diagram is simpler and is usually easier to read. Figures 19-18 and 19-27 depict actual wiring schematic diagrams.

FIGURE 19-23
Adjusting the air shutter on the oven burner.

Bake burner air shutter

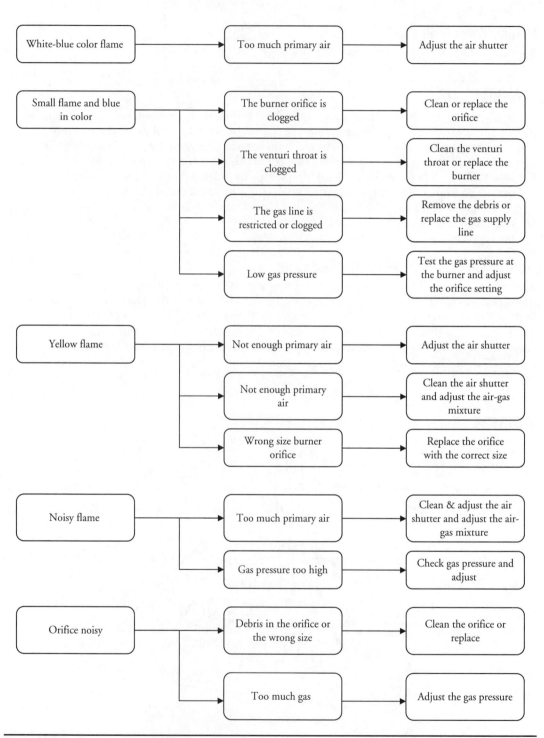

FIGURE 19-24   Flow chart identifying the causes of various flame characteristics. (*continued*)

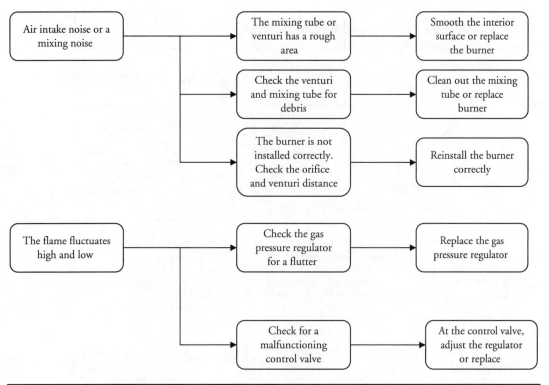

Figure 19-24   Flow chart identifying the causes of various flame characteristics.

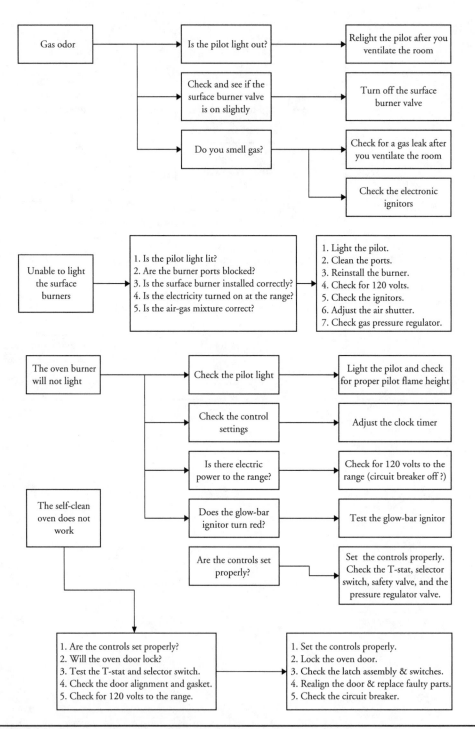

FIGURE 19-25 Diagnostic flow chart: Gas odor; unable to light the surface burners; the oven burner will not light; the self-cleaning feature does not work.

FIGURE 19-26   The strip circuit diagram is easy to read and understand.

FIGURE 19-27 A typical wiring schematic for a self-cleaning range with electronic controls.

Microwave Ovens

Microwave ovens have been around since the late 1940s, and their popularity increased in the mid-1970s. Over the years, the cost to purchase a microwave oven has decreased. Today, you can purchase a microwave oven for under $50, as compared to the 1940s price of around $3000.

Microwave ovens produce electromagnetic non-ionizing radiation to cook the food, operating at approximately 3000 volts. They operate in the microwave region in the electromagnetic spectrum at a frequency of 2450 MHz. For domestic use, the microwave oven (see Figures 1-6, 20-1, 20-2, 20-3, and 20-4) is manufactured in different sizes and styles, countertop, hanging, and built-in models with electromechanical or electronic controls.

Microwave energy can travel at the speed of light in straight lines; it is produced in the oven by a device called a magnetron tube and reflected by metal channels to enter the oven cavity. Microwaves can pass through non-metallic objects and be absorbed into the food.

This chapter will provide the basics needed to diagnose and repair the components in a microwave oven. Figure 20-5 identifies where various components are typically located within the microwave oven. However, this illustration is used as an example only. The actual construction and features might vary, depending on the brand and model you are servicing.

Safety First

Any person who cannot use basic tools or follow written instructions should *not* attempt to install, maintain, or repair any microwave oven. Any improper installation, preventive maintenance, or repairs could create a risk of death, personal injury, death to others, or property damage.

If you do not fully understand the installation, preventive maintenance, or repair procedures in this chapter, or if you doubt your ability to complete the task on the microwave, do not attempt to make the repairs. Call your service manager instead.

The following precautions should also be followed:

- Do not operate the microwave with the door open.
- Do not operate the microwave if the glass in the door is broken.
- Do not allow children under the age of 10 to operate a microwave oven.
- Do not jump out or bypass any components to operate the microwave oven.

FIGURE 20-1
A countertop
microwave oven
with electronic
controls.

- Do not operate the microwave oven if the door hinges and door assembly are loose.

- Always perform a microwave leakage test when repairs are completed.

- Before you begin to service any high-voltage components within the microwave oven, you must discharge the high-voltage capacitors and inverter first.

- Never touch any wires or components with your hands or tools when the microwave oven is operating.

- Never run the microwave oven with the oven cavity empty. The microwave energy needs to be absorbed into food or a liquid; otherwise, the energy will bounce around, making its way to the high-voltage system and causing damage.

Before continuing, take a moment to refresh your memory of the safety procedures in Chapter 2.

FIGURE 20-2
A countertop
microwave
oven with
electromechanical
controls.

FIGURE 20-3
Over-the-range
microwave ovens
are available in
various colors
and features.

Principles of Operation

The microwave oven circuitry consists of two primary circuits divided by the high-voltage transformer. The high-voltage circuit consists of the secondary winding of the high-voltage transformer (or inverter on some models), magnetron, high-voltage capacitor, and the high-voltage diode. The low-voltage circuit consists of the primary side of the high-voltage transformer, fans, lights, switches, timer, motors, fuses, and a low-voltage transformer that controls the electronic components in some models.

When the microwave oven is turned on and running, the low-voltage alternating current (120 VAC) is stepped up to a high-voltage alternating current (1800 VAC) and converted into an even higher direct current (3000 to 4500 VDC).

FIGURE 20-4
An under-the-
counter microwave
oven.

A. Oven cavity thermostat	H. Cavity thermostat	O. High-voltage capacitor	V. Touch panel
B. Humidity sensor	I. Fuse holder	P. High-voltage diode	W. Interlock switch (secondary)
C. Exhaust hood fan motor assembly	J. Resistor	Q. Lamp holder	X. Interlock switch (monitor)
D. Cavity thermostat	K. Magnetron	R. Cooktop lamp	Y. Interlock switch (primary)
E. Fuse	L. Magnetron thermostat	S. Exhaust fan thermostat	Z. Turntable motor
F. Fuse	M. High-voltage transformer	T. Fuse	AA. Cavity lamp
G. Motor capacitor	N. Cooling fan motor	U. Electronic control	BB. Stirrer motor

FIGURE 20-5   An exploded view of an over-the-range microwave oven, illustrating the various component locations.

WARNING *Extreme care must be taken when handling high-voltage components.*

The direct current travels to the magnetron. The output of the magnetron produces microwave energy out of the antenna into the waveguide. From the waveguide, the non-ionizing energy will then be channeled to the stirrer blade. The stirrer blade will begin to spin around, dispersing the microwaves into the oven cavity evenly. Then the microwaves will reflect off the interior walls, ceiling, and floor of the oven cavity, penetrating the food.

Cooking the food is achieved by three methods: conduction, convection, and radiation. The microwave energy will penetrate the food approximately one inch, cooking the food from the outside toward the center. The molecules within the food will begin to resonate violently, producing heat to cook the food. When the cooking cycle is completed, the oven will turn off and the magnetron will stop producing non-ionizing energy.

Location and Installation of Microwave Ovens

Locate the microwave oven where it will be well lighted and have access to proper ventilation. The oven must be level for proper baking and cooking results. The microwave oven must also have the proper electrical and grounding requirements, as described in the manufacturer's installation instructions.

The proper installation instructions for your model are included with your purchase. These instructions will assist you with the installation requirements (dimensions, cutout dimensions, venting, etc.) needed to complete the installation according to the manufacturer's specifications.

Microwave Oven Cookware, Utensils, and Coverings

The following "do not use" list for microwave cooking should be followed for safety reasons:

- Do not use metal pots, pans, and bakeware.
- Do not use dishes with metallic trim.
- Do not use non-heat-resistant glass.
- Do not use non-microwave-safe plastics.
- Do not use recycled paper products.
- Do not use brown paper bags.
- Do not use food storage bags.
- Do not use metal twist-ties.

Any of these items can cause arcing, fire, and damage to the microwave oven. If you are uncertain if a dish is safe for microwaving, place the dish in the microwave on high for 30 seconds. If the dish is hot, it is unsafe for microwaving and should not be used.

The following items are safe to use in microwave ovens:

- Glass ceramic manufactured out of PyroCeram (for example, CorningWare products)
- Pyrex heat-resistant glass products
- Any plastic product with "microwave-safe" stamped on it
- Paper plates, paper towels, wax paper
- Any pottery, porcelain, or stoneware that has "microwave-safe" stamped on it
- A browning dish (follow the manufacturer's directions)
- Wood, straw, and wicker for short-term reheating purposes only
- Plastic wrap that is approved for microwave use only
- Oven cooking bags approved for microwave cooking

PART V

If you are not sure if the item can be used in the microwave oven, check your use and care instructions or go to the manufacturer's Web site.

Can aluminum foil be used in the microwave? Yes, it can. Small, flat pieces of aluminum foil can be placed on the food in areas that cook too quickly to prevent the food from overcooking. When using aluminum foil, you must make sure to keep it from contacting any part of the oven's interior surfaces (about one inch clearance to prevent arcing). For best cooking results, always follow the manufacturer's recommendations in the use and care guide.

Microwave Ovens in General

Much of the troubleshooting information in this chapter covers microwave ovens in general, rather than specific models, in order to present a broad overview of service techniques. The illustrations that are used in this chapter are for demonstration purposes only to clarify the description of how to service these appliances. They in no way reflect on a particular brand's reliability.

Step-by-Step Troubleshooting by Symptom Diagnosis

When servicing an appliance, don't overlook the simple things that might be causing the problem. Step-by-step troubleshooting by symptom diagnosis is based on diagnosing malfunctions, with their possible causes arranged into categories relating to the operation of the microwave oven. This section is intended only to serve as a checklist to aid you in diagnosing a problem. Look at the symptom that best describes the problem that you are experiencing with the oven, and then proceed to correct the problem.

Microwave Oven Dead in the "Off" Position

- Check for the proper voltage and grounding to the oven.
- Check the fuses in the microwave oven.
- Test the thermal fuses within the oven.
- Test the electronic control board or electromechanical timer.
- Check the power supply.
- Check the door alignment and the door interlock switches.
- Check the service cord, wiring, and wiring connectors.
- Test all of the high-voltage components.

Microwave Oven Will Not Operate After the Controls Are Set

- Check for proper voltage and grounding to the oven.
- Check the door interlock switches.
- Make sure that the door closes properly, with the latch assembly engaging the interlock switches.
- Test the electronic control board or electromechanical timer.
- Check the power supply.

- Check all the fuses.
- Check the start button.
- Try reprogramming the control panel. The programming may have timed out.
- Check the wiring and wiring connectors.
- Test the service cord for an open or short circuit.
- On electronic models, run the diagnostic test mode. Check for fault codes.

Unable to Program the Touchpad Controls

- Does the oven door close properly?
- Disconnect the voltage and wait two minutes. Can you reprogram the touchpad control?
- Check the oven door interlock switches.
- Test the electronic control board and the touchpad panel.

Microwave Oven Works, but There Is No Display

- Check the electronic control board.
- Check the display board and power supply.
- Check the wiring and wiring connectors.
- Disconnect the electricity and wait two minutes.
- Try resetting the microwave controls.

Microwave Oven Still Runs with the Door Open

- Check the oven door interlock switches.
- Check the electromechanical timer.
- Check the electronic controls for a stuck relay contact.
- Check the electromechanical timer.
- Check for a bad sensor.

Microwave Oven Starts Running as Soon as the Door Is Closed

- Check the microwave relay or triac.
- Check the electronic control board or electromechanical timer.
- Test the power supply.
- Test the touchpad panel.

Microwave Oven Is Running but Will Not Cook

- Check the magnetron.
- Test the diode.
- Check the triac or relay.
- Check the thermal fuse.

- Check the high-voltage transformer fuse.
- Check the door interlock switches.
- Check the high-voltage capacitor.
- Check the electronic control board or electromechanical timer.
- On electronic models, run the diagnostic test mode. Check for fault codes.

Microwave Oven Is Running, but It Takes Too Long to Cook the Food

- Check for proper voltage to the oven.
- Check the thermostats.
- Check the high-voltage transformer or inverter.
- Check the magnetron.
- Check the diode.
- Check the triac or relay.
- Check the electronic control board.
- Check the fans.
- Check the stirrer blade. Is it turning?
- Check the turntable. Does it rotate?
- Check the vents for blockage.
- On electronic models, run the diagnostic test mode. Check for fault codes.

Microwave Oven Is Running, but the Operation Is Erratic

- Test the voltage to the microwave oven.
- Check the magnetron performance.
- Check the electronic control board or electromechanical timer.
- Is the stirrer blade turning?
- Check the high-voltage transformer or inverter.
- Check the wiring and wiring connections.
- Check the diode.
- Check the thermostats.
- Check the turntable operation.
- Is the oven cavity clean?
- Check the door alignment and check for microwave leakage.
- On electronic models, run the diagnostic test mode. Check for fault codes.

Microwave Oven Keeps Blowing Fuses

- Check the door interlock switches.
- Check the oven door and the door latch assembly for proper operation.

- Check the diode.
- Check for a shorted magnetron.
- Check the high-voltage transformer.
- Check the high-voltage capacitor.
- Check the triac or relay.
- Check the incoming voltage supply for power surges.

Microwave Oven Is Running, but Will Not Cook and Makes a Loud Humming Noise

- Check the triac or relay.
- Check the diode.
- Check the magnetron.
- Check the high-voltage transformer or inverter.
- Check the wiring or wiring connections.

Microwave Oven Has a Burning Smell

- Ask the customer what was cooking in the oven.
- Inspect in and above the oven cavity for carbonized food or grease.
- Inspect the wiring and wiring connections.
- Inspect the waveguide and cover for grease.
- Check the high-voltage system.

Microwave Oven Is Arcing

- Ask the customer if aluminum foil or any other metal product was used for cooking.
- Inspect the oven cavity.
- Check for arcing in the waveguide.
- Check the magnetron.
- Check for any loose screws in the door.
- Inspect for sharp exposed metal edges.

Microwave Oven Is Very Noisy

- Check the fan blade(s) and assembly.
- Check the stirrer blade. Is it turning?
- Check for unusual noises coming from the high-voltage transformer.
- Check the turntable motor.
- Check the electromechanical timer motor.

Microwave Oven Door Will Not Close Properly

- Check the door latch assembly and catch.
- Check the door hinges.

- Check the door alignment.
- Check for loose or missing screws in the door.

Electromechanical Timer Will Not Advance in the Cooking Cycle

- Check for voltage to the timer motor.
- Check wiring and wiring connections.
- Test the motor on the timer for an open winding.
- Does the timer knob turn easily or is it jammed?

Microwave Oven Performs the Wrong Functions

- Ask the customer if the cycle was interrupted and then someone else tried to run the oven.
- Disconnect the electricity to the oven; then turn on the electricity and try to reprogram the controls. Does it operate okay?
- Check the electronic control board.
- Check the power supply to the electronic control board.
- Test the touchpad control panel and display board for proper operation.
- Check all sensors.
- On electronic models, run the diagnostic test mode. Check for fault codes.

Touchpad Does Not Function or the Wrong Commands Appear

- Turn off the electricity to the oven, wait two minutes, turn the electricity back on, and try to reprogram the electronic controls.
- Check the wiring and wiring connections from the touchpad membrane to the electronic control board.
- Test the touchpad membrane circuitry.
- Check the electronic control board.
- On electronic models, run the diagnostic test mode. Check for fault codes.

Microwave Oven Turntable Is Not Working

- Check the wiring and wiring connections.
- Is the glass tray installed properly?
- Check the turntable motor.
- Check the thermostat.
- Check the electronic control board.
- On older models, check the turntable bearings.

Microwave Oven Runs Okay, but Randomly Shuts Off During the Cooking Cycle

- Check the cooling fan motor.
- Check the stirrer blade. Is it turning?

- Check the stirrer blade motor.
- Check the air vents for a blockage.
- Check the thermostats and check the sensors.
- Check the door alignment and the door interlock switches.
- Check for proper voltage to the oven.
- Check the electronic control board.
- Check the wiring and wiring connections.
- Check the magnetron.
- On electronic models, run the diagnostic test mode. Check for fault codes.

Stirrer Blade Will Not Rotate

- Check for obstructions.
- Check the stirrer blade. Is it bent or broken?
- Check the stirrer motor.
- Check for a loose or broken belt.
- Check the wiring and wiring connections.

Microwave Oven Is in Cook Mode and Running, but the Light Is Out in the Oven

- Check the light bulb.
- Check the light socket.
- Check the wiring and wiring connections.

Microwave Oven Only Heats on High Power

- Check the control settings.
- Check the electronic control board.
- Check the electromechanical timer.
- On older models, check for a stuck switch.
- Check the triac or relay.
- On electronic models, run the diagnostic test mode. Check for fault codes.

Microwave Oven Maintenance

The maintenance on a microwave oven is simple: keep the interior and exterior clean. Make sure that the air vents are clear of debris for proper ventilation. The interior can be cleaned with a damp cloth and a mild detergent. If food particles are allowed to collect in the oven cavity, on the stirrer blade or cover, or waveguide, the food will become carbonized, causing arcing, improper heating, and possible damage to the components. When cleaning the control panel, use a damp cloth only, and do not use a spray cleaner or harsh chemical on the touch panel. This will prevent the touch panel from shorting out or preventing some of the keypad functions from working properly. Inspect the service cord and outlet; check the prongs of the plug for a tight fit. A loose connection can cause the service cord to heat up and short out. This is especially important on built-in models, given the amount of heat that some of these ovens can generate. It is also recommended that the consumer keep the area around the microwave oven clean.

After cooking in the microwave oven, the smell of food lingers for quite a while. There have been cases where cockroaches have penetrated into the microwave oven, traveling to the electronics area, and eating the wires and the metal traces off the printed circuit boards, causing expensive repairs. This can be prevented by placing a fine mesh over the intake and discharge vent areas. You might have to clean the mesh more often, however, because dust and dirt will build up on the mesh much faster.

Recipe for Microwave Oven Odor Removal

To remove cooking odors from the microwave, combine one cup of water, grated lemon peels, lemon juice, and several slices of lemon. Stir and place in a microwave-approved glass container. Cook on high power, and let the mixture boil for several minutes. When the oven shuts off, let the mixture cool, remove the container from the oven, and wipe the interior with a damp cloth.

Microwave Leakage Testing

On every microwave oven service call, you must perform a radio frequency (RF) leakage test, before and after the repairs, to determine that the energy emission level is below 4 mw/cm^2 at a distance of five centimeters, with a scan rate not to exceed 2.54 cm/second. Make sure that you use an approved microwave leak detector for testing.

Take a glass and fill it with water, about 275 ml (±15—just under 9.5 ounces). Place it in the center of the floor in the oven cavity. Set the microwave power level to high, set the timer for three minutes, and press the start button. Take the RF leakage detector probe, and scan the following areas at a rate of one inch per second:

- Scan around the front of the cabinet
- Scan around the oven door
- Scan horizontally across the oven door
- Scan vertically across the oven door
- Scan across the control panel
- Scan diagonally across the oven door
- Scan across the oven intake and discharge vents
- Scan all seams and welds
- Scan the bottom plate (for hanging microwave ovens)
- Scan the microwave oven feet

WARNING *Never let a consumer operate a microwave oven if it fails the RF leakage test.*

Microwave Oven Magnetron Power Output Test

The magnetron's performance can be tested for proper operation by performing a power output test in the following manner:

1. The oven cavity must be cool and clean. Inspect the waveguide and stirrer blade.

2. The voltage to the oven should be around 120 volts (±10 percent).

3. Place a 8-ounce glass of water heated to about 75 degrees Fahrenheit in the middle of the cavity floor (remove the metal rack).

4. Take a thermometer and stir the water for 30 seconds. Record the water temperature.

5. Set the oven controls to 100 percent power and the timer to one minute, and press the start button.

6. When the oven turns off, remove the glass, and stir the water with the thermometer for 30 seconds. Record the water temperature.

7. Subtract the two recorded water temperatures to determine the temperature rise.

The temperature rise should be between 19 and 26 degrees Fahrenheit at 120 volts (±10 percent), indicating that the magnetron output power is operating properly.

Repair Procedures

Each of the following repair procedures is a complete inspection and repair process for a single microwave oven component, containing the information you need to test a component that might be faulty and to replace it, if necessary.

Electronic Control Board, Touchpad Panel, or Display Board

The typical complaints associated with the electronic control board, touchpad panel, or display board are:

- The microwave oven will not run or power up.
- Unable to program the touchpad panel functions.
- The display board will not display anything.
- One or more key pads will not accept commands.
- Unusual display readouts.

To prevent electrostatic discharge (ESD) from damaging expensive electronic components, follow the steps in Chapter 9.

To handle these problems, perform the following steps:

1. **Verify the complaint.** Verify the complaint by operating the microwave oven. Turn off the electricity to the appliance and wait for two minutes before turning it back on. If a fault code appears, look up the code. If the microwave oven will not power up, locate the technical data sheet behind the control panel or cabinet for diagnostics information.

2. **Check for external factors.** You must check for external factors not associated with the appliance. Is there electricity to the microwave oven?

3. **Disconnect the electricity.** Before working on the microwave oven, disconnect the electricity. This can be done by pulling the plug out of the wall receptacle. Or disconnect the electricity at the fuse panel or circuit breaker panel. Turn off the electricity.

WARNING *Before you begin to service any high-voltage components within the microwave oven, you must discharge the high-voltage capacitors and inverter first.*

WARNING *Some diagnostic tests will require you to test the components with the power turned on. When you disassemble the control panel or remove the outer cabinet, you can position the panel in such a way that the wiring will not make contact with metal. This will allow you to test the components without electrical mishaps.*

4. **Remove the outer cabinet or control panel to gain access.** You can gain access to the electrical components, touchpad panel, or electronic control board (some models have multiple circuit boards) by removing the access panel or the outer cabinet on a microwave oven or combination oven. If you are repairing a wall oven or an eye-level combination range, the control panel can be removed (see Figures 9-1 and 18-9) by opening the door and removing the screws that secure the panel. These might be underneath the front of the exhaust hood or just below the control panel. Some control panels are hinged; just tilt the control panel toward you for servicing.

5. **Test the electronic control board, touchpad panel, or display board.** If you are able to run the microwave oven's diagnostic test mode, check the different functions of the oven. Use the technical data sheet for the model you are servicing to locate the test points on the wiring schematic. Check all wiring connections and wiring. Using the technical data sheet, test the electronic control and display board, input voltages, and output voltages. On some models, fuses are soldered to the printed circuit board (PCB). These fuses must be tested before condemning the component. The touchpad panel can also be tested for proper operation by checking the continuity for each keypad function.

6. **Remove the defective component.** To remove the defective component, remove the screws that secure the boards to the control panel or oven frame. Disconnect the connectors from the electronic control board, touchpad membrane, or display board.

7. **Install the new component.** To install a new component, just reverse the disassembly procedure, and reassemble. Reinstall all panels or the console panel, and restore the electricity to the oven. Test the microwave oven operation. Perform the microwave performance and leakage test.

Magnetron

The typical complaints associated with the magnetron are:

- Unable to cook the food.
- Failed magnetron power output test.
- Noisy operation.

FIGURE 20-6
Place a glass of water in the microwave cavity along with the microwave test light bar.

To handle these problems, perform the following steps:

1. **Verify the complaint.** Verify the complaint by operating the microwave oven with a glass of water and the microwave test light bar centered on the oven floor (Figures 20-6 and 20-7). The lights in the light bar will light up if the magnetron and high-voltage system are working correctly. At the same time, place an ammeter on one wire of the service cord—the meter should indicate approximately 15 amps at 100 percent power. It may be necessary to turn off the electricity to the appliance and wait for two minutes before turning it back on. If a fault code appears, look up the code. If the microwave oven will not power up, locate the technical data sheet behind the panel or cabinet for diagnostics information.

2. **Check for external factors.** You must check for external factors not associated with the appliance. Is there electricity to the microwave oven? With over-the-range models, is the vent system blocked?

3. **Disconnect the electricity.** Before working on the microwave oven, disconnect the electricity. This can be done by pulling the plug out of the wall receptacle. Or disconnect the electricity at the fuse panel or circuit breaker panel. Turn off the electricity.

FIGURE 20-7
The microwave oven door is closed, the controls are set, and the oven is on. This illustration shows that the microwave test lights are on, indicating that the magnetron and the high-voltage system are functioning correctly.

WARNING *Before you begin to service any high-voltage components within the microwave oven, you must discharge the high-voltage capacitors and inverter first.*

WARNING *Some diagnostic tests will require you to test the components with the power turned on. When you disassemble the control panel or remove the outer cabinet, you can position the panel in such a way that the wiring will not make contact with metal. This will allow you to test the components without electrical mishaps.*

4. **Remove the outer cabinet or control panel to gain access.** You can gain access to the magnetron by removing the access panel or the outer cabinet on a microwave oven or combination oven. If you are repairing a wall oven or an eye-level combination range, the control panel can be removed (see Figures 9-1 and 18-9) by opening the door and removing the screws that secure the panel. These might be underneath the front of the exhaust hood or just below the control panel. Some control panels are hinged; just tilt the control panel toward you for servicing. On some combination models, you might have to uninstall the appliance to gain access.

5. **Test the magnetron.** Locate the magnetron tube within the high-voltage section of the microwave oven (see Figure 20-5). Set your ohmmeter on the R × 1 scale. Carefully remove the wires from the magnetron filament terminals (Figure 20-8). Inspect the terminal seal for any damage. Take your ohmmeter probes, and place one probe on each filament terminal. The ohmmeter should read less than 1 ohm. Next, leave one probe on the filament terminal, and place the other probe on the chassis. The meter should read infinite ohms. Do the same procedure for the other filament terminal. The meter should read infinite ohms also. If the testing results differ, replace the magnetron.

6. **Remove the defective magnetron.** To remove the defective magnetron, remove the screws that secure the magnetron to the oven frame. Carefully remove the magnetron from the oven so as not to damage any other components. Inspect the magnetron tube antenna and embossed ring; replace the magnetron if there are signs of arcing, burn marks, or a broken antenna.

7. **Install the new magnetron.** To install a new magnetron, just reverse the disassembly procedure, and reassemble. Reinstall all panels or the console panel, and restore the electricity to the oven. Test the microwave oven operation. Perform the microwave performance and leakage test.

FIGURE 20-8   The magnetron.

High-Voltage Transformer

The typical complaints associated with the high-voltage transformer are:

- The microwave oven operates but is unable to cook the food.
- Erratic cooking.
- The oven turns off the electricity or blows a fuse in the circuitry when the oven door is closed.
- You hear a constant buzzing noise coming from the high-voltage section.

To handle these problems, perform the following steps:

1. **Verify the complaint.** Verify the complaint by operating the microwave oven with a glass of water and the microwave test light bar centered on the oven floor (see Figures 20-6 and 20-7). The lights in the light bar will light up if the magnetron and high-voltage system are working correctly. At the same time, place an ammeter on one wire of the service cord; the meter should indicate approximately 15 amps at 100 percent power. It may be necessary to turn off the electricity to the appliance and wait for two minutes before turning it back on. If a fault code appears, look up the code. If the microwave oven will not power up, locate the technical data sheet behind the panel or cabinet for diagnostics information.

2. **Check for external factors.** You must check for external factors not associated with the appliance. Is there electricity to the microwave oven?

3. **Disconnect the electricity.** Before working on the microwave oven, disconnect the electricity. This can be done by pulling the plug out of the wall receptacle. Or disconnect the electricity at the fuse panel or circuit breaker panel. Turn off the electricity.

WARNING *Before you begin to service any high-voltage components within the microwave oven, you must discharge the high-voltage capacitors first.*

WARNING *Some diagnostic tests will require you to test the components with the power turned on. When you disassemble the control panel or remove the outer cabinet, you can position the panel in such a way that the wiring will not make contact with metal. This will allow you to test the components without electrical mishaps.*

4. **Remove the outer cabinet or control panel to gain access.** You can gain access to the high-voltage transformer by removing the access panel or the outer cabinet on a microwave oven or combination oven. If you are repairing a wall oven or an eye-level combination range, the control panel can be removed (see Figures 9-1 and 18-9) by opening the door and removing the screws that secure the panel. These might be underneath the front of the exhaust hood or just below the control panel. Some control panels are hinged; just tilt the control panel toward you for servicing. On some combination models, you might have to uninstall the appliance to gain access.

Figure 20-9
The high-voltage
transformer.

Filament
winding
terminals

Secondary
winding
terminals

Primary
winding
terminals

Oven chassis

5. **Test the high-voltage transformer.** To test the high-voltage transformer (Figure 20-9), disconnect the wires from the primary winding terminals on the transformer (label them). Set your ohmmeter on the R × 1 scale, and measure the resistance at the primary winding terminals on the transformer. The ohmmeter scale should indicate resistance in the winding of less than 1 ohm. Now remove the wires from the secondary winding terminals on the transformer (label them as well), and measure the resistance of the winding. The ohmmeter scale should indicate resistance in the secondary winding from 50 to 130 ohms, depending on the type of transformer used. Next, test the resistance of the filament winding—the ohmmeter should indicate zero to 1 ohm. Finally, take the ohmmeter probes and test each winding to ground: you should have an infinite reading for each winding. If the readings differ, replace the transformer.

6. **Remove the defective high-voltage transformer.** To remove the defective transformer, remove the screws that secure it to the oven frame. Be careful: the high-voltage transformer is heavy.

7. **Install the new high-voltage transformer.** To install a new high-voltage transformer, just reverse the disassembly procedure, and reassemble. Reinstall all panels or the console panel, and restore the electricity to the oven. Test the microwave oven operation. Perform the microwave performance and leakage test.

Inverter

The typical complaints associated with the inverter are:

- The microwave oven operates, but is unable to cook the food.
- Erratic cooking.
- The oven turns off the electricity or blows a fuse in the circuitry when the oven door is closed.
- You hear a constant buzzing noise coming from the high-voltage section.

To handle these problems, perform the following steps:

1. **Verify the complaint.** Verify the complaint by operating the microwave oven with a glass of water and the microwave test light bar centered on the oven floor (see Figures 20-6 and 20-7). The lights in the light bar will light up if the magnetron and high-voltage system are working correctly. At the same time, place an ammeter on one wire of the service cord; the meter should indicate approximately 15 amps at 100 percent power. It may be necessary to turn off the electricity to the appliance and wait for two minutes before turning it back on. If a fault code appears, look up the code. If the microwave oven will not power up, locate the technical data sheet behind the panel or cabinet for diagnostics information.

2. **Check for external factors.** You must check for external factors not associated with the appliance. Is there electricity to the microwave oven?

3. **Disconnect the electricity.** Before working on the microwave oven, disconnect the electricity. This can be done by pulling the plug out of the wall receptacle. Or disconnect the electricity at the fuse panel or circuit breaker panel. Turn off the electricity.

WARNING *Before you begin to service any high-voltage components within the microwave oven, you must discharge the high-voltage capacitors first.*

WARNING *Some diagnostic tests will require you to test the components with the power turned on. When you disassemble the control panel or remove the outer cabinet, you can position the panel in such a way that the wiring will not make contact with metal. This will allow you to test the components without electrical mishaps.*

4. **Remove the outer cabinet or control panel to gain access.** You can gain access to the inverter by removing the access panel or the outer cabinet on a microwave oven or combination oven. If you are repairing a wall oven or an eye-level combination range, the control panel can be removed (see Figures 9-1 and 18-9) by opening the door and removing the screws that secure the panel. These might be underneath the front of the exhaust hood or just below the control panel. Some control panels are hinged; just tilt the control panel toward you for servicing. On some combination models, you might have to uninstall the appliance to gain access.

5. **Test the inverter.** The inverter replaces the high-voltage transformer, high-voltage capacitor, and high-voltage diode. To test the inverter (Figure 20-10), locate the service cord wires, and connect an ammeter to the white wire. Then place a glass of water in the oven cavity, turn on the electricity, and run the microwave oven at 100 percent power. If the reading on the ammeter is above 0.5 amps, check the magnetron and the wiring. If the reading is below 0.5 amps, there is no 120-volt input voltage to the inverter. In this case, you will have to check the cooking relay, electronic control board, or power relay board. Under normal operating conditions the high-voltage circuit can be measured with an the ammeter and it will read 15 amps.

Some models use a smaller inverter (about 40 watts) to provide 12 volts DC to the hood lamp and the cooling fan motor. To test this type of inverter, set your

FIGURE 20-10
The high-voltage
inverter.

voltmeter to the AC voltage scale, and test for 120 volts AC to the input connector on the inverter. Then set your voltmeter to the DC voltage scale, and check for 12 volts DC from the hood lamp and cooling fan motor connectors on the inverter.

WARNING *Do not measure the high-voltage side of the circuit with your multimeter. Only test the 120-volt input voltage.*

6. **Remove the defective inverter.** Before you begin to remove the inverter, you must use a 20,000-ohm 2-watt resistor to discharge the inverter. Hold the resistor body with a pair of insulated pliers (see Figure 3-8f), and discharge the inverter at the discharge points, as shown in Figure 20-10. To remove the inverter, remove the inverter cover, wire connectors, grounding wire, and screws; then slide the inverter out of its holding container. On some models, you might have to remove other components to gain access to the inverter.

7. **Install the new inverter.** To install a new inverter, just reverse the disassembly procedure, and reassemble. Reinstall all panels or the console panel, and restore the electricity to the oven. Test the microwave oven operation. Perform the microwave performance and leakage test.

High-Voltage Diode

The typical complaint associated with the high-voltage diode is that the microwave oven operates but is unable to cook the food. To handle this problem, perform the following steps:

1. **Verify the complaint.** Verify the complaint by operating the microwave oven with a glass of water and the microwave test light bar centered on the oven floor (see Figures 20-6 and 20-7). The lights in the light bar will light up if the magnetron and high-voltage system are working correctly. At the same time, place an ammeter on one wire of the service cord; the meter should indicate approximately 15 amps at 100 percent power. It may be necessary to turn off the electricity to the appliance and wait for two minutes before turning it back on. If a fault code appears, look up the code. If the microwave oven will not power up, locate the technical data sheet behind the panel or cabinet for diagnostics information.

2. **Check for external factors.** You must check for external factors not associated with the appliance. Is there electricity to the microwave oven?

3. **Disconnect the electricity.** Before working on the microwave oven, disconnect the electricity. This can be done by pulling the plug out of the wall receptacle. Or disconnect the electricity at the fuse panel or circuit breaker panel. Turn off the electricity.

WARNING *Before you begin to service any high-voltage components within the microwave oven, you must discharge the high-voltage capacitors first.*

WARNING *Some diagnostic tests will require you to test the components with the power turned on. When you disassemble the control panel or remove the outer cabinet, you can position the panel in such a way that the wiring will not make contact with metal. This will allow you to test the components without electrical mishaps.*

4. **Remove the outer cabinet or control panel to gain access.** You can gain access to the high-voltage diode by removing the access panel or the outer cabinet on a microwave oven or combination oven. If you are repairing a wall oven or an eye-level combination range, the control panel can be removed (see Figures 9-1 and 18-9) by opening the door and removing the screws that secure the panel. These might be underneath the front of the exhaust hood or just below the control panel. Some control panels are hinged; just tilt the control panel toward you for servicing. On some combination models, you might have to uninstall the appliance to gain access.

5. **Test the diode.** To test a diode, you must first disconnect one side of the diode to isolate it from the remainder of the circuit. Set your ohmmeter to the R × 1K scale, and connect the leads to the diode. The meter may or may not read continuity. Then reverse the leads on the diode and check for continuity. You should only read continuity in one direction only. If continuity is not indicated in either direction, the diode is open (see Figure 5-26a). If continuity is indicated in both directions, the diode is shorted (see Figure 5-26b).

6. **Remove the high-voltage diode.** To remove the diode from the oven, disconnect the wires and remove.

7. **Install the new high-voltage diode.** To install a new high-voltage diode, just reverse the disassembly procedure, and reassemble. Reinstall all panels or the console panel, and restore the electricity to the oven. Test the microwave oven operation. Perform the microwave performance and leakage test.

High-Voltage Capacitor

The typical complaint associated with the high-voltage capacitor is that the microwave oven operates but is unable to cook the food. To handle this problem, perform the following steps:

1. **Verify the complaint.** Verify the complaint by operating the microwave oven with a glass of water and the microwave test light bar centered on the oven floor (see Figures 20-6 and 20-7). The lights in the light bar will light up if the magnetron and high-voltage system are working correctly. At the same time, place an ammeter on one wire of the service cord; the meter should indicate approximately 15 amps at 100 percent power. It may be necessary to turn off the electricity to the appliance

and wait for two minutes before turning it back on. If a fault code appears, look up the code. If the microwave oven will not power up, locate the technical data sheet behind the panel or cabinet for diagnostics information.

2. **Check for external factors.** You must check for external factors not associated with the appliance. Is there electricity to the microwave oven?

3. **Disconnect the electricity.** Before working on the microwave oven, disconnect the electricity. This can be done by pulling the plug out of the wall receptacle. Or disconnect the electricity at the fuse panel or circuit breaker panel. Turn off the electricity.

WARNING *Before you begin to service any high-voltage components within the microwave oven, you must discharge the high-voltage capacitor first.*

WARNING *Some diagnostic tests will require you to test the components with the power turned on. When you disassemble the control panel or remove the outer cabinet, you can position the panel in such a way that the wiring will not make contact with metal. This will allow you to test the components without electrical mishaps.*

4. **Remove the outer cabinet or control panel to gain access.** You can gain access to the high-voltage capacitor by removing the access panel or the outer cabinet on a microwave oven or combination oven. If you are repairing a wall oven or an eye-level combination range, the control panel can be removed (see Figures 9-1 and 18-9) by opening the door and removing the screws that secure the panel. These might be underneath the front of the exhaust hood or just below the control panel. Some control panels are hinged; just tilt the control panel toward you for servicing. On some combination models, you might have to uninstall the appliance to gain access.

5. **Test the high-voltage capacitor.** Do not touch the high-voltage capacitor until it's discharged.

WARNING *A capacitor will hold a charge indefinitely, even when it is not currently in use. A charged capacitor is extremely dangerous. Discharge the capacitor immediately any time that work is being conducted in its vicinity. Recharge after repowering the microwave oven if further work must be done.*

Use a screwdriver with an insulated handle to discharge the capacitor by shorting it across both terminals. Remove the wire leads, one at a time, with needle nose pliers. Set the ohmmeter on the highest scale, and place one probe on one terminal and the other probe on the other terminal (see Figure 5-22). Observe the meter action. While the capacitor is charging, the ohmmeter will read nearly zero ohms for a short period of time. Then the ohmmeter reading will slowly return toward infinity. If the ohmmeter reading deflects to zero and does not return to infinity, the capacitor is shorted and should be replaced with a duplicate of the original. If the ohmmeter reading remains at infinity and does not dip toward zero, the capacitor is open and should be replaced with a duplicate of the original.

6. **Remove the high-voltage capacitor.** To remove the capacitor, remove the wires from the capacitor terminals. Then remove the screws from the capacitor-mounting bracket.

7. **Install the new high-voltage capacitor.** To install a new high-voltage capacitor, just reverse the disassembly procedure, and reassemble. Reinstall all panels or the console panel, and restore the electricity to the oven. Test the microwave oven operation. Perform the microwave performance and leakage test.

Microwave Oven Thermostats

The typical complaints associated with the thermostats are:

- The microwave oven operates but turns off after a while.
- The microwave oven will not cook.
- The microwave oven takes too long to cook the food.
- Erratic operation.

To handle these problems, perform the following steps:

1. **Verify the complaint.** Verify the complaint by operating the microwave oven with a glass of water and the microwave test light bar centered on the oven floor (see Figures 20-6 and 20-7). The lights in the light bar will light up if the magnetron and high-voltage system are working correctly. At the same time, place an ammeter on one wire of the service cord; the meter should indicate approximately 15 amps at 100 percent power. Check the operation of the cooling fans and stirrer blade. It may be necessary to turn off the electricity to the appliance and wait for two minutes before turning it back on. If a fault code appears, look up the code. If the microwave oven will not power up, locate the technical data sheet behind the panel or cabinet for diagnostics information.

2. **Check for external factors.** You must check for external factors not associated with the appliance. Is there electricity to the microwave oven? Is the vent system blocked?

3. **Disconnect the electricity.** Before working on the microwave oven, disconnect the electricity. This can be done by pulling the plug out of the wall receptacle. Or disconnect the electricity at the fuse panel or circuit breaker panel. Turn off the electricity.

WARNING *Before you begin to service any high-voltage components within the microwave oven, you must discharge the high-voltage capacitors or inverter first.*

WARNING *Some diagnostic tests will require you to test the components with the power turned on. When you disassemble the control panel or remove the outer cabinet, you can position the panel in such a way that the wiring will not make contact with metal. This will allow you to test the components without electrical mishaps.*

4. **Remove the outer cabinet or control panel to gain access.** You can gain access to the thermostats by removing the access panel or the outer cabinet on a microwave oven or combination oven. If you are repairing a wall oven or an eye-level

combination range, the control panel can be removed (see Figures 9-1 and 18-9) by opening the door and removing the screws that secure the panel. These might be underneath the front of the exhaust hood or just below the control panel. Some control panels are hinged; just tilt the control panel toward you for servicing. On some combination models, you might have to uninstall the appliance to gain access.

5. **Test the thermostats.** The microwave oven can have up to four thermostats (see Figure 20-5) in the oven that operate as a safety switch to turn the oven off if the oven temperature exceeds its limits. The cavity thermostats monitor the airflow temperatures in the oven. The magnetron thermostat monitors the operating temperature of the magnetron. Some models have an exhaust fan thermostat that will monitor the temperature from beneath the oven and the airflow temperature (this is common in over-the-range models).

   The cavity and magnetron thermostats are normally closed—when the temperature exceeds its rated limits, it will open the circuit. When the temperature drops below its rating, it will reset and close the circuit. The exhaust fan thermostat is normally open; it closes on temperature rise, turning on the exhaust fan. When the temperature drops below the rated temperature, the thermostat will reset to open the circuit.

   To test the thermostat, set the ohmmeter on R × 1, remove the wires from the thermostat terminals, and test for continuity.

6. **Remove the thermostat.** Locate the thermostat (see Figure 20-5), remove the wires from the terminals, and remove the screws that secure the thermostat to the cavity or magnetron.

7. **Install the new thermostat.** To install a new thermostat, just reverse the disassembly procedure, and reassemble. Reinstall all panels or the console panel, and restore the electricity to the oven. Test the microwave oven operation. Perform the microwave performance and leakage test.

Microwave Oven Door Switches

The typical complaints associated with the door switches are:

- The microwave oven will not operate.
- The oven goes off when the oven door is closed.

To handle these problems, perform the following steps:

1. **Verify the complaint.** Verify the complaint by operating the microwave oven with a glass of water and the microwave test light bar centered on the oven floor (see Figures 20-6 and 20-7). The lights in the light bar will light up if the magnetron and high-voltage system are working correctly. At the same time, place an ammeter on one wire of the service cord; the meter should indicate approximately 15 amps at 100 percent power. It may be necessary to turn off the electricity to the appliance and wait for two minutes before turning it back on. If a fault code appears, look up the code. If the microwave oven will not power up, locate the technical data sheet behind the panel or cabinet for diagnostics information.

2. **Check for external factors.** You must check for external factors not associated with the appliance. Is there electricity to the microwave oven?

3. **Disconnect the electricity.** Before working on the microwave oven, disconnect the electricity. This can be done by pulling the plug out of the wall receptacle. Or disconnect the electricity at the fuse panel or circuit breaker panel. Turn off the electricity.

WARNING *Before you begin to service any high-voltage components within the microwave oven, you must discharge the high-voltage capacitors or inverter first.*

WARNING *Some diagnostic tests will require you to test the components with the power turned on. When you disassemble the control panel or remove the outer cabinet, you can position the panel in such a way that the wiring will not make contact with metal. This will allow you to test the components without electrical mishaps.*

4. **Remove the outer cabinet or control panel to gain access.** You can gain access to the door switches by removing the access panel or the outer cabinet on a microwave oven or combination oven. If you are repairing a wall oven or an eye-level combination range, the control panel can be removed (see Figures 9-1 and 18-9) by opening the door and removing the screws that secure the panel. These might be underneath the front of the exhaust hood or just below the control panel. Some control panels are hinged; just tilt the control panel toward you for servicing. On some combination models, you might have to uninstall the appliance to gain access.

5. **Test the microwave door switches.** Locate the primary interlock, monitor, and secondary interlock switches (see Figures 8-3, 20-5, and 20-11). Identify each switch using the technical data sheet for the model you are servicing. The switch housing is usually marked with the terminal identification numbers that correspond to the wiring diagram. It identifies the contacts by number: normally open (NO) contacts, normally closed (NC) contacts, or common (COM). Internally, the switch can house many contact points for controlling more than one circuit.

   In Figure 20-11a, the wiring diagram illustrates the door switches in the closed and open positions. Figure 20-11b illustrates the location of the door switches in a countertop model. Set your ohmmeter to the R × 1 scale, remove the wires from the primary interlock switch terminals, and place your meter probes on the terminals. With the door closed, the ohmmeter will indicate continuity. When the door is opened, the ohmmeter reads no continuity. Next, test the monitor switch. Remove the wires from the switch terminals, and place your meter probes on the terminals. With the door closed, the ohmmeter will indicate no continuity. When the door is opened, the ohmmeter reads continuity. Finally, test the secondary interlock switch. Remove the wires from the switch, and place the meter probes on the terminals. With the door closed, the ohmmeter will indicate continuity. When the door is opened, the ohmmeter reads no continuity. If any of the readings differ from what is stated here, replace all of the switches in the microwave.

6. **Remove the door switches.** Remove the wires from the door switches and label them. Depending on the model you are servicing, remove the door switch assembly as a complete unit or remove each switch separately.

FIGURE 20-11   (a) The strip circuits for the microwave oven door interlock switches with the oven door closed and open. (b) A typical view of one oven model with the door interlock switches. The latch heads and the door switch assembly can be adjusted.

7. **Install the new door switch.** To install a new door switch, just reverse the disassembly procedure, and reassemble. Check the in and out play in the door: it should be less than 0.5 mm when the door is closed. Reinstall all panels or the console panel, and restore the electricity to the oven. Test the microwave oven operation. Perform the microwave performance and leakage test.

Temperature Probe

The typical complaints associated with the temperature probe are:

- Erratic cooking of the food.
- The food will not cook.
- The food burns.

The temperature probe is a separate component that plugs into a receptacle located in the oven cavity. The consumer inserts the probe into the food product being cooked, programs the controls, and starts the oven. The temperature probe will sense the temperature of the food cooking and sends a signal back to the electronic control to turn off the microwave oven when the food is fully cooked. To test the temperature probe, set your ohmmeter on R × 1K, and place the meter probes on the male end that goes into the receptacle (Figure 20-12). With some models at room temperature, the ohmmeter scale will read between 19,000 and 75,000 ohms. Next, with the ohmmeter probes still attached to the male jack end, insert the probe tip in a glass of hot water, and measure the resistance. The resistance reading should decrease. If the reading does not decrease, replace the temperature probe.

Temperature probe

FIGURE 20-12   Testing the resistance in the temperature probe.

Humidity Sensor

The typical complaints associated with the humidity sensor are:

- The microwave oven will not cook the food properly.
- The microwave oven takes too long to cook the food.
- Erratic operation.

To handle these problems, perform the following steps:

1. **Verify the complaint.** Verify the complaint by operating the microwave oven with a glass of water and the microwave test light bar centered on the oven floor (see Figures 20-6 and 20-7). The lights in the light bar will light up if the magnetron and high-voltage system are working correctly. At the same time, place an ammeter on one wire of the service cord; the meter should indicate approximately 15 amps at 100 percent power. Check the operation of the cooling fans and stirrer blade. It may be necessary to turn off the electricity to the appliance and wait for two minutes before turning it back on. If a fault code appears, look up the code. If the microwave oven will not power up, locate the technical data sheet behind the panel or cabinet for diagnostics information.

2. **Check for external factors.** You must check for external factors not associated with the appliance. Is there electricity to the microwave oven? Is the vent system blocked? On countertop models, is there enough open space surrounding the microwave oven for proper ventilation?

3. **Disconnect the electricity.** Before working on the microwave oven, disconnect the electricity. This can be done by pulling the plug out of the wall receptacle. Or disconnect the electricity at the fuse panel or circuit breaker panel. Turn off the electricity.

WARNING *Before you begin to service any high-voltage components within the microwave oven, you must discharge the high-voltage capacitors or inverter first.*

WARNING *Some diagnostic tests will require you to test the components with the power turned on. When you disassemble the control panel or remove the outer cabinet, you can position the panel in such a way that the wiring will not make contact with metal. This will allow you to test the components without electrical mishaps.*

4. **Remove the outer cabinet or control panel to gain access.** You can gain access to the humidity sensor (see Figure 12-5) by removing the access panel or the outer cabinet on a microwave oven or combination oven. If you are repairing a wall oven or an eye-level combination range, the control panel can be removed (see Figures 9-1 and 18-9) by opening the door and removing the screws that secure the panel. These might be underneath the front of the exhaust hood or just below the control panel. Some control panels are hinged; just tilt the control panel toward you for servicing. On some combination models, you might have to uninstall the appliance to gain access.

5. **Test the humidity sensor.** The humidity sensor can be tested at the electronic control board behind the control panel. Locate the sensor wires, trace them to the electronic control board, and disconnect the three-wire connector from the board. Set your ohmmeter on the R × 1K scale. Place your meter probes on pins 1 and 3 (check the technical data sheet for the pin designations), located on the humidity sensor wire connector. The ohmmeter scale should read approximately 2800 ohms at 77 degrees Fahrenheit (±2 percent). Next, place the meter probes on the connector pins 2 and 3—the ohmmeter scale should read approximately 2800 ohms at 77 degrees Fahrenheit (±2 percent). If the ohmmeter readings differ, replace the humidity sensor.

6. **Remove the humidity sensor.** Disconnect the sensor wire connector from the electronic control board and trace it back to the sensor. Remove the screws that secure the sensor to the oven cavity.

7. **Install the new humidity sensor.** To install a new humidity sensor, just reverse the disassembly procedure, and reassemble. Make sure that you place the wires in the same route that you removed them and do not pinch the wires. Reinstall all panels or the console panel, and restore the electricity to the oven. Test the microwave oven operation. Perform the microwave performance and leakage test.

Stirrer Motor

The typical complaints associated with the stirrer motor are:

- The microwave oven will not cook the food properly.
- The microwave oven takes too long to cook the food.
- Erratic operation.

To handle these problems, perform the following steps:

1. **Verify the complaint.** Verify the complaint by operating the microwave oven with a glass of water and the microwave test light bar centered on the oven floor (see Figures 20-6 and 20-7). The lights in the light bar will light up if the magnetron and high-voltage system are working correctly. At the same time, place an ammeter on one wire of the service cord; the meter should indicate approximately 15 amps at 100 percent power. Check the operation of the cooling fans and stirrer blade. It may be necessary to turn off the electricity to the appliance and wait for two minutes before turning it back on. If a fault code appears, look up the code. If the microwave oven will not power up, locate the technical data sheet behind the panel or cabinet for diagnostics information.

2. **Check for external factors.** You must check for external factors not associated with the appliance. Is there electricity to the microwave oven? Is the vent system blocked? On countertop models, is there enough open space surrounding the microwave oven for proper ventilation?

3. **Disconnect the electricity.** Before working on the microwave oven, disconnect the electricity. This can be done by pulling the plug out of the wall receptacle. Or disconnect the electricity at the fuse panel or circuit breaker panel. Turn off the electricity.

WARNING *Before you begin to service any high-voltage components within the microwave oven, you must discharge the high-voltage capacitors or inverter first.*

WARNING *Some diagnostic tests will require you to test the components with the power turned on. When you disassemble the control panel or remove the outer cabinet, you can position the panel in such a way that the wiring will not make contact with metal. This will allow you to test the components without electrical mishaps.*

4. **Remove the outer cabinet or control panel to gain access.** You can gain access to the stirrer motor (see Figure 12-5) by removing the access panel or the outer cabinet on a microwave oven or combination oven. From inside the oven cavity area, remove the stirrer cover and stirrer blade first in order to remove the stirrer motor if necessary.

5. **Test the stirrer motor.** To test the stirrer motor, locate the stirrer motor leads, disconnect the wires, and attach your ohmmeter leads to the motor wires. Set your ohmmeter on the R × 1K scale. The ohmmeter scale should indicate 2.5K to 4.0K ohms. If the reading is different from this, replace the stirrer motor. On some older models, the stirrer blade is belt-driven—check for a broken belt. On other models, the stirrer blade turns with the airflow from the cooling fan motor that circulates the microwave energy. Make sure the stirrer blade spins freely.

6. **Remove the stirrer motor.** On some models, the stirrer cover and stirrer blade will have to be removed first. The stirrer motor is mounted to the oven cavity with screws—remove the screws.

7. **Install the new stirrer motor.** To install a new stirrer motor, just reverse the disassembly procedure, and reassemble. Reinstall all panels and restore the electricity to the oven. Test the microwave oven operation. Perform the microwave performance and leakage test.

Turntable Motor

The typical complaints associated with the turntable motor are:

- The microwave oven will not cook the food evenly.
- The microwave oven takes too long to cook the food.
- Erratic operation.

To handle these problems, perform the following steps:

1. **Verify the complaint.** Verify the complaint by operating the microwave oven with a glass of water and the microwave test light bar centered on the oven floor (see Figures 20-6 and 20-7). The lights in the light bar will light up if the magnetron and high-voltage system are working correctly. At the same time, place an ammeter on one wire of the service cord; the meter should indicate approximately 15 amps at 100 percent power. Check the operation of the cooling fans and stirrer blade. It may be necessary to turn off the electricity to the appliance and wait for two minutes before turning it back on. If a

fault code appears, look up the code. If the microwave oven will not power up, locate the technical data sheet behind the panel or cabinet for diagnostics information.

2. **Check for external factors.** You must check for external factors not associated with the appliance. Is there electricity to the microwave oven? Is the vent system blocked? On countertop models, is there enough open space surrounding the microwave oven for proper ventilation?

3. **Disconnect the electricity.** Before working on the microwave oven, disconnect the electricity. This can be done by pulling the plug out of the wall receptacle. Or disconnect the electricity at the fuse panel or circuit breaker panel. Turn off the electricity.

WARNING *Before you begin to service any high-voltage components within the microwave oven, you must discharge the high-voltage capacitors or inverter first.*

WARNING *Some diagnostic tests will require you to test the components with the power turned on. When you disassemble the control panel or remove the outer cabinet, you can position the panel in such a way that the wiring will not make contact with metal. This will allow you to test the components without electrical mishaps.*

---

4. **Gain access to the turntable motor.** To gain access to the turntable motor, lay the oven on its back, and remove the bottom panel. On older models, the turntable motor might be belt-driven—inspect the belt.

5. **Test the turntable motor.** To test the turntable motor, locate the motor leads, disconnect the wires, and attach your ohmmeter leads to the motor wires. Set your ohmmeter on the R × 1K scale. The ohmmeter scale should indicate 25,000 to 40,000 ohms. If the reading differs from this, replace the turntable motor.

6. **Remove the turntable motor.** To remove the turntable motor, remove the screws that secure the motor to the cavity. On some older models, you might have to remove the gearbox before removing the motor.

7. **Install the new turntable motor.** To install a new turntable motor, just reverse the disassembly procedure, and reassemble. Reinstall all panels and restore the electricity to the oven. Test the microwave oven operation. Perform the microwave performance and leakage test.

Cooling Fan Motor

The typical complaints associated with the cooling fan motor are:

- The microwave oven operates but is unable to cook the food.

- The microwave oven will shut off.

1. **Verify the complaint.** Verify the complaint by operating the microwave oven with a glass of water and the microwave test light bar centered on the oven floor (see Figures 20-6 and 20-7). The lights in the light bar will light up if the magnetron and high-voltage system are working correctly. At the same time, place an ammeter on one wire of the service cord; the meter should indicate approximately 15 amps at

100 percent power. It may be necessary to turn off the electricity to the appliance and wait for two minutes before turning it back on. If a fault code appears, look up the code. If the microwave oven will not power up, locate the technical data sheet behind the panel or cabinet for diagnostics information.

2. **Check for external factors.** You must check for external factors not associated with the appliance. Is there electricity to the microwave oven? Is the vent system blocked? On countertop models, is there enough open space surrounding the microwave oven for proper ventilation?

3. **Disconnect the electricity.** Before working on the microwave oven, disconnect the electricity. This can be done by pulling the plug out of the wall receptacle. Or disconnect the electricity at the fuse panel or circuit breaker panel. Turn off the electricity.

WARNING *Before you begin to service any high-voltage components within the microwave oven, you must discharge the high-voltage capacitors or inverter first.*

WARNING *Some diagnostic tests will require you to test the components with the power turned on. When you disassemble the control panel or remove the outer cabinet, you can position the panel in such a way that the wiring will not make contact with metal. This will allow you to test the components without electrical mishaps.*

4. **Remove the outer cabinet or control panel to gain access.** You can gain access to the cooling fan motor by removing the access panel or the outer cabinet on a microwave oven or combination oven. If you are repairing a wall oven or an eye-level combination range, the control panel can be removed (see Figures 9-1 and 18-9) by opening the door and removing the screws that secure the panel. These might be underneath the front of the exhaust hood or just below the control panel. Some control panels are hinged; just tilt the control panel toward you for servicing. On some combination models, you might have to uninstall the appliance to gain access.

5. **Test the cooling fan motor.** The cooling fan motor serves two purposes in some models. The fan motor cools the magnetron and high-voltage transformer, and circulates the air through the waveguide to the stirrer blade and out the discharge vent (Figure 20-13). Over-the-range models use two fan motors to cool the microwave oven. Other models with stirrer motors circulate the air only in the high-voltage section of the microwave oven.

   To test the cooling fan motor, locate the technical data sheet for the model you are servicing, and look up the motor's operating voltage. Microwave oven manufacturers use a 120-volt AC or 12-volt DC motor. Check and see if the fan blade spins freely. Then disconnect the wires from the motor, isolating it from the rest of the circuit, and place your ohmmeter probes on the motor wire leads. Set your ohmmeter on the R × 1 scale. For direct current (DC) motors, the ohmmeter scale should indicate 15 to 30 ohms. For alternating current (AC) motors, the ohmmeter scale should indicate 80 to 140 ohms. If the readings are different from this, replace the cooling fan motor.

6. **Remove the cooling fan motor.** To remove the cooling fan motor, disconnect the wiring, and remove the screws that secure the motor assembly to the oven frame. Depending on the model you are servicing, you may have to remove other components to access the motor.

Figure 20-13 A typical airflow pattern. This model shows the cooling fan circulating the air through the magnetron, waveguide, stirrer duct, and stirrer blade; through the oven cavity; and then out through the discharge vent.

7. **Install the new cooling fan motor.** To install a new cooling fan motor, just reverse the disassembly procedure, and reassemble. Reinstall all panels and restore the electricity to the oven. Test the microwave oven operation. Perform the microwave performance and leakage test.

Exhaust Fan Motor

The typical complaints associated with the exhaust fan motor are:

- The exhaust fan motor will not come on.
- The microwave oven overheats and shuts off.

To handle these problems, perform the following steps:

1. **Verify the complaint.** Verify the complaint by operating the microwave oven with a glass of water and the microwave test light bar centered on the oven floor (see Figures 20-6 and 20-7). The lights in the light bar will light up if the magnetron and high-voltage system are working correctly. At the same time, place an ammeter on one wire of the service cord; the meter should indicate approximately 15 amps at

100 percent power. It may be necessary to turn off the electricity to the appliance and wait for two minutes before turning it back on. If a fault code appears, look up the code. If the microwave oven will not power up, locate the technical data sheet behind the panel or cabinet for diagnostics information.

2. **Check for external factors.** You must check for external factors not associated with the appliance. Is there electricity to the microwave oven? Is the vent system blocked? On countertop models, is there enough open space surrounding the microwave oven for proper ventilation?

3. **Disconnect the electricity.** Before working on the microwave oven, disconnect the electricity. This can be done by pulling the plug out of the wall receptacle. Or disconnect the electricity at the fuse panel or circuit breaker panel. Turn off the electricity.

WARNING *Before you begin to service any high-voltage components within the microwave oven, you must discharge the high-voltage capacitors or inverter first.*

WARNING *Some diagnostic tests will require you to test the components with the power turned on. When you disassemble the control panel or remove the outer cabinet, you can position the panel in such a way that the wiring will not make contact with metal. This will allow you to test the components without electrical mishaps.*

4. **Remove the outer cabinet or control panel to gain access.** You can gain access to the exhaust fan motor by removing the access panel or the outer cabinet on a microwave oven. If you are repairing a wall oven or an eye-level combination range, the control panel can be removed (see Figures 9-1 and 18-9) by opening the door and removing the screws that secure the panel. These might be underneath the front of the exhaust hood or just below the control panel. Some control panels are hinged; just tilt the control panel toward you for servicing. On some combination models or over-the-range models, you might have to uninstall the appliance to gain access (see Figure 20-5).

5. **Test the exhaust fan motor.** Disconnect the motor wire leads from the wiring harness. Check the motor windings for continuity. To check for a grounded winding in the motor, take the ohmmeter probes, and check from each motor wire lead terminal to the motor housing. The ohmmeter will indicate continuity if the windings are grounded.

6. **Remove the exhaust fan motor.** Remove the fan blade and the screws that secure the fan motor to the oven frame. On some models, the entire fan motor housing can be removed from the oven for easy accessibility. On other models, the motor is located on top of the oven.

7. **Install the new exhaust fan motor.** To install a new exhaust fan motor, just reverse the disassembly procedure, and reassemble. Reinstall all panels or reinstall the oven, and restore the electricity to the oven. Test the microwave oven operation. Perform the microwave performance and leakage test.

Diagnostic Chart and Wiring Diagrams

The following diagnostic chart will help you to pinpoint the likely causes of various problems associated with microwave ovens (Figure 20-14).

The wiring diagrams in this chapter are only examples. You must refer to the actual technical data sheet or the wiring diagrams for the microwave oven that you are servicing. Figure 20-15 depicts an actual wiring schematic diagram, and Figure 20-16 depicts an actual pictorial electrical wiring diagram of the same model. Figures 20-17 and 20-18 depict the strip circuits for the same model. The strip circuits are easier to read.

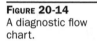

FIGURE 20-14
A diagnostic flow chart.

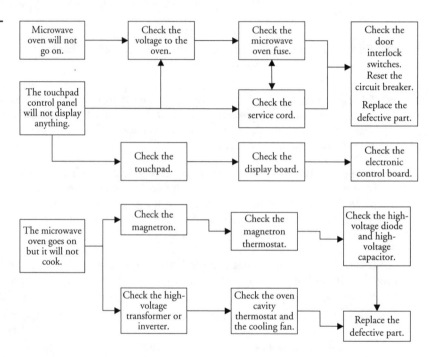

FIGURE 20-15
A wiring schematic.

BK: Black   W: White   BR: Brown   BU: Blue   G: Green   O: Orange   R: Red   Y: Yellow   GY: Gray   TR: Transparent

FIGURE 20-16   A pictorial wiring diagram.

Microwave cooking

Blower fan on automatic

Blower fan on variable

Microwave time-of-day displayed

FIGURE 20-17 Microwave oven strip circuits for various functions.

Door open—cavity lamp on

Cooktop lamp on (variable light)

FIGURE 20-18   Microwave oven strip circuits for various functions.

Refrigerators and Freezers

One of the most important applications of refrigeration (which was invented in the early 1900s) was for the preservation of food. When different types of food[1] are kept at room temperature, some of them will spoil rapidly. When foods are kept cold, they will last longer. Refrigerators/freezers prevent food spoilage by keeping the food cold. The refrigerator/freezer consists of three parts:

- The cabinet.
- The sealed system, which consists of the evaporator coil, the condenser coil, the compressor, and the connecting tubing.
- The electrical circuitry, including fan motors and other electrical components.

This chapter covers the electrical components and how to diagnose the sealed system. The actual repair or replacement of any sealed-system component is not included in this chapter. It is recommended that you acquire refrigerant certification (or call an authorized service company) to repair or replace any sealed-system component. The refrigerant in the sealed system must be recovered properly.

The Refrigeration Cycle

The sealed system (Figure 21-1) in a refrigerator or freezer consists of a compressor, a condenser coil, an evaporator coil, a capillary tube, and a heat exchanger and its connecting tubing. This is the heart of the refrigerator or freezer that keeps the food cold inside the cabinet.

Starting at the compressor, refrigerant gas is pumped out of the compressor, through the discharge tubing, and into the condenser coil. When the gas is in the condenser coil, the temperature and pressure of the refrigerant gas greatly increase because of the capillary tube at the discharge end of the condenser coil. From the surface of the condenser coil, the heat spreads out into the room via air moving over the condenser coil. The condenser coil cools the hot refrigerant gas. As the refrigerant gas gives up the heat it obtained from inside the refrigerator or freezer cabinet, the refrigerant gas changes into a liquid. This liquid then leaves the condenser coil and enters the capillary tube.

Figure 21-1
The sealed-system components in a side-by-side refrigerator.

Capillary tube

Evaporator

Suction line

Compressor

Discharge line

Condenser

Heat exchanger

This capillary tube is carefully made with regard to its length and inside diameter to meter out the exact amount of liquid through the sealed system (this is designed by the manufacturer for a particular size and model). As the liquid refrigerant leaves the capillary tube and enters the larger tubing of the evaporator coil, the sudden increase of tubing size causes a low-pressure area. It is here that the liquid refrigerant changes from a liquid to a mixture of liquid and gas.

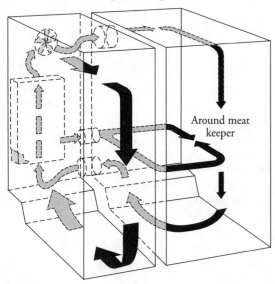

Around meat keeper

As this mixture passes through the evaporator coil, the refrigerant absorbs heat from the warmer items (food) within the refrigerator or freezer cabinet, slowly changing any liquid back to gas. As the refrigerant gas leaves the evaporator coil, it returns to the compressor through the suction line.

This entire procedure is called a *cycle*. Depending on where the cold control is set, the thermostat can show how cold it is inside the cabinet and then control the actuation of the cooling cycle. It will determine whether to turn the system on or off to maintain the temperature within the cabinet.

Inside the cabinet, the cold air is circulated by convection and/or by means of an electrical fan. In Figure 21-2, the arrows show the airflow patterns in this type of side-by-side refrigerator. Figure 21-3 shows the air patterns in a two-door refrigerator with a top freezer.

Figure 21-2   The airflow pattern in a side-by-side refrigerator.

Storage Requirements for Perishable Products

Table 21-1 lists the recommended storage temperatures, relative humidity, and the approximate storage life for perishable products. These values are used in designing commercial refrigeration systems, which house large quantities of perishable products. Large warehouses are usually equipped to store foods at those temperatures best adapted to prolonging the safe storage period for each type of food. In the domestic refrigerator, most foods are kept at 35 to 45 degrees Fahrenheit and the humidity is kept around 50 percent. The freezer temperature is between zero and 10 degrees Fahrenheit.

It can be difficult to maintain these temperatures and humidity for each individual product. Therefore, refrigerator manufacturers have designed separate compartments within the refrigerated cabinet to maintain a variable temperature and humidity selected by the consumer. The storage life of various products will vary in a domestic refrigerator/freezer. This period will be influenced by many factors, such as the storage temperature, the type of container, the condition of the food, and the kind of food.

FIGURE 21-3 The airflow pattern in a two-door refrigerator with top freezer.

Product	Storage Temp.° F	Relative Humidity %	Approximate Storage Life
Apples	30–40	90	3–8 months
Apricots	31–32	90	1–2 weeks
Artichokes	31–32	95	2 weeks
Asparagus	32–36	95	2–3 weeks
Avocados	45–55	85–90	2–4 weeks
Bananas	55–65	85–95	—
Beans (green or snap)	40–45	90–95	7–10 days
Beans, lima	32–40	90	1 week
Blackberries	31–32	95	3 days
Blueberries	31–32	90–95	2 weeks
Broccoli	32	95	10–14 days
Cabbage	32	95–100	3–4 months
Carrots	32	98–100	5–9 months
Cauliflower	32	95	2–4 weeks
Celery	32	95	1–2 months
Cherries, sour	31–32	90–95	3–7 days
Cherries, sweet	30–31	90–95	2–3 weeks
Collards	32	95	10–14 days
Corn, sweet (fresh)	32	95	4–8 days
Cranberries	36–40	90–95	2–4 months
Cucumbers	50–55	90–95	10–14 days

TABLE 21-1 Storage Requirements for Perishable Products. (*continued*)

Product	Storage Temp.° F	Relative Humidity %	Approximate Storage Life
Dairy products			
Cheddar cheese	40	65–70	6 months
Processed cheese	40	65–70	12 months
Butter	40	75–85	1 month
Cream	35–40	—	2–3 weeks
Ice cream	–20 to –15	—	3–12 months
Milk, fluid whole			
Pasteurized, grade A	32–34	—	2–4 months
Condensed, sweetened	40	—	15 months
Evaporated	40	—	24 months
Dates (dried)	0 or 32	75 or less	6–12 months
Dried fruits	32	50–60	9–12 months
Eggplant	45–50	90–95	7–10 days
Eggs, shell	29–31	80–85	5–6 months
Figs, dried	32–40	50–60	9–12 months
Figs, fresh	31–32	85–90	7–10 days
Fish, fresh	30–35	90–95	5–15 days
Haddock, cod	30–35	90–95	15 days
Salmon	30–35	90–95	15 days
Smoked	40–50	50–60	6–8 months
Shellfish, fresh	30–33	86–95	3–7 days
Tuna	30–35	90–95	15 days
Grapefruit	50–60	85–90	4–6 weeks
Grapes, American type	31–32	85–90	2–8 weeks
Grapes, European type	30–31	90–95	3–6 months
Greens, leafy	32	95	10–14 days
Guavas	45–50	90	2–3 weeks
Honey	38–50	50–60	1 year, plus
Horseradish	30–32	95–100	10–12 months
Lemons	32 or 50–58	85–90	1–6 months
Lettuce, head	32–34	95–100	2–3 weeks
Limes	48–50	85–90	6–8 weeks
Maple sugar	75–80	60–65	1 year, plus
Mangoes	55	85–90	2–3 weeks
Meat			
Bacon, cured (farm style)	60–65	85	4–6 months
Game, fresh	32	80–85	1–6 weeks
Beef, fresh	32–34	88–92	1–6 weeks
Hams & shoulders, fresh	32–34	85–90	7–12 days
Cured	60–65	50–60	0–3 years
Lamb, fresh	32–34	85–90	5–12 days
Livers, frozen	–10–0	90–95	3–4 months
Pork, fresh	32–34	85–90	3–7 days
Smoked sausage	40–45	85–90	6 months
Fresh	32	85–90	1–2 weeks
Veal, fresh	32–34	90–95	5–10 days

TABLE 21-1   Storage Requirements for Perishable Products. (*continued*)

Product	Storage Temp.° F	Relative Humidity %	Approximate Storage Life
Melons, Cantaloupe	36–40	90–95	5–15 days
Honeydew & Honey Ball	45–50	90–95	3–4 weeks
Watermelons	40–50	80–90	2–3 weeks
Mushrooms	32	90	3–4 days
Milk	34–40	—	7 days
Nectarines	31–32	90	2–4 weeks
Nuts (dries)	32–50	65–75	8–12 months
Okra	45–50	90–95	7–10 days
Olives, fresh	45–50	85–90	4–6 weeks
Onions (dry) & onion sets	32	65–70	1–8 months
Oranges	32–48	85–90	3–12 weeks
Orange juice, chilled	30–35	—	3–6 weeks
Papayas	45	85–90	1–3 weeks
Parsley	32	95	1–2 months
Parsnips	32	98–100	4–6 months
Peaches	31–32	90	2–4 weeks
Pears	29–31	90–95	2–7 months
Peas, green	32	95	1–3 weeks
Peppers, sweet	45–50	90–95	2–3 weeks
Pineapples, ripe	45	85–90	2–4 weeks
Plums, including fresh prunes	31–32	90–95	2–4 weeks
Popcorn, unpopped	32–40	85	4–6 months
Potatoes, early crop	50–55	90	0–2 months
Potatoes, late crop	38–50	90	5–8 months
Poultry			
Fresh chicken	32	85–90	1 week
Fresh goose	32	85–90	1 week
Fresh turkey	32	85–90	1 week
Pumpkins	50–55	70–75	2–3 months
Radishes—spring, prepacked	32	95	3–4 weeks
Raisins (dried)	40	60–70	9–12 months
Rabbits, fresh	32–34	90–95	1–5 days
Raspberries, black	31–32	90–95	2–3 days
Raspberries, red	31–32	90–95	2–3 days
Rhubarb	32	95	2–4 weeks
Spinach	32	95	10–14 days
Squash, summer	32–50	85–95	5–14 days
Squash, winter	50–55	70–75	4–6 months
Strawberries, fresh	31–32	90–95	5–7 days
Sugar, maple	75–80	60–65	1 year, plus
Sweet potatoes	55–60	85–90	4–7 months
Syrup, maple	31	60–70	1 year, plus
Tangerines	32–38	85–90	2–4 weeks
Tomatoes, mature green	55–70	85–90	1–3 weeks
Tomatoes, firm ripe	45–50	85–90	4–7 days
Turnips, roots	32	95	4–5 months
Vegetables (mixed)	32–40	90–95	1–4 weeks
Yams	60	85–90	3–6 months

TABLE 21-1 Storage Requirements for Perishable Products.

Safety First

Any person who cannot use basic tools or follow written instructions should *not* attempt to install, maintain, or repair any refrigerators or freezers. Any improper installation, preventive maintenance, or repairs could create a risk of personal injury or property damage.

If you do not fully understand the installation, preventive maintenance, or repair procedures in this chapter, or if you doubt your ability to complete the task on your refrigerator or freezer, please call your service manager.

Before continuing, take a moment to refresh your memory of the safety procedures in Chapter 2.

Refrigerators/Freezers in General

Much of the troubleshooting information in this chapter covers refrigerators and freezers in general, rather than specific models, in order to present a broad overview of service techniques. The illustrations in this chapter are for demonstration purposes only, to clarify the description of how to service these appliances. They in no way reflect on a particular brand's reliability.

Location and Installation of Refrigerator/Freezer

Thoroughly read the installation instructions that are provided with every new refrigerator/freezer. These instructions will provide you with the information you need to properly install the refrigerator or freezer. The following are some general principles that should be followed when performing the installation:

- The refrigerator/freezer must be installed on a solid floor capable of supporting the product up to 1000 pounds.
- For proper air circulation around the refrigerator/freezer, some models require a one-inch clearance at the rear and top of the cabinet and adequate clearance near the front grille at the bottom of the refrigerator/freezer.
- Do not leave the refrigerator/freezer on its side longer than necessary to remove the shipping base.
- When removing or reversing the doors on a refrigerator, always reinstall them according to the installation instructions, and remember to realign the doors properly.
- Level the refrigerator/freezer cabinet so that the doors close properly.

What's that "Different" Sound in Your Kitchen?

If you have bought or serviced a new refrigerator within the past few years, you've probably noticed that it sounds "different." Here's why: New refrigerators use only half as much electricity as the older models. In fact, a new 20.6-cubic foot refrigerator with top freezer uses no more electricity than a 75-watt light bulb. Most new refrigerators are also larger, and they have such added conveniences as automatic defrost systems, ice makers, and perhaps even a "built-in" look. These new features result in different sounds, such as:

- High-pitched whine: This is due to the more energy-efficient compressors that have smaller, higher-speed motors.
- Soft hum: This is from the evaporator fan in the freezer and/or from the condenser fan under the refrigerator.
- Clicks: These can be from the automatic defrost timer switching on and off, the thermostat turning the refrigerator on and off, or the water valve refilling the ice maker.
- Boiling and/or gurgling or trickling water: When the refrigerator stops running, the refrigerant continues to circulate within the system or the defrost water runs into the drain pan.
- Running water and "thuds": These sounds occur as the ice cube tray fills and as the ice cubes drop into the storage bin.

To help mute these new sounds:

- Be sure that the refrigerator is level and that the defrost water collection pan is in position (usually reachable behind the bottom front "toe plate").
- Put a piece of carpet or a sound-absorbing ceiling tile on the wall behind the refrigerator.
- Allow enough space between the back of the refrigerator and the wall, unless it is designed as a "built-in" unit. Check your use and care manual for the needed space.

To reduce the compressor run time:

- Vacuum the condenser coils at least twice a year—more often if you have pets.
- Keep your freezer at least three-fourths full. Use partially filled water jugs to fill any large empty space.

Many of the consumers filing "refrigerator sound" complaints with the Major Appliance Consumer Action Panel (MACAP) represent one- or two-family households; they have recently moved to smaller retirement homes; or they have remodeled, and the kitchen is now open to a family living area. Sounds are more noticeable in quieter surroundings. Consumers with hearing aids are especially sensitive.

Some consumers report to MACAP that their refrigerators are "louder" than an identical model in a friend's or a relative's house. This might be because of the number of people in the house, as well as different furnishings and room arrangements. Carpeting, drapery, upholstered furniture, and wall coverings can help to muffle refrigerator sounds.[2]

Are Refrigerators/Freezers Snowbirds?

Putting that extra refrigerator or freezer in any area in which the temperature falls below 60 degrees Fahrenheit might be a problem in colder areas when the winter months are approaching. Combination refrigerator/freezers and freezers with automatic defrost systems are sensitive to the ambient air temperature surrounding them. As this ambient temperature rises, the compressor runs more to maintain the storage temperature in the fresh food and frozen food compartments, thus wasting energy. As the ambient temperature falls, compressor operation decreases.

When temperatures fall below 60 degrees Fahrenheit, the compressor will not operate long enough to maintain low storage temperatures in the freezer compartment. This is because the fresh food compartment contains the primary sensor, and it is satisfied quickly at a low ambient temperature. The lower the temperature goes, the worse this condition becomes. At about 38 to 42 degrees Fahrenheit, the compressor will not run at all. The freezer compartment temperature will increase to the ambient room temperature, and the frozen food in the compartment will defrost and spoil.

Combination refrigerator/freezers and automatic-defrost freezers should not be operated in unheated places, like garages or porches, where room temperatures are likely to drop below 60 degrees Fahrenheit, unless they are specifically designed for operation in low temperatures. Check the manufacturer's use and care manual for the lowest safe operating temperature.

At any time the temperature will be less than 60 degrees Fahrenheit, it is best to empty the freezer compartment of the refrigerator/freezer in order to prevent defrosting and the possible spoilage of frozen foods. You might want to consider removing all of the food from the unit, turning it off, and propping the door open if you will be absent for an extended period of time. The door must stay open to prevent mold and odor. *Don't* do this if your refrigerator has a latch-type handle (pre-1958 model) because of the potential for child entrapment.

Manual-defrost freezers can usually be operated in an unheated garage or porch without affecting the unit or the frozen food. However, check the use and care manual to determine if your unit needs special care.[3]

"Freezer" in Single-Door Refrigerator Has Limited Function

"I just bought a single-door refrigerator and my ice cream won't harden. It's like cold soup," a consumer recently complained to MACAP. "If I turn the temperature control to a colder setting, the food in the refrigerator section freezes, but the ice cream remains soup. Something's wrong with the freezer."

Approximately 6 percent of all refrigerators sold each year are "single-door" models. That is, the model has only one outside door. Inside (usually across the top or to one side) is a small freezer compartment with its own door. Most combination refrigerator/ freezers have two or more doors on the outside, providing independent access to separate freezer or refrigerated sections.

In the course of investigating this consumer's complaint, MACAP found that consumers have reason to be confused about the capabilities of "single-door" refrigerators. Many manufacturers refer to the separate frozen food compartment as a "freezer section," as a "freezer compartment," or as a "freezer" in their specification literature. MACAP found that a majority of consumers expect to be able to keep such hard-to-freeze items as ice cream and orange juice in this "freezer" compartment. These items have a high sugar content, and they freeze at lower temperatures than water.

The Association of Home Appliance Manufacturers (AHAM), in its nationally accepted standard, calls such a unit a "basic refrigerator" and specifies that it is "intended for short-term storage of foods at temperatures below 32 degrees Fahrenheit and normally above 8 degrees

Fahrenheit." However, most newer models have temperatures at (or near) freezing level, which is not adequate to freeze foods. Distilled water freezes at 32 degrees Fahrenheit, but all frozen foods must be stored at a temperature lower than that to freeze. Vegetables begin to freeze at 29 to 31 degrees Fahrenheit, meats at 25 to 29 degrees Fahrenheit, and orange juice concentrate at about 8 degrees Fahrenheit. Ice cream begins to stiffen at 27 degrees Fahrenheit, but is considered at an ideal hardness for scooping at 8 degrees Fahrenheit.

If reference is made to the compartment being a "food freezer" or a "frozen food storage compartment" (as in double-door units), the consumer can expect that it will store already frozen foods for several days, or even months, without deterioration. However, a "freezer section" or "freezer compartment," as found in single-door refrigerators, might only freeze ice cubes.

MACAP recommends that consumers determine their food freezing needs and carefully read available literature before making a purchasing decision.[4]

Refrigerator/Freezer Maintenance

The inside of the cabinet should be cleaned at least once a month to help prevent odors from building up. Of course, any spills that might happen should be wiped up immediately. Wash all removable parts by hand with warm water and a mild detergent; then rinse and dry the parts. The inside walls of the cabinet, the door liners, and the gaskets should also be washed using warm water and a mild detergent, rinsed, and dried. Never use cleaning waxes, concentrated detergents, bleaches, or cleansers containing petroleum products on plastic parts. On the outside of the cabinet, use a sponge with warm water and a mild detergent to clean dust and dirt. Then rinse off and dry thoroughly.

At least two times a year, the outside cabinet should be waxed with an appliance wax or with a good auto paste wax. Waxing painted metal surfaces provides rust protection. The defrost pan, which is located behind the toe plate or behind the cabinet, should be cleaned out once a month. The condenser coil should also be cleaned of dust and lint at least once a month. The floor should be free of dirt and debris when the cabinet is rolled out away from the wall. After the cabinet is rolled back into place, you must check to be sure that the cabinet is level.

Step-by-Step Troubleshooting by Symptom Diagnosis

When servicing an appliance, don't overlook the simple things that might be causing the problem. Step-by-step troubleshooting by symptom diagnosis is based on diagnosing malfunctions, with possible causes arranged into categories relating to the operation of the refrigerator/freezer. This section is intended only to serve as a checklist to aid you in diagnosing a problem. Look at the symptom that best describes the problem you are experiencing with the refrigerator/freezer, and then correct the problem.

Compressor Will Not Run

- Is there voltage at the wall receptacle? Check this with the voltmeter.
- Check for loose electrical connections.
- Is the condenser coil dirty? A dirty condenser coil will overheat the compressor.

PART V

- Check the condenser fan motor.
- Test the cold control for continuity.
- Test the compressor, the relay, and the overload switch.

Compressor Kicks Out on Overload

- Check for high or low voltage when the compressor tries to start. High voltage will overheat the compressor. Low voltage will try to run the compressor with the start winding. A compressor is designed to start and run within a 10 percent tolerance of the rated voltage.
- Test the capacitor. A shorted or open capacitor will overheat the compressor.
- Test the compressor relay.
- Test the overload for continuity.
- Test the compressor windings for a short.

Refrigerator Is Too Cold

- Check the damper control setting. Check to see if the damper is stuck open (thermostatically controlled dampers only).
- Test the cold control switch contacts for continuity. Test for stuck contacts.
- Check the location of the refrigerator. If outside in the winter, the ambient temperature may be too cold.

Refrigerator Is Too Warm

- Check for restricted air circulation around the condenser coil.
- Check the location of the refrigerator.
- Check the door gaskets for proper sealing.
- Check to see if the cabinet light is staying on when the door is closed.
- Check the defrost heaters. Use a clamp-on ammeter (or wattmeter) to test the heaters if they are coming on when the refrigeration cycle is running.
- Check the cold control setting.
- Check the compressor. Is it operating properly?
- Is the evaporator fan running?
- Check the air duct for restriction.
- Check for a leaking air duct.
- Check the evaporator coil for excessive frost buildup.
- Check the defrost cycle. Is it working properly?
- Check the damper control setting. Check to see if the damper is stuck closed (thermostatically controlled dampers only).

Refrigerator/Freezer Is Too Noisy

- Check for loose parts.
- Check for rattling pipes.
- Check the fan assembly, the evaporator, and the condenser.
- Check the compressor.
- If these are normal operational noises, inform the consumer.
- Refrigerator/freezer not properly leveled.
- Check the floor—it may not be structurally sound.

Sweating on the Outside of the Cabinet

- Check the location of the refrigerator/freezer. If located in an area of high humidity, it will begin to sweat.
- Check for a void in the insulation between the cabinet and the inner liner.
- Test the mullion and/or stile heaters for continuity.
- Is the energy saver switch in the "on" position?
- On older models, check for wet insulation.
- Check for suction line or any low side tubing touching the cabinet.
- Check for water leaks from the ice maker.
- Check chilled water supply lines and connections.
- Check for a kinked, misaligned, or blocked drain system.
- Check the defrost drain pan for misalignment or for leaking cracks.
- Are the doors aligned properly?

Sweating on the Inside of the Cabinet

- Check for any abnormal usage. Instruct the consumer on proper usage.
- Check the door gaskets for proper sealing.
- Check for defrost drain water leaking into the cabinet.
- Is the condensate drain blocked?
- Are the doors aligned properly?
- Inspect all access holes where tubing or wires enter the refrigerator/freezer. Seal with permagum if necessary.
- Inspect cabinet outer walls and seams for any openings. Seal with Permagum if necessary.
- Are there excessive door openings on hot, humid days?
- Check for improper food storage.

Incomplete Defrosting of the Evaporator Coil or High Temperature During the Defrost Cycle

- Test the defrost thermostat.
- Check for loose wiring in the defrost electrical circuit.
- Test the defrost timer for continuity.
- Test for defective defrost heaters.

Odor in Cabinet

- Check for spoiling food in the cabinet.
- Check the defrost water drain system.
- Check the defrost heaters.

Excessive Frost Buildup on Evaporator Coil in Freezer

- Check the defrost cycle.
- Check for loose wiring.
- Is the heater making contact with the evaporator coil?
- Check for proper door alignment.
- Check the door gaskets.

Freezer Run Time Is Too Long

- Check the thermostat setting.
- Check for excessive loading of unfrozen food.
- Check for incorrect wiring.

Temperature in Freezer Is Higher Than Normal

- Check the thermostat for proper temperature calibration.
- Test evaporator fan motor and blade.
- Check the defrost timer.
- Check for excessive loading of unfrozen food.
- Check door gasket for proper sealing.

Refrigerator/Freezer Runs Excessively or Continuously

- Check if the interior lights are staying on continuously.
- Check condenser coil for air restriction.
- Check door gaskets.
- On models with automatic ice makers, make sure the ice maker is operating properly.

Troubleshooting Sealed-System Problems

If you suspect a sealed-system malfunction, be sure to check all external factors first. These include:

- Thermostats
- Compressor
- Relay and overload on the compressor
- Interior lights
- Evaporator and condenser fans
- Timers
- Refrigerator/freezer getting good air circulation
- Food loaded in the refrigerator/freezer properly
- Check if heat exchanger has separated
- Check the wiring harness

After eliminating all of these external factors, you will systematically check the sealed system. This is accomplished by comparing the conditions found in a normally operating refrigerator/freezer. These conditions are:

- Refrigerator/freezer storage temperature
- Wattage
- Condenser temperature
- Evaporator inlet sound (gurgle, hiss, etc.)
- Evaporator frost pattern
- High-side pressure[5]
- Low-side pressure[6]
- Pressure equalization time

One thing to keep in mind is that no single indicator is conclusive proof that a particular sealed-system problem exists. Rather, it is a combination of findings that must be used to definitely pinpoint the exact problem.

Refrigerant Leak

The following symptoms may indicate there is a refrigerant leak in the sealed system:

- Temperatures in the storage area are below normal.
- The wattage and amperage are below normal, as indicated on the model/serial plate.
- The condenser coil will be cool to the touch at the last pass or even as far as midway through the coil.
- At the evaporator coil, you will hear a gurgling noise, a hissing noise, or possibly an intermittent hissing or gurgling noise.

- When the evaporator coil cover is removed, the evaporator coil will show a receded frost pattern.
- The high- and low-side pressures will be below normal.[6]
- The pressure equalization time might be normal or shorter than normal.

Overcharged Refrigerator/Freezer

If the sealed system is overcharged, the symptoms are:

- The storage temperature will be higher than normal.
- The wattage and amperage are above normal, as indicated on the model/serial plate.
- The temperature of the condenser coil will be above normal.
- At the evaporator coil, you will hear a constant gurgling noise. Generally, this is a higher sound level than normal.
- When the evaporator coil cover is removed, the evaporator coil will show a full frost pattern. If you remove the back cover, located behind the refrigerator/freezer, you will possibly see the suction line frosted back to the compressor.
- The high- and low-side pressures will be above normal.[6]
- The pressure equalization time will be normal.

Slight Restriction

The symptoms of a slight restriction in the sealed system are:

- The storage temperature will be below normal.
- The wattage and amperage are below normal, as indicated on the model/serial plate.
- The temperature of the condenser coil will be slightly below normal.
- At the evaporator coil, you will hear a constant gurgling noise and a low sound level.
- When the evaporator coil cover is removed, the evaporator coil pattern will be receded.
- The high- and low-side pressures will be below normal.[6]
- The pressure equalization time will be longer than normal.

Partial Restriction

The symptoms of a partial restriction in the sealed system are:

- The storage temperature will be higher than normal.
- The wattage and amperage are below normal, as indicated on the model/serial plate.
- The temperature of the condenser coil will be below normal more than halfway on the coil.
- At the evaporator coil, you will hear a constant gurgling noise and a considerably low sound level.

- When the evaporator coil cover is removed, the evaporator coil will be considerably receded.
- The high- and low-side pressures will be below normal.[6]
- The pressure equalization time will be longer than normal.

Complete Restriction
The symptoms of a complete restriction in the sealed system are:

- The storage temperature will be warm.
- The wattage and amperage will be considerably below normal, as indicated on the model/serial plate.
- The temperature of the condenser coil will be cool or at room temperature.
- At the evaporator coil, you will hear no sounds.
- When the evaporator coil cover is removed, the evaporator coil will not have any frost on it or the frost will be melting.
- The high-side pressure will be equal to the pressure of refrigerant at room temperature.[6]
- The low-side pressure will be in a deep vacuum.[6]
- There will be no pressure equalization time.

Moisture Restriction
The symptoms of a moisture restriction in the sealed system are:

- The storage temperature will be above normal.
- The wattage and amperage will be considerably below normal, as indicated on the model/serial plate.
- The temperature of the condenser coil will be below normal.
- At the evaporator coil, you will hear a constant gurgle, low sound level, or no sound at all.
- When the evaporator coil cover is removed, the evaporator coil might have some frost on the evaporator inlet.
- The high-side pressure will be below normal.[6]
- The low-side pressure will be below normal or in a deep vacuum.[6]
- The pressure equalization time will be longer than normal or there will be no equalization at all.

Low-Capacity Compressor
The symptoms of a low-capacity compressor in the sealed system are:

- Temperatures in the storage area will be above normal.
- The wattage and amperage will be below normal, as indicated on the model/serial plate.
- The temperature of the condenser coil will be below normal.

- At the evaporator coil, you will hear a slightly reduced gurgling noise.
- When the evaporator coil cover is removed, the evaporator coil will show a normal frost pattern.
- The high-side pressure will be below normal; the low-side pressure will be above normal.[6]
- The pressure equalization time might be normal or shorter than normal.

Repair Procedures

Each repair procedure is a complete inspection and repair process for a single refrigerator/freezer component. It contains the information you need to test the components and replace them, if necessary.

Electronic Components

The typical complaints associated with the electronic components are:

- Unable to program the touchpad panel functions.
- The display board will not display anything.
- Unusual display readouts.
- Unable to control the temperatures.
- The compressor will not run.

To prevent electrostatic discharge (ESD) from damaging expensive electronic components, follow the steps in Chapter 9.

To handle these problems, perform the following steps:

1. **Verify the complaint.** Verify the complaint by operating the refrigerator controls. Turn off the electricity to the appliance and wait for two minutes before turning it back on. If a fault code appears, look up the code. If the refrigerator will not power up, locate the technical data sheet behind the panel or cabinet for diagnostics information.

2. **Check for external factors.** You must check for external factors not associated with the appliance. Is there electricity to the refrigerator? Is the electrical receptacle polarized and properly grounded?

3. **Disconnect the electricity.** Before working on the refrigerator, disconnect the electricity. This can be done by pulling the plug out of the wall receptacle. Or disconnect the electricity at the fuse panel or circuit breaker panel. Turn off the electricity.

4. **Gain access to the electronic components.** You can gain access to the electronic components by removing the screws on the access panel. Depending on the model you are servicing, the electronic components can be in the refrigerator or freezer compartment, on the outside of the freezer door, or located under the cabinet.

5. **Test the electronic components.** If you are able to run the refrigerator's diagnostic test mode, check the different functions of the refrigerator. Use the technical data sheet for the model you are servicing to locate the test points on the wiring schematic. Check all

wiring connections and wiring. Using the technical data sheet, test the electronic control and display board, input voltages, and output voltages. Some high-end refrigerator models have variable-speed compressors. These compressors are controlled by an inverter, which controls the compressor speed.

6. **Remove the defective component.** To remove the defective component, remove the screws that secure the PCB or inverter to the refrigerator. Disconnect the connectors from the electronic control board, display board, or inverter.

7. **Install the new component.** To install a new component, just reverse the disassembly procedure, and reassemble. Reinstall all panels or the console panel, and restore the electricity to the refrigerator. Test the refrigerator operation.

Door Gasket

The typical complaints associated with door gasket failure are:

- Sweating inside the cabinet.
- Temperatures inside the cabinet are warmer than normal.
- Ice forming on the freezer walls.
- Door gaskets not maintained properly.

To handle these problems, perform the following steps:

1. **Verify the complaint.** Verify the complaint by checking the door gasket for proper sealing and alignment. Inspect the gaskets for any damage.

2. **Check for external factors.** You must check for external factors not associated with the appliance. Is the appliance installed properly? Were the doors reinstalled correctly?

3. **Disconnect the electricity.** Before working on the refrigerator/freezer, disconnect the electricity. This can be done by pulling the plug from the receptacle. Or disconnect the electricity at the fuse panel or at the circuit breaker panel. Turn off the electricity.

4. **Gain access to the door gaskets.** To access the door gaskets, open the refrigerator/freezer door. The gaskets are located on the door.

5. **Test the door gaskets.** To test the gaskets for proper sealing, take a dollar bill and place it between the gasket and the flange of the outer cabinet (Figure 21-4). Pull on the dollar bill. When pulling on the dollar bill, you should feel some tension as the bill is gripped. This means that the gasket is making good contact with the refrigerator/freezer flange. Repeat this test in other areas where you suspect problems with the gasket. If the gasket fails this test, the next step is to replace the gasket. For the doors to close and seal properly, the refrigerator should tilt backwards 1/4 of an inch. This is accomplished by raising the front legs or wheels

Figure 21-4 The gasket must make full contact with the cabinet flange.

FIGURE 21-5 Warped or sagging doors must be corrected so that the refrigerator/freezer will operate properly.

according to the installation instructions. If you are still unable to get the doors to close properly, check the doors for sagging or warping (Figure 21-5). Also, check the floor to see if it is level under the refrigerator/freezer. Check from front to back and from side to side.

6. **Remove the door gasket.** Before you get started, remove all of the food from the door. To remove the door gasket, pull back on the gasket—this exposes the retaining strip and screws. Loosen the screws about halfway, but do not remove them (Figure 21-6). Gently remove the gasket from around the door (Figure 21-7).

FIGURE 21-6 Peel back the gasket to gain access to the retaining strip and screws.

FIGURE 21-7 Carefully remove the gasket so as not to damage the inner door liner.

FIGURE 21-8 Have someone help you hold the door straight when tightening the screws.

7. **Install a new door gasket.** Before installing the new gasket, soak it in warm water for about 15 to 20 minutes. This will make the gasket soft and easier to install. Starting at either top corner, insert the flange of the gasket behind the retaining strip and/or door liner. Proceed all the way around the door. When the gasket is in place, begin to tighten the screws slightly all around the door. Now close the door; the gasket should make contact with the cabinet flange evenly and all around the door (Figure 21-8). The door gasket might be adjusted by aligning the door panel, as shown in Figure 21-9. To align the door, twist the door in the opposite direction of the warp. Close the door, and check that the gasket is sealed against the cabinet. Now that the door gasket is sealing properly, tighten the screws completely. If the gasket is distorted, or if it has wrinkles in it, use a hair dryer to heat the gasket and mold it to its original form.

Then recheck to be sure the gasket seats against the flange properly. Next, check the gap between the door and the cabinet on the hinge side. Use a penny, which is about 3/4 of an inch in diameter, to check the gap. Slide the penny from the top hinge to the bottom hinge. The door might be adjusted by moving the top hinge and by adding or removing shims to the center and bottom hinges. Be sure that the doors line up evenly with the sides of the cabinet and evenly with each other. Also, check that when both doors are opened simultaneously (top-mount refrigerators only), they do not hit against each other.

FIGURE 21-9
The correct gap between the door and cabinet will allow the door to close and seal properly. The gap should also be checked as the gasket starts rolling inward when the door is closed.

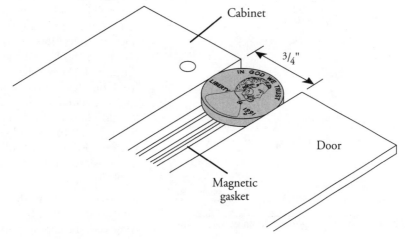

Cabinet

3/4"

Door

Magnetic gasket

Thermostat (Cold Control)

The typical complaints associated with failure of the thermostat (cold control) are:

- The refrigerator/freezer is not cold enough.
- The refrigerator/freezer is too cold.
- The refrigerator/freezer runs all the time.
- The refrigerator/freezer doesn't run at all.

To handle these problems, perform the following steps:

1. **Verify the complaint.** Verify the complaint by checking the control setting. Turn the control off; then turn it on again and see if the refrigerator/freezer starts.

2. **Check for external factors.** You must check for external factors not associated with the appliance. Is the appliance installed properly? Explain to the user how to set the controls.

3. **Disconnect the electricity.** Before working on the refrigerator/freezer, disconnect the electricity to the refrigerator/freezer. This can be done by pulling the plug from the receptacle. Or disconnect the electricity at the fuse panel or at the circuit breaker panel. Turn off the electricity.

4. **Gain access to the thermostat.** To access the thermostat, open the fresh food door. Look for the control dial that has the word "off" printed on it. This is the control that turns the compressor on and off. Remove the dial (Figure 21-10). Next, remove the two screws that secure the control (Figure 21-11). Remove the wires from the terminals. On some models, the capillary tube is inserted in the air duct; on other models, the capillary tube might be attached to the evaporator coil. At this point, if the capillary tube is attached to the coil, do not remove the capillary tube yet. Test the control first.

5. **Test the thermostat.** If the capillary tube lost its charge, the refrigerator/freezer might not cool or it could freeze the food in the fresh food compartment or keep the temperatures warmer than normal. To test the thermostat, place the ohmmeter probes on the terminals (Figure 21-12). Set the range scale on R × 1, and test for continuity. With the control set in the "off" position, you should not read continuity. When the control is set to the highest position, you should read continuity. If the thermostat is good, the problem must be elsewhere.

6. **Remove the thermostat.** With the thermostat control housing already removed, the capillary tube must now

Thermostat
control dial

FIGURE 21-10 Pull the dial off gently to gain access to the thermostat.

FIGURE 21-11   Be sure that the electricity is disconnected before attempting to remove the thermostat. Note how the capillary tube is routed and if it is secured to anything. Also note in which direction the terminal end of the thermostat is pointed when removing it. If you install the thermostat upside down, the dial indicators will be incorrect. The thermostat inside the cabinet could be set on the wrong position.

FIGURE 21-12   Check thermostat switch contacts for continuity and inspect the capillary tube for damage.

be removed. Because there are many different models on the market today, Figure 21-13 represents only a few types. The capillary tube might be routed through the control housing (Figure 21-13a). The capillary tube might be secured to the evaporator by means of a clamp (Figure 21-13b). The capillary tube might be inserted into a housing that senses how cold the air is (Figure 21-13c). Whichever way the capillary tube is installed, remove it carefully so as not to damage the other components.

7. **Install a new thermostat.** To install the new thermostat, just reverse the order of disassembly, and reassemble. Then test the control. Remember to reinstall the capillary tube in the same location from which it was removed. If you do not, the refrigerator/ freezer will not work properly.

Defrost Timer

The purpose of the defrost timer is to regulate the frequency of the defrost cycles and their duration. The typical complaints associated with failure of the defrost timer are:

- The refrigerator/freezer does not defrost.
- The storage temperature in the cabinet is too warm.
- The compressor will not run.

To handle these problems, perform the following steps:

1. **Verify the complaint.** Verify the complaint by asking the customer to describe what the refrigerator/freezer is doing or did.

2. **Check for external factors.** You must check for external factors not associated with the appliance. Is the appliance installed properly? Are the doors aligned properly?

Figure 21-13 (a) The capillary tube is attached to the control panel. It senses the airflow temperature. (b) The capillary tube is attached to the evaporator plate. Note exactly where the capillary tube is attached. If you remove it for any reason, you must reattach it in the same position. (c) Capillary tube location within the housing.

3. **Disconnect the electricity.** Before working on the refrigerator/freezer, disconnect the electricity to the appliance. This can be done by pulling the plug from the receptacle. Or disconnect the electricity at the fuse panel or at the circuit breaker panel. Turn off the electricity.

4. **Gain access to the defrost timer.** To access the defrost timer, you must first locate it. On some models, the defrost timer is located on the bottom, behind the toe plate; or it might be behind the temperature control housing, in the fresh food section (Figure 21-14); or it might be in the back of the refrigerator, behind the rear leg (Figure 21-15).

5. **Remove the defrost timer.** In order to test the defrost timer, it must be removed from its mounting position. Remove the two mounting screws from the defrost timer (see Figure 21-15). Next, remove the wire harness plug from the defrost timer (Figure 21-16).

FIGURE 21-14
Defrost timer and
thermostat located
together in one
housing.

Defrost timer

Lamp
socket

Temperature
control

Housing

FIGURE 21-15
Remember to always
reconnect the green
ground wire to the
defrost timer and
ground.

Rear leg

Defrost timer

Quick
disconnect
plug

Ground
wire

Timer
mounting
screws (2)

FIGURE 21-16
Not all defrost timers
have harness plug
connectors. Some are
wired with individual
wires. If you forget
how to reconnect the
wires properly, read
the wiring diagram.

Harness
plug

Defrost
timer

FIGURE 21-17 Always rotate the timer cam clockwise.

6. **Test the defrost timer.** To test the defrost timer, place a screwdriver in the timer cam slot (Figure 21-17), and turn it clockwise until you hear the first "snap." The defrost timer is now in the defrost cycle. At this point, you must read the wiring diagram to determine which numbered terminals are for the defrost circuit. For the purpose of demonstrating how to check for continuity of the switch contacts, Figure 21-18a illustrates the internal components of this sample timer. Set the ohmmeter scale on R × 1, and place the probes on the terminals marked 2 and 3 (Figure 21-19). You should measure continuity. Next, rotate the timer cam until you hear the second "snap." The meter will show no continuity, indicating that the defrost cycle is over and that the refrigeration cycle begins.

Now place the meter probes on the terminals marked 3 and 4. The ohmmeter will show continuity, indicating the refrigeration cycle is activated. Turn the timer cam once again, until you hear the first "snap." The meter will show no continuity. At no time should there be continuity between terminals 2 and 4. (If so, the switch contacts are burned and welded together and the defrost timer must be replaced.) If the defrost timer passes this portion of the test, you must determine if the timer mechanism is functioning. Place the ohmmeter probes on the timer motor leads and read the resistance. The resistance can be between 800 and 4000 ohms, depending on the type of timer used by the manufacturer. If you are unable to read resistance, the timer motor is defective.

If the defrost timer passes this portion of the test, rotate the timer cam until you hear the first "snap." Advance the timer cam again, counting the number of clicks until you hear the second "snap." Write down the number of clicks on a piece of paper. Now rotate the

FIGURE 21-18 If you have continuity between terminals 2, 3, and 4, this indicates the switch contact points are all welded together. Symptoms in the refrigerator/freezer will be warmer temperatures than normal.

Figure 21-19   Check switch contacts for continuity.

timer cam again until the marks line up (Figure 21-18b), which indicates the beginning of the defrost cycle, and the "snap" is heard. Advance the timer cam and count the clicks until there is one click left before the end of the defrost cycle. Take the timer and reconnect it to the wiring harness (see Figure 21-16). Place the defrost timer on a nonmetallic surface.

Reconnect the voltage supply to the refrigerator/freezer. *Note:* Be cautious when working with live wires. Avoid getting shocked. Place the ammeter jaws around the wire attached to the number 4 terminal. The meter should indicate no amperage. Next, place the jaws on the number 2 terminal wire. The ammeter should indicate some amperage. Wait for approximately 10 to 15 minutes: you should hear a "snap," indicating that the timer has completed the defrost cycle. At this point, the ammeter will show no amperage on number 2, but will indicate current flow at number 4. If not, replace the timer.

7. **Install a new defrost timer.** To install the new defrost timer, just reverse the order of disassembly, and resemble. Remember to reconnect the ground wire to the defrost timer.

Evaporator Fan Motor

The typical complaints associated with failure of the evaporator fan motor are:

- The refrigerator/freezer temperature is warm.
- The evaporator fan motor runs slower than normal.
- The evaporator fan motor does not run at all.
- The evaporator fan motor is noisy.

To handle these problems, perform the following steps:

1. **Verify the complaint.** Verify the complaint by asking the customer to describe what the refrigerator/freezer is doing or did. Is the evaporator fan motor running? Is it noisy?

2. **Check for external factors.** You must check for external factors not associated with the appliance. Is the appliance installed properly? Is there something hitting the fan blade?

3. **Disconnect the electricity.** Before working on the refrigerator/freezer, disconnect the electricity. This can be done by pulling the plug from the receptacle. Or disconnect the electricity at the fuse panel or at the circuit breaker panel. Turn off the electricity.

4. **Gain access to the evaporator fan motor.** To access the evaporator fan motor, the evaporator cover must be removed (Figure 21-20). Remove the screws that secure the cover in place. On some models, the evaporator fan assembly is located on the rear wall of the interior freezer compartment.

PART V

5. **Test the evaporator fan motor.** The evaporator fan motor should be tested for proper resistance, as indicated on the wiring diagram. To test the evaporator fan motor, remove the wires from the motor terminals. Next, place the probes of the ohmmeter on the motor terminals (Figure 21-21). Set the scale on R × 1. The meter should show resistance. If no reading is indicated, replace the motor. If the fan blade does not spin freely, replace the motor.

6. **Remove the evaporator fan motor.** To remove the evaporator fan motor, you must first remove the fan blade. On most models, just pull the blade off the motor shaft. Be careful not to break the blade. On other models, the fan blade is held on the motor shaft with screws. Remove the screws. Then remove the screws that secure the fan assembly to the cabinet (Figure 21-22). On some models, you must remove the fan shroud (Figure 21-23) by removing the shroud screws.

7. **Install a new evaporator fan motor.** To install the new evaporator fan motor, just reverse the order of disassembly, and reassemble. When reinstalling the fan blades onto the motor shaft, the fan blades should be positioned on the shaft so that one-third of its depth (approximately 1/4 inch) protrudes through the fan orifice in the direction of airflow. When reinstalling any shrouds, grilles, ducts, or gaskets, always position them correctly to ensure the proper airflow through the evaporator and within both compartments of the refrigerator/freezer. Remember to reconnect the ground wire to the motor. Reconnect the wires to the motor terminals, and test.

Condenser Fan Motor

The typical complaints associated with failure of the condenser fan motor are:

- The refrigerator/freezer temperature is warm.

- The condenser fan motor runs slower than normal.

- The condenser fan motor does not run at all.

- The compressor is sometimes noisier than normal.

FIGURE 21-21 Check the evaporator fan motor for resistance. Also check the motor for grounded windings.

FIGURE 21-22   Exploded view of an evaporator fan motor assembly.

To handle these problems, perform the following steps:

1. **Verify the complaint.** Verify the complaint by asking the customer to describe what the refrigerator/freezer is doing. Is the condenser fan motor running?

2. **Check for external factors.** You must check for external factors not associated with the appliance. Is the appliance installed properly? Is there any foreign object blocking the condenser fan blade?

FIGURE 21-23   Removing the fan shroud to gain access to the motor terminals.

FIGURE 21-24
Removing the rear panel will expose the components that need to be serviced.

3. **Disconnect the electricity.** Before working on the refrigerator/freezer, disconnect the electricity. This can be done by pulling the plug from the receptacle. Or disconnect the electricity at the fuse panel or at the circuit breaker panel. Turn off the electricity.

4. **Gain access to the condenser fan motor.** Pull the refrigerator/freezer out and away from the wall. Remove the back panel, which is located at the bottom of the refrigerator/freezer. This will expose the compressor, the condenser fan assembly, and the condenser coil (Figure 21-24).

5. **Test the condenser fan motor.** A condenser fan motor should be tested for proper resistance, as indicated on the wiring diagram. Check the fan blade for obstructions. The blade should turn freely. Next, rotate the fan blade and check for bad bearings. If you hear any unusual noises coming from the motor, or if the fan blade is sluggish when spinning, replace the motor.

To test the condenser fan motor, remove the wires from the motor terminals. Next, place the probes of the ohmmeter on the motor terminals (Figure 21-25). Set the scale on R × 1. The meter should show resistance. If no reading is indicated, replace the motor.

6. **Remove the condenser fan motor.** To remove the condenser fan motor, you must first remove the fan blades. Unscrew the nut that secures the blades to the motor. Remove the blades from the motor. Then remove the motor assembly by removing the mounting bracket screws (Figure 21-26).

7. **Install a new condenser fan motor.** To install the new condenser fan motor, just reverse the order of disassembly, and reassemble. Remember to reconnect the ground wire to the motor. Reconnect the wires to the motor terminals, and test.

FIGURE 21-25 Remember to set the meter on the ohm scale when testing for resistance in the condenser fan motor.

Figure 21-26   Remove the fan blade first. This will prevent the blades from bending out of shape and becoming off balance. On some models, the compressor is located within inches of the motor. Be careful! The compressor might be hot, and you could burn yourself.

Defrost Heater

The typical complaints associated with failure of the defrost heater are:

- The refrigerator temperature is warm.
- The freezer temperature is warm.
- The refrigerator/freezer does not defrost.
- Food is spoiling.
- No ice cubes.

To handle these problems, perform the following steps:

1. **Verify the complaint.** Verify the complaint by asking the customer to describe what the refrigerator/freezer is doing. Is food spoiling? Check the temperature in the compartments. Check for ice buildup on the evaporator cover.

2. **Check for external factors.** You must check for external factors not associated with the appliance. Is the appliance installed properly?

3. **Disconnect the electricity.** Before working on the refrigerator/freezer, disconnect the electricity. This can be done by pulling the plug from the receptacle. Or disconnect the electricity at the fuse panel or at the circuit breaker panel. Turn off the electricity.

4. **Gain access to the defrost heater.** To access the defrost heater, the evaporator cover must be removed (see Figure 21-20). Remove the screws that secure the cover in place.

5. **Test the defrost heater.** A defrost heater should be tested for proper resistance, as indicated on the wiring diagram. To test the defrost heater, remove the wires from the heater terminals. Next, place the probes of the ohmmeter on the heater terminals (Figure 21-27). Set the scale on R × 1. The meter should show resistance. If no reading is indicated, replace the defrost heater.

6. **Remove the defrost heater.** To remove the defrost heater in this type of model, you must first remove the reflector shield (Figure 21-28). Bend the clip up and lift the shield. Do the same for the other end of the heater. Once the shield is removed, you can lift the defrost heater from its brackets.

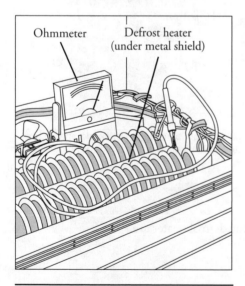

Figure 21-27 If the model you are repairing has glass defrost heaters in it that look black, dark smoky-gray, or burned, the heater is either defective or soon will be.

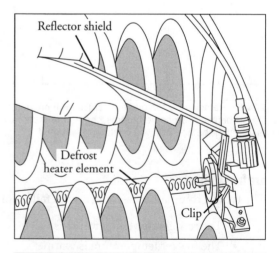

Figure 21-28 When removing this type of heater, be careful not to break the glass. The clip must be straight so the heater can slide out of the brackets.

FIGURE 21-29   After the heater is installed and the shield is back in place, bend back the clip on the bracket. When you reinstall the wires on the terminals, use one hand to hold the end of the heater. This will prevent the glass cracking if you push down too hard.

7. **Install a new defrost heater.** To install the new defrost heater (Figure 21-29), just reverse the order of disassembly, and reassemble. *Note:* On some models, do not touch the glass because it will shorten the life of the heater. Remember to reconnect the wires to the heater. When reinstalling any shrouds, grilles, ducts, or gaskets, always position them correctly to ensure the proper airflow through the evaporator and within both compartments of the refrigerator/freezer.

Defrost Thermostat

The typical complaints associated with failure of the defrost thermostat are:

- The refrigerator temperature is warm.
- The freezer temperature is warm.
- The refrigerator/freezer does not defrost.
- Food is spoiling.
- No ice cubes.

To handle these problems, perform the following steps:

1. **Verify the complaint.** Verify the complaint by asking the customer to describe what the refrigerator/freezer is doing. Is food spoiling? Check the temperature in the compartments. Check for ice buildup on the evaporator cover.

2. **Check for external factors.** You must check for external factors not associated with the appliance. Is the appliance installed properly?

FIGURE 21-30   Test the defrost thermostat when it is connected to the evaporator coil.

3. **Disconnect the electricity.** Before working on the refrigerator/freezer, disconnect the electricity. This can be done by pulling the plug from the receptacle. Or disconnect the electricity at the fuse panel or at the circuit breaker panel. Turn off the electricity.

4. **Gain access to the defrost thermostat.** To access the defrost thermostat, the evaporator cover must be removed (see Figure 21-20). Remove the screws that secure the cover in place.

5. **Test the defrost thermostat.** Failure of a defrost thermostat usually results in a frost-blocked evaporator. To test the defrost thermostat, disconnect the wires to isolate the thermostat from the rest of the defrost circuit. Next, place the probes of the ohmmeter on the defrost thermostat wire leads (Figure 21-30).

Part number	Temperature setting °F	
	Open	Close
L - 45	45	25
L - 50	50	30
L - 55	55	35
L - 60	60	40
L - 70	70	50
L - 80	80	30
L - 90	90	35

Rating stamped
on side of thermostat

Figure 21-31 A temperature rating chart for common defrost thermostats.

Set the meter scale on R × 1. The meter will show continuity when the thermostat is either frozen or very cold, indicating the defrost thermostat is good. The defrost thermostat switch contacts close when the temperature is colder than its temperature rating (Figure 21-31). If no reading is indicated, replace the defrost thermostat. At ambient temperature, you will read no continuity, which will indicate the thermostat might be good.

6. **Remove the defrost thermostat.** To remove the defrost thermostat, you must remove the hold-down clamp. On some models, the defrost thermostat and clamp are one assembly. On other models, the defrost thermostat clamps around the evaporator tubing. As shown in Figure 21-32, remove this type of defrost thermostat by squeezing in on the clip and lifting the thermostat up.

7. **Install a new defrost thermostat.** To install the new defrost thermostat, just reverse the order of disassembly, and reassemble. Remember to reconnect the wires to the thermostat. When reinstalling any shrouds, grilles, ducts, or gaskets, always position them correctly to ensure the proper airflow through the evaporator and within both compartments of the refrigerator/freezer. On models that have the defrost thermostat attached to the evaporator coil, you must reinstall the defrost thermostat in the same location from which it was removed.

Wire cap

Figure 21-32 When replacing the defrost thermostat, be sure you reinstall it in the same position from which you removed it. Otherwise, the defrost cycle will not function properly.

Compressor, Relay, and Overload Protector

The typical complaints associated with failure of the compressor are:

- The refrigerator temperature is warm.
- The freezer temperature is warm.
- The refrigerator/freezer does not run at all.
- Food is spoiling.

To handle these problems, perform the following steps:

1. **Verify the complaint.** Verify the complaint by asking the customer to describe what the refrigerator/freezer is doing.

2. **Check for external factors.** You must check for external factors not associated with the appliance. Is the appliance installed properly? Does the refrigerator/freezer have the correct voltage? Check for a voltage drop during refrigerator startup.

3. **Disconnect the electricity.** Before working on the refrigerator/freezer, disconnect the electricity. This can be done by pulling the plug from the receptacle. Or disconnect the electricity at the fuse panel or at the circuit breaker panel. Turn off the electricity.

4. **Gain access to the compressor.** To access the compressor, pull the refrigerator/ freezer out and away from the wall. Remove the back panel, which is located at the bottom of the refrigerator/freezer. This will expose the compressor, the condenser fan assembly, and the condenser coil (see Figure 21-24). Next, remove the compressor terminal cover (Figure 21-33) by removing the retaining clip that secures the cover. Remove the terminal cover.

5. **Test the compressor relay.** To test the compressor relay, remove the relay by pulling it off the compressor terminals without twisting it (Figure 21-34a). Remove the wires from the relay and label them. On the relay body is stamped the word TOP. Hold the relay so that TOP is in the up position.

Next, place the probes of the ohmmeter on the relay terminals marked S and M. Set the meter scale on R × 1. The reading will show no continuity. Then remove the probe from the terminal marked M, and place it on the side terminal marked L. The reading will show no continuity. Now, move the probe from terminal S, and place it on the terminal marked M. The reading will show continuity. With the probes

Figure 21-33 Removing the terminal cover to gain access to the relay and overload protector.

<figure>
(a)   (b)

FIGURE 21-34   Pull the relay off the terminals without twisting it. This will prevent you from breaking the compressor terminals.
</figure>

still attached, turn the relay upside down (Figure 21-34b), and perform the same tests. By turning the relay over, the switch contacts in the relay will close. When you retest the relay, you should get the opposite results: You should have continuity between terminals S and M and between S and L; however, the meter will not read continuity between M and L. If the relay fails this test, replace it.

6. **Test the overload protector.** To test the overload protector, remove the wires from the overload and compressor terminals. Then remove the overload protector from the compressor by removing the retaining clip that secures the overload protector to the compressor (Figure 21-35a). Next, place the probes of the ohmmeter on the overload terminals (Figure 21-35b). Set the meter scale on R × 1. The reading will show continuity. If not, replace the overload protector.

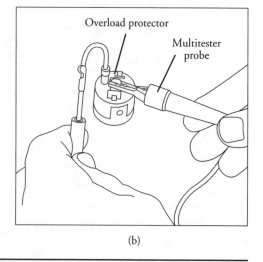

(a)   (b)

FIGURE 21-35   Testing the overload protector for continuity between the terminals.

(a)   (b)

FIGURE 21-36   Testing the compressor motor windings for resistance.

7. **Test the compressor.** To test the compressor, remove the relay and the overload protector. This will expose the compressor terminals. The compressor terminals are marked C, S, and R. C indicates the common winding terminal; S indicates the start winding terminal; and R indicates the run winding terminal. (Refer to the actual wiring diagram for the model that you are servicing.) Set the meter scale on R × 1, touch the probes together, and adjust the needle setting to indicate a zero reading. Next, place the probes of the ohmmeter on the terminals marked S and R (Figure 21-36a). The meter reading will show continuity. Now place the meter probes on the terminals marked C and S. The meter reading will show continuity. Finally, place the meter probes on the terminals marked C and R. The meter reading will show continuity. The total number of ohms measured between S and R is equal to the sum of C to S plus C to R. The compressor should be tested for proper resistance, as indicated on the wiring diagram.

To test the compressor for ground, place one probe on a compressor terminal, and attach the other probe to the compressor housing or any good ground (Figure 21-36b). Set the meter scale to R × 1000. The meter reading will show no continuity. Repeat this for the remaining two terminals. The meter reading will show no continuity. If you get a continuity reading from any of these terminals to ground, the compressor is grounded. Replace it.

Upright Freezers

Upright freezers are similar to refrigerator/freezers in design and operation. They share some of the same features:

- Fan motors
- Compressor
- Automatic defrost system
- Door gasket
- Thermostat
- Interior lighting

FIGURE 21-37 Refrigerant flow of an upright freezer with manual defrost.

FIGURE 21-38 The airflow pattern of an upright automatic-defrost freezer.

On manual-defrost models,[7] the evaporator coils are the shelves inside the cabinet. Figure 21-37 illustrates the refrigerant flow on this type of manually defrosted upright freezer. The condenser coils are embedded between the cabinet liners and are secured to the inside wall of the outer cabinet. This provides for even heat removal and it eliminates the need for a condenser fan motor.

Automatic-defrost models[8] use a fan motor to circulate the air inside the cabinet through air ducts. The evaporator coil is mounted on the inside back wall of the inner liner. Figure 21-38 illustrates the airflow pattern in an upright automatic-defrost freezer. Figure 21-39 illustrates the refrigerant flow in this type of upright freezer.

To diagnose and repair the upright freezer, consult the troubleshooting and the repair procedures sections of this chapter.

FIGURE 21-39 The refrigerant flow of an upright automatic-defrost freezer.

FIGURE 21-40
The component
location of a chest
freezer.

FIGURE 21-40
The component
location of a chest
freezer.

Chest Freezers

The chest freezer (Figure 21-40) has the evaporator coils and the condenser coils embedded between the inner liner and the outer cabinet. These coils are inaccessible for replacement or repair if a refrigerant leak occurs. The differences between the upright freezer and the chest freezer are:

- Door hinges.
- Gasket.
- Location and access of temperature controls.
- Location and access of the compressor, relay, and overload protector. Most models have a power indicator light. This light stays on as long as the freezer is plugged into the wall receptacle. The light alerts the consumer when the power is off to the freezer, but it does not tell you what the temperature is inside the cabinet.

Chest freezers must be defrosted once or twice a year to remove the ice buildup from the inside. To gain access to the components, remove the side access panel (Figure 21-41).

FIGURE 21-41
Removing the side
access panel to gain
access to the
components.

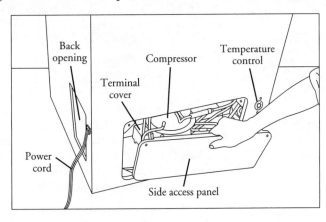

Repair Procedures for Chest Freezers

Each repair procedure in this section is a complete inspection and repair process for a single freezer component, containing the information you need to test a component that might be faulty and to replace it, if necessary.

Thermostat (Cold Control)

The typical complaints associated with failure of the thermostat (cold control) are:

- The freezer is not cold enough.
- The freezer is too cold.
- The freezer runs all the time.
- The freezer doesn't run at all.

To handle these problems, perform the following steps:

1. **Verify the complaint.** Verify the complaint by checking the control setting. Turn the control off and on again to see if the freezer starts up. Is the power indicator light on?

2. **Check for external factors.** You must check for external factors not associated with the appliance. Is the appliance installed properly? Explain to the user how to set the controls.

3. **Disconnect the electricity.** Before working on the freezer, disconnect the electricity. This can be done by pulling the plug from the receptacle. Or disconnect the electricity at the fuse panel or at the circuit breaker panel. Turn off the electricity.

4. **Gain access to the thermostat.** To access the thermostat, remove the access panel (see Figure 21-41). Next, remove the two screws that secure the control. Remove the wires from the terminals. The capillary tube is inserted into a channel. Do not remove capillary tube yet. Test the control first.

5. **Test the thermostat.** To test the thermostat, place the ohmmeter probes on the terminals (Figure 21-42). Set the range scale on R × 1, and test for continuity. With the control set in the "off" position, you should not read continuity. When the control is set to the highest position, you should read continuity. If the thermostat is good, the problem must be elsewhere.

6. **Remove the thermostat.** With the thermostat control housing already removed, the capillary tube must now be removed. Remove the capillary tube from the channel.

7. **Install a new thermostat.** To install the new thermostat, just reverse the order of disassembly, and reassemble. Then test the control. Remember to reinstall the capillary tube in the same location from which it was removed. Be careful not to kink the tube. If you do, the freezer will not work properly.

FIGURE 21-42 Testing the thermostat with a multitester with the range set on the ohms scale. If you read no continuity, replace the thermostat.

Power Indicator Light

The typical complaints associated with the failure of the power indicator light are:

- The light is not on.
- The light flickers.
- The light is dim.

To handle these problems, perform the following steps:

1. **Verify the complaint.** Verify the complaint by checking if the freezer is plugged into the wall receptacle. Is the power indicator light on?

2. **Check for external factors.** You must check for external factors not associated with the appliance. Is the appliance installed properly? Check the voltage to the freezer.

3. **Disconnect the electricity.** Before working on the freezer, disconnect the electricity. This can be done by pulling the plug from the receptacle. Or disconnect the electricity at the fuse panel or at the circuit breaker panel. Turn off the electricity.

4. **Gain access to the power indicator light.** To access the power indicator light, use a screwdriver to pry out the power indicator light from the front of the freezer cabinet. Next, remove the wires from the indicator light (Figure 21-43).

5. **Test the power indicator light.** To test the power indicator light, place the ohmmeter probes on its terminals (Figure 21-44). Set the range scale on R × 1, and test for continuity. The meter should show continuity; if not, replace the component.

6. **Install a new power indicator light.** To install the new power indicator light, just reverse the order of disassembly, and reassemble (Figure 21-45). Then test the control.

FIGURE 21-43 Pry gently to remove the power indicator light. Be careful not to scratch the cabinet.

FIGURE 21-44 Testing the power indicator light.

Figure 21-45 When installing the power indicator light, you must first reconnect the wires and then insert the light in the cabinet.

Gasket

The typical complaints associated with failure of the door gasket are:

- Sweating inside and/or outside of the cabinet.
- Temperatures inside the cabinet are warmer than normal.
- Ice is forming on the freezer walls.
- Door gasket not maintained properly.

To handle these problems, perform the following steps:

1. **Verify the complaint.** Verify the complaint by checking the door gasket for proper sealing and alignment. Inspect the gaskets for any damage.

2. **Check for external factors.** You must check for external factors not associated with the appliance. Is the appliance installed properly?

3. **Disconnect the electricity.** Before working on the refrigerator/freezer, disconnect the electricity. This can be done by pulling the plug from the receptacle. Or disconnect the electricity at the fuse panel or at the circuit breaker panel. Turn off the electricity.

4. **Test the door gasket.** To test the gasket for proper sealing, take a dollar bill and place it between the gasket and the flange of the outer cabinet. Pull on the dollar bill. When pulling on the dollar bill, you should feel some tension as the gasket and flange grip the bill. Repeat this test in other areas where you suspect problems with the gasket. If the gasket fails this test, the next step is to replace the gasket.

5. **Gain access to the door gasket.** To access the door gasket, the door must be removed (Figure 21-46). Turn the door over on its back.

Figure 21-46
Removing the screws from the hinges and lifting the door from the freezer.

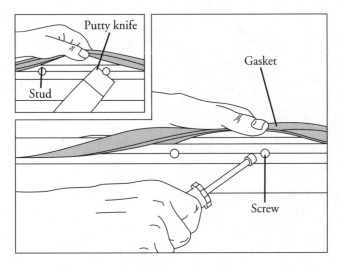

FIGURE 21-47
Removing the studs (inset). On some models, you must remove the screws that secure the gasket to the door.

6. **Remove the gasket.** Remove the gasket, either by prying the studs out or by removing the screws (Figure 21-47).

7. **Install a new door gasket.** Before installing the new gasket, soak it in warm water for 15 to 20 minutes. This will make the gasket soft and easier to install. Starting at either top corner, insert the flange of the gasket behind the retaining strip and/or door liner. Proceed all the way around the door. When the gasket is in place, begin to tighten the screws slightly all around the door or reinstall the studs. If the gasket is distorted, or if it has wrinkles in it, use a hair dryer to heat the gasket and mold it to its original form. Then be sure the gasket seats against the flange properly. Next, check the gap between the door and the cabinet; adjust it, if necessary (see step 4).

Diagnostic Charts

The following refrigerator/freezer diagnostic charts will help you to pinpoint the likely causes of the particular problem associated with these appliances (Figures 21-48, 21-49, 21-50, and 21-51).

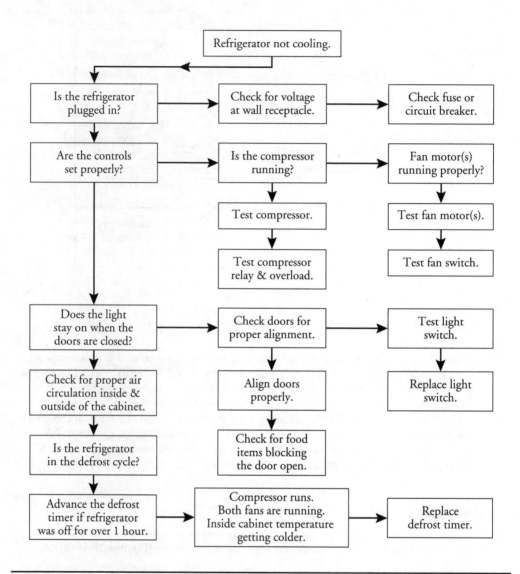

FIGURE 21-48   The diagnostic flow chart: Refrigerator not cooling.

FIGURE 21-49 The diagnostic flow chart: Refrigerator temperature too cold.

Refrigerator temperature too cold

Check temperature control setting → Adjust thermostat / Adjust damper control → Replace if faulty

Refrigerator has odor →
Food spoiled
Drain pan dirty
Food not properly stored
Poor maintenance on refrigerator

Refrigerator noisy →
Refrigerator not level
Drain pan rattling
Check condenser fan motor
Check evaporator fan motor
Check compressor
Check for rattling tubing
Jars rattling in refrigerator

FIGURE 21-50 The diagnostic flow chart.

Freezer temperature too cold →
Adjust temperature control
Test temperature control

Moisture around door →
Is food properly stored?
Is the door opened too frequently?
Is the freezer located in a humid location?
Check the freezer door gasket for proper sealing

Freezer not running Power indicator light off →
Is the freezer plugged in?
Check fuses or circuit breakers
Check service cord
Check compressor, relay, overload

Compressor on and off rapidly →
Test compressor relay
Test overload protector
Check for low voltage
Test compressor

Freezer runs constantly →
Manual defrost freezers must be defrosted once or twice a year
Check door gasket for proper sealing

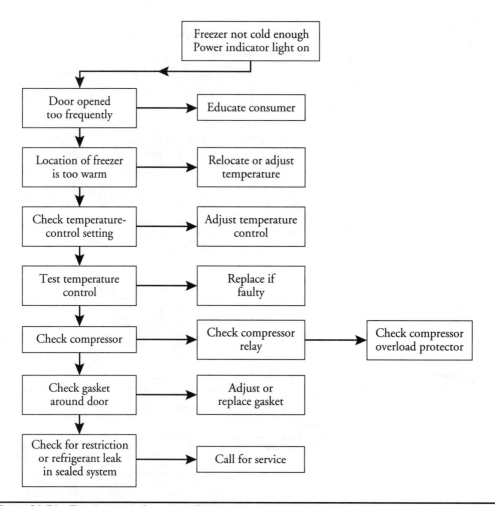

FIGURE 21-51 The diagnostic flow chart: Freezer not cold enough/power indicator light on.

Refrigerator/Freezer Electrical Schematics

The wiring diagrams in this chapter are examples only. You must refer to the actual wiring diagram for the refrigerator/freezer you are servicing. Figure 21-52 depicts an actual wiring schematic for a side-by-side refrigerator, which includes an automatic ice maker and ice and water dispensers. On models with the rapid electrical diagnosis (R-E-D) feature, a technician can make a quick and accurate diagnosis of electrical faults without disassembling the refrigerator.[9] To perform this test, a special adapter is connected to the wiring harness through the multicircuit connector, which is located behind the front grill. Upon separating the multicircuit connector, the parallel circuits in the wiring harness will be isolated. This process will permit you to test all of the electrical components and the related wiring within the main wiring harness. If an R-E-D test adapter is not available, you can still check the circuits with an ohmmeter.

FIGURE 21-52 Side-by-side refrigerator wiring schematic.

Caution *Disconnect the electricity from the refrigerator/freezer before measuring resistances.*

A basic understanding of the symbols used in the schematic diagram is essential (see Figure 21-52). The numbered terminals, located in the multicircuit connector, are shown on the schematic diagram. The component circuits on the schematic diagram are indicated by an arrow and a number. The point of the arrow indicates a male terminal, and the tail of the arrow indicates a female terminal. The number identifies the terminal location in the connector (Figure 21-53).

Remember how to read a wiring schematic. Give the following examples a try.

FIGURE 21-53 Side-by-side refrigerator wiring schematic of multicircuit connector circuitry. (a) Defrost heater circuit. (b) Defrost thermostat circuit. (c) Defrost control motor circuit. (d) Defrost control switch contacts circuit.

Example #1

The customer explains that the food in the freezer is thawing and the food in the refrigerator is warm. The customer states the compressor runs constantly. You made your observation and confirmed that the food is thawing and the compressor is operating. You also noticed frost on the evaporator cover, which indicates a defrost problem. This indicates to you the possibility of three components malfunctioning: the defrost heaters, the defrost thermostat, or the defrost timer. Also, there is a possibility of a broken or loose wiring connection. Set the refrigerator in the defrost mode first. Advance the defrost timer until you hear the first "snap" sound coming from the timer.

Using the ohmmeter, set the range on R × 1, and adjust the meter scale to read zero. Separate the multicircuit connector. *Did you turn off the electricity?* To test the defrost heaters, insert one meter probe into the number 5 male connector pin and the other meter probe into the number 1 male connector pin (see Figure 21-53a). The ohmmeter should show the combined resistance of the heaters. This resistance is 16 ohms. If the test is okay, continue on to the defrost thermostat. If not, replace the heaters.

When testing the defrost thermostat, insert one meter probe into the number 1 male connector pin and the other meter probe into the number 4 male connector pin (see Figure 21-53b). The meter reading should indicate zero ohms. If not, replace the defrost thermostat.

Now test the defrost timer motor. Insert one meter probe into the number 4 male connector pin and the other meter probe into the number 4 female connector pin (see Figure 21-53c). The meter should show between 800 and 4000 ohms. Adjust the ohmmeter range accordingly. To test the defrost timer switch contacts, insert one meter probe into the number 2 female connector pin and the other meter probe into the number 5 female connector pin (see Figure 21-53d). The meter should show zero ohms. If not, replace the defrost timer control.

Figure 21-54 depicts the active defrost circuits. If you want to check the entire defrost circuit, insert one meter probe into the number 3 male connector pin and the other meter probe into the number 4 male connector pin. The meter should read continuity. Be sure that the cold control (thermostat) is set for maximum cooling and that the defrost timer is set on defrost.

Example #2

The wiring diagrams in Figures 21-55 and 21-57 are examples only. You must refer to the actual wiring diagram for the refrigerator that you are servicing. Figure 21-55 depicts an actual wiring schematic for a cycle-defrost refrigerator.

The refrigerator door is closed, the thermostat is calling for cooling, and the compressor is running. What circuits are active? Trace the active circuits in Figure 21-55. The active components are the thermostat, overload, relay, compressor, stile heater, and the mullion heater. Check the end result with Figure 21-56.

The thermostat in Figure 21-57 has cycled off, and the switch contacts are open. Also, the compressor stopped running. What circuits are active? Trace the active circuits in Figure 21-57. The active components are the evaporator heater, drain heater, stile heater, and the mullion heater. Check the end result with Figure 21-58. Current will flow through the overload, the relay coil, and the compressor running winding, but there is not enough current to energize the start relay and to run the compressor.

FIGURE **21-54**
Illustration of the
active circuits when
the refrigerator is
operating in the
defrost mode.

FIGURE 21-55
A typical cycle-defrost refrigerator electrical schematic.

Typical cycle-defrost refrigerator
electrical schematic

FIGURE 21-56
The active circuits when the refrigerator is operating in the cooling mode.

Typical cycle-defrost refrigerator
electrical schematic

FIGURE 21-57
The cycle-defrost
refrigerator wiring
schematic, showing
that the refrigerator
has cycled off.

Typical cycle-defrost refrigerator
electrical schematic

FIGURE 21-58
The active circuits
when the refrigerator
is operating in the
defrost mode.

Typical cycle-defrost refrigerator
electrical schematic

Example #3

The wiring diagram in Figure 21-59 is an example only. It depicts a no-frost refrigerator. The refrigerator is in the cooling mode, and the compressor is running. What circuits are active? Trace the active circuits. Check the end result with Figure 21-60.

FIGURE 21-59 A typical no-frost refrigerator electrical schematic.

FIGURE 21-60 The active circuits when the refrigerator is operating in the cooling mode.

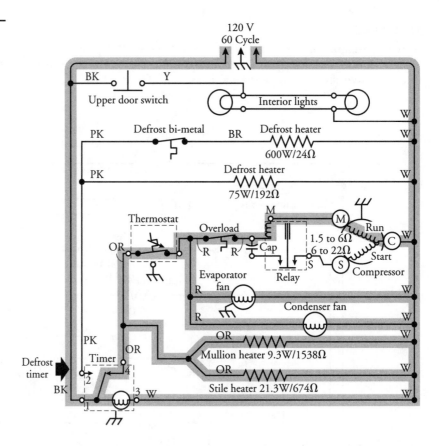

Endnotes

1. Dairy products, meats, seafood, fruits, and vegetables will all spoil rapidly if not kept cold or frozen. The colder temperatures of the refrigeration compartment should be between 35 and 40 degrees Fahrenheit to prevent the spoiling of foods. Most foods will last from three to seven days at that temperature. If the foods were frozen and packaged properly, they could last for several weeks in a domestic refrigerator with a temperature of zero degrees Fahrenheit to –10 degrees Fahrenheit.

2. Reprinted from MACAP Consumer Bulletin, Issue no. 15, December 1992.

3. Reprinted from MACAP Consumer Bulletin, Issue no. 12, November 1989.

4. Reprinted from MACAP Consumer Bulletin, Issue no. 6, February 1984.

5. If you open up the sealed system, you will void your warranty. The sealed system must be repaired by an authorized service company.

6. If you open up the sealed system, you will void your warranty. The sealed system must be repaired by an authorized service company.

7. Manual-defrost freezers must be defrosted once or twice a year, depending upon usage.

8. Automatic-defrost freezers will self-defrost the evaporator coil. The defrost components are the defrost timer, the defrost thermostat, and the defrost heater.

9. Robinair part # 14442, purchased at your local parts dealer.

Automatic Ice Makers

The automatic ice maker is an independent appliance, installed in the freezer or in the freezer compartment of a refrigerator. The ice maker has three distinct operations: fill, freeze, and harvest the ice cubes. It is the refrigerator/freezer cooling system that allows the water to freeze into whatever shape is designed by the ice mold or tray. The amount of time it takes for the ice maker to produce and harvest ice cubes will depend on:

- Freezer temperature
- Food load conditions
- Number of door openings
- Ambient temperature

Safety First

Any person who cannot use basic tools or follow written instructions should *not* attempt to install, maintain, or repair any automatic ice makers. Any improper installation, preventive maintenance, or repairs could create a risk of personal injury or property damage.

If you do not fully understand the installation, preventive maintenance, or repair procedures in this chapter, or if you doubt your ability to complete the task on your automatic ice maker, please call your service manager.

The following precautions should also be followed:

- Never place fingers or hands on the automatic ice maker mechanism while the refrigerator/freezer is plugged in.
- Disconnect the electrical supply to the freezer or refrigerator before servicing the automatic ice maker.
- Be careful of any sharp edges on the automatic ice maker, which might result in personal injury.
- Do not attempt to operate the automatic ice maker unless it has been properly reinstalled, including the grounding and electrical connections.

Before continuing, take a moment to refresh your memory on safety procedures in Chapter 2.

Automatic Ice Makers in General

Much of the troubleshooting information in this chapter covers automatic ice makers in general, rather than specific models, in order to present a broad overview of service techniques. The illustrations that are used in this chapter are for demonstration purposes only, to clarify the description of how to service ice makers. They in no way reflect on a particular brand's reliability.

Principles of Operation for Type-1 and Type-2 Automatic Ice Makers

The freezer temperature determines the efficiency of the automatic ice maker. The colder the temperature in the freezer, the faster ice will freeze in the mold or tray. In order for the ice maker to harvest the ice cubes, the temperature in the freezer should be colder than 12 degrees Fahrenheit.

The water supply (Figure 22-1) is metered into the ice maker mold or tray by a water valve, usually located in the compressor compartment. The water freezes in the automatic ice maker mold or tray, as it would in a manual ice cube tray.

FIGURE 22-1
A typical ice maker water supply installation. Note the direction of the water flow.

Type-1 Ice Maker

Type-1 ice makers, as illustrated in Figure 22-2, have a thermostat that is mounted on the ice maker mold, which will sense the temperature of the ice. When the water and ice have cooled down to 17 degrees Fahrenheit (±3 degrees), the thermostat will activate the ice maker motor and heater circuit. The motor will rotate the ejector blade, allowing it to rest on the ice cubes and, at the same time, exerting pressure. The ice maker mold heater will warm up the ice cubes just enough to release them from the mold. This procedure will take approximately three to five minutes to release the ice cubes from the mold.

When the ice cubes become free from the ice maker mold, the ejector blade will continue to rotate, scooping the ice cubes out of the ice maker mold and depositing them into the ice bucket. The shutoff arm moves up and down as the ice maker cycles.

At the end of the cycle, the arm will lie on top of the ice cubes. As the cubes fill the ice bucket, the shutoff arm rises to a designated point and turns off the ice maker, halting ice production. As the ice level in the bucket falls from use, the ice maker cycle will resume. Then the water valve opens again, allowing the water to enter the ice maker mold to be frozen into ice cubes, thus beginning the cycle all over again. The fill time for the type-1 ice maker is approximately seven seconds. This type of ice maker can produce up to eight pounds of ice within a 24-hour period, providing the conditions are ideal. Some customers will have ice production between three to five pounds per day, depending on usage and door openings.

Type-2 Ice Maker

The type-2 ice maker (Figure 22-3) has the components in the ice maker head necessary for the ice-making operation. The temperature in the freezer compartment must be below 15 degrees Fahrenheit before the ice maker will begin to operate. Inside the ice maker head assembly is a thermostat that senses the temperature in the freezer compartment. When the thermostat is satisfied, it will energize the timer motor. When the motor is energized, it will begin to turn a timing gear and the shutoff arm will also begin to move. The shutoff arm will descend into the ice bucket to determine how much ice is there. If the bucket is full, the

FIGURE 22-2   Type-1 ice maker.

FIGURE 22-3   Type-2 ice maker.

Figure 22-4
The ice tray will twist, releasing the ice cubes.

Tray in twist mode

Tray stop engaging corner of tray

shutoff arm will rest against the ice cubes and the ice maker will not cycle. If the shutoff arm continues to its normal position, the ice bucket is not full and the cycle will continue.

If the cycle continues, the tray will begin to rotate. When the tray has rotated about 140 degrees, one corner of the tray engages the tray stop, which prevents that part of the tray from rotating further (Figure 22-4). The rest of the tray will continue to rotate and twist the tray to about 40 degrees. At this point, the twisting of the tray will loosen the ice cubes. As the shaft continues to turn, the tray stop will begin to retract. As this happens, the tray will rapidly release the ice cubes into the ice bucket. After the ice has been released, the tray will continue to rotate until it has completed its 360-degree rotation. Near the end of the tray rotation, the water valve fill switch is energized, allowing the water to enter and fill the tray for the next cycle (Figure 22-5). The fill time for type-2 ice makers is approximately 12 seconds.

Testing the Type-1 Ice Maker

The type-1 ice maker is designed to allow all of the components to be tested without removing the ice maker and without moving the refrigerator away from the wall to test the water valve. The ice maker in Figure 22-6a is the old-style ice maker, and it is not in production any more. Parts for this ice maker are still available, however. Figure 22-6b is the new-style modular ice maker. The components that make up this ice maker are illustrated in Figure 22-7.

Figure 22-5
The ice cube tray must return to a level position after the filling cycle is complete.

Water fill valve solenoid energized

Tray rotating toward level position

Ice cubes harvested

FIGURE 22-6
(a) Old-style type-1 ice maker. (b) New-style type-1 ice maker.

(a)   (b)

The test holes that are on the ice maker head module (see Figure 22-6b) are identified as "N," "M," "H," "T," "L," and "V." These are the test points for testing this type of ice maker. The letters indicate the following:

- N = neutral side of line voltage
- M = ice maker motor connection
- H = mold heater connection
- L = L1 side of line voltage
- V = ice maker water valve connection

This test requires the electricity to be turned on.

FIGURE 22-7
Exploded view of new-style type-1 ice maker.

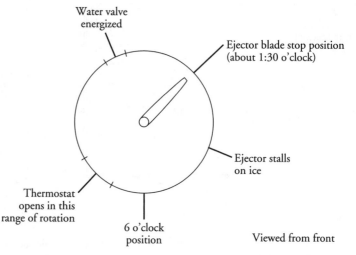

FIGURE 22-8   The jumper wire is used to short out the test points on the ice maker module.

FIGURE 22-9   Indicating the ejector blade position when ice maker is cycling.

NOTE   *Caution should be used when working with live wires. Avoid getting shocked. Stay clear of live wires. Only handle the meter probes by the insulated handles.*

To test this ice maker, the unit must be installed and plugged into the freezer ice maker receptacle, the shutoff arm must be placed in the down position, and the temperature in the freezer should be colder than 12 degrees Fahrenheit. Set your multimeter for voltage, and place one probe in test point "L" and the other probe in test point "N." Be sure that the test probes go into the test points about 1/2 inch. You should have a reading of 120 volts, which indicates line voltage to the ice maker. Next, place the meter probes in test points "T" and "H." This will verify if the bi-metal thermostat is open or closed. If the thermostat is open, you will read 120 volts. If it is closed, you will read no line voltage. At this point, you are going to use an insulated jumper wire (Figure 22-8) to short the test points "T" and "H."[1] This procedure will run the motor. If the motor doesn't run, replace the ice maker module assembly. If the ice maker motor runs, replace the bi-metal thermostat. If you leave the jumper wire in for half of a revolution, you will feel the mold heater heating up. This means that the mold heater is working. Now remove the jumper wire, and the water valve will be energized in the last half of the revolution. The ejector blade shown in Figure 22-9 is rotating clockwise. This illustration shows what is about to happen as the cycle begins (from the stop position) and rotates 360 degrees back again.

Another way to test the water valve without cycling the ice maker is to place one end of the jumper wire in test point "N" and the other end in test point "V." Water will immediately enter the ice maker mold, so be ready to disconnect the jumper wire. If no water enters the mold, check the water supply, the water valve, and the connecting tubing.

Testing the Type-2 Ice Maker

In testing the operation of the type-2 ice maker head, you will have to initiate the harvest cycle. Do not remove the ice maker for this test. The shutoff arm will not operate during the

FIGURE 22-10
Manually twist the ice tray clockwise. When the tray reaches 30 degrees, release the switch actuator.

Manually twist tray to start cycle

Push toward tray shaft to unlock tray

Switch actuator must be down for motor to operate

manual harvest. When performing the following procedures, refer to Figure 22-10. To cycle the ice maker, do the following:

1. Place the shutoff arm in the "on" position.
2. Push the switch actuator tab down, and hold it. This will activate the ice maker motor.
3. Push the tray lock tab toward the tray shaft. This will unlock the tray.
4. Twist the tray clockwise. This will start the cycle.
5. As the tray turns past 30 degrees, you can release the switch actuator tab.

Tray stop

Motor shaft

Lift

OFF position

ON position

Tray lock

Shutoff arm

FIGURE 22-11 To shut off the ice maker, lift the shutoff arm and rest it on the tab, as indicated by the broken lines.

6. After the harvest cycle is complete, empty the water from the ice tray. This will prevent the next automatic harvest cycle that occurs from dumping water into the ice bucket. When you perform a manual harvest, you interrupt the timing sequence only. A manual harvest will not reset the timing mechanism.

If you are unable to start the harvest cycle, check the motor shaft to see if it turns while you depress the switch actuator (Figure 22-11). Also check the temperature in the freezer. If the temperature is above 15 degrees Fahrenheit, the ice maker will not operate. If the motor shaft still does not rotate, remove the ice maker and test the unit on the work bench. Use a test cord to test the ice maker motor (Figure 22-12).[2]

FIGURE 22-12
A test cord to test a
type-2 ice maker.

Water/Temperature Problems

The water quality can cause the ice maker to malfunction or produce poor-quality ice cubes. If the minerals in the water or sand particles become lodged in the water valve fill screen, it can restrict or stop the water from entering the ice maker mold. If sand particles bypass the fill screen in the water valve, it can possibly keep the water valve from closing completely, thus causing the water to enter the ice maker mold continuously. If this condition happens, the ice maker will produce small cubes or may even flood the ice maker compartment and freezer. Another indication of a defective water valve is that the ice maker fill tube will be completely frozen shut, possibly causing water to flow onto the floor every time the ice maker cycles. The minerals in the water can also cause lime deposits to build up on the ice maker mold. If this happens, the ice cubes will not be released easily from the mold.

Ice production can be slowed down if the temperature in the freezer compartment is above normal. To correct this problem, adjust the temperature controls to a colder position. The more the doors are opened, the colder the temperature setting must be.

Step-by-Step Troubleshooting Type-1 Ice Makers by Symptom Diagnosis

When servicing an appliance, don't overlook the simple things that might be causing the problem. Step-by-step troubleshooting by symptom diagnosis is based on diagnosing malfunctions, with possible causes arranged in categories relating to the operation of the automatic ice maker. This section is intended only to serve as a checklist to aid you in diagnosing a problem. Look at the symptom that best describes the problem you are experiencing with the automatic ice maker, and then correct the problem.

Ice Maker Producing too Much Ice

- Check if the shutoff arm is not connected to the actuator in the ice maker head.
- Check the shutoff arm. Is it bent out of its original shape?
- Check the shutoff linkage in the ice maker head. Is it broken?

Ice Maker Will Not Make Ice/Low Ice Production

- Check freezer temperature.
- Is the shutoff arm in the "off" position?
- Check motor operation. Did it stall? Are the gears stripped?

- Test the voltage at the ice maker. Is the ice maker plugged into the receptacle?
- Test the thermostat for continuity. Bypass the thermostat. Will the ice maker run?
- Check the shutoff linkage in the ice maker head. Is it broken?
- Is the water supply turned on? Does water enter the ice maker mold when it cycles?
- Check the ice maker fill tube. Is it frozen?
- Check for an ice jam.
- Check for defective wiring.

Ice Cubes Are Too Small

- Check the ice maker. Is it level?
- Check the fill tube. Is it frozen?
- Check the water supply.
- Check the ice maker water valve.
- Check the water pressure. Is it between 20 and 120 psi?
- Check for a self-piercing saddle valve. Mineral deposits will restrict the opening.
- Cycle the ice maker, and catch the fill water in a glass. Measure the amount of water. Are there at least 140 cc of water in the glass?
- Test for openings in the mold heater.
- Check the ice maker thermostat. There might be insufficient thermal bond between the thermostat and the ice maker mold.

Ice Maker Is Producing Hollow Ice Cubes

- Cycle the ice maker, and catch the fill water in a glass. Measure the amount of water. Are there at least 140 cc of water in the glass?
- Check for improper airflow in the freezer compartment. Direct the airflow away from the ice maker thermostat.
- Check the ice maker thermostat.
- Check the temperature in the freezer compartment.

Ice Maker Is Flooding the Freezer Compartment or Ice Bucket

- Check the thermostat.
- Check for an ice jam when the ice maker is in the fill position.
- Check for a leaky water valve.
- Check for proper ice maker water fill. Too much water will spill over the mold, causing the ice in the bucket to freeze together.
- Check the ejector blade position. If the blade is in the 12 o'clock position, the ice maker motor has stalled.
- Check the ice maker module for contamination and/or burned switch contacts.

- Check the linkage for proper operation.
- Is the refrigerator level? Is the ice maker level?
- Check water pressure. Ice makers fill according to time, not volume. Water pressure should be between 20 and 120 psi.
- Is the fill tube located in the fill cup?

Step-by-Step Troubleshooting Type-2 Ice Makers by Symptom Diagnosis

When servicing an appliance, don't overlook the simple things that might be causing the problem. Step-by-step troubleshooting by symptom diagnosis is based upon diagnosing malfunctions, with possible causes arranged into categories relating to the operation of the automatic ice maker. This section is intended only to serve as a checklist to aid you in diagnosing a problem. Look at the symptom that best describes the problem you are experiencing with the automatic ice maker, and then proceed to correct the problem.

Ice Maker Will Not Run

- Test the freezer temperature.
- Check the shutoff arm. Is it in the "off" position?
- Run an ice maker test. Does the ice maker motor run?
- Check for defective wiring.
- Is the ice maker plugged into the receptacle?
- Test for voltage at the ice maker receptacle.

Ice Cubes Are Stuck Together

- Test temperature of freezer.
- Check for proper fill.
- Check ice tray for mineral deposits. Mineral deposits will cause the ice to stick to the tray, which can cause the ice cubes to stick together on the next cycle.

Ice Maker Spills Water from the Tray

- Check the ice tray. When the harvest cycle is complete, the ice cube tray should return to its starting position and the tray should be level.
- Check the inlet water fill tube. Be sure that the fill tube and fill trough fit together properly.
- Check for a leaking water inlet valve.

Water Will Not Enter Ice Tray

- Check for proper water supply to ice maker.
- Check water valve strainer for restrictions.
- Check the water valve.

- Check for proper water pressure.
- Check wiring circuit.

Ice Cubes too Small, or Some of the Ice Cube Compartments Are Empty

- Check the ice maker. Is it level?
- Check the fill tube. Is it frozen?
- Check the water supply.
- Check the ice maker water valve.
- Check the water pressure. Is it between 20 and 120 psi?
- Check for a self-piercing saddle valve. Mineral deposits will restrict the opening.
- Cycle the ice maker and catch the fill water in a glass. Measure the amount of water. Are there at least 200 cc of water in the glass? The fill time will be between 12 and 13 seconds.

Diagnostic Chart

The diagnostic chart (Figure 22-13) will help you to pinpoint the likely cause of the problem with an ice maker.

FIGURE 22-13
The diagnostic flow chart: Ice maker not making ice.

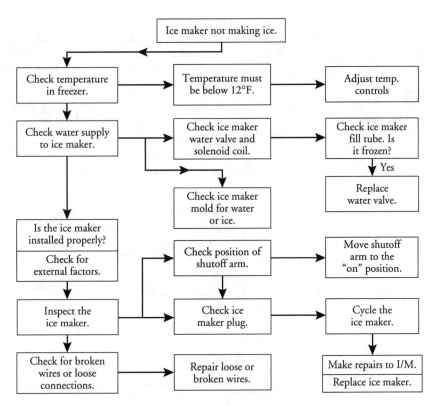

Repair Procedures

Each repair procedure is a complete inspection and repair process for a single ice cube maker component, containing the information you need to test and replace components.

Type-1 Ice Maker: Module

The typical complaints associated with failure of the ice maker module are:

- Ice maker is not making any ice.
- Ice maker stalls in the middle of cycling.

To handle these problems, perform the following steps:

1. **Verify the complaint.** Verify the complaint by checking the ice maker and the temperature in the freezer.

2. **Check for external factors.** You must check for external factors not associated with the ice maker. For example, is the ice maker installed properly?

3. **Disconnect the electricity.** Before working on the ice maker, disconnect the electricity to the refrigerator/freezer. This can be done by pulling the plug from the receptacle. Or disconnect the electricity at the fuse panel or at the circuit breaker panel. Turn off the electricity.

4. **Gain access to the ice maker.** Open the freezer door to access the ice maker. Remove the screws that secure the ice maker to the wall of the freezer (Figure 22-14). Next, unplug the ice maker from the receptacle and remove the ice maker (Figure 22-15a). Now depress the retaining tab, and pull the wiring harness from the ice maker head (Figure 22-15b).

5. **Disassemble and remove the ice maker module.** After testing the ice maker and determining that you must replace the ice maker module, it is time to disassemble it. Insert a flat-blade screwdriver between the shutoff arm and the white elongated hole on the ice maker head (Figure 22-16). Remove the shutoff arm by prying the

FIGURE 22-14 Support the ice maker when removing the screws.

FIGURE 22-15
(a) Pull the ice maker plug out of the receptacle. (b) Remove the wire harness by depressing the tab, and then pull it.

(a)

Plug

(b)

Depress tab and pull harness out

FIGURE 22-16
Removing the shutoff arm by prying it out of the slot. When installing the shutoff arm, be sure that the end is pressed in all the way.

Fill cup

Straight end

Shutoff arm

Shutoff arm

White elongated hole

PART V

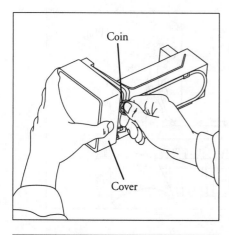

FIGURE 22-17   Gain access to the ice maker module.

FIGURE 22-18   Removing the three screws to remove the module.

arm out of the ice maker head. Pry off the ice maker cover (Figure 22-17) with a coin or a screwdriver. This will expose the ice maker module. To remove the ice maker module, remove the three screws (Figure 22-18), and pull the module out of the ice maker head. Inspect the module linkage and switch contacts (Figure 22-19).

6. **Install a new ice maker module.** To install the new ice maker module, just reverse the order of disassembly, and reassemble. To reinstall the shutoff arm on the ice maker (see Figure 22-16), you must first insert the straight end into the round hole in the fill cup. Be sure the flat side on the arm goes through the fill cup. This will prevent the shutoff arm from coming out of the fill cup hole. Next, insert the other end of the shutoff arm into the white elongated hole in the ice maker housing. Do not install the arm into any of the round holes in the ice maker housing. Push on the arm so that it will be completely in place and even with the ice maker surface. Reinstall the ice maker in the freezer, and test it.

FIGURE 22-19
Inspecting the shutoff arm linkage, cam follower, and switch contacts. Note: The ground terminal is slightly longer than the other three terminals. The ground connection is either made first or breaks last when the wiring harness is removed or plugged in.

Type-1 Ice Maker: Thermostat

The typical complaints associated with failure of the ice maker thermostat are:

- Ice maker is not making any ice cubes.
- Ice maker is producing hollow ice cubes.

To handle these problems, perform the following steps:

1. **Verify the complaint.** Verify the complaint by checking the ice maker and the temperature in the freezer.

2. **Check for external factors.** You must check for external factors not associated with the ice maker. Is the ice maker installed properly?

3. **Disconnect the electricity.** Before working on the ice maker, disconnect the electricity to the refrigerator/freezer. This can be done by pulling the plug from the receptacle. Or disconnect the electricity at the fuse panel or at the circuit breaker panel. Turn off the electricity.

4. **Gain access to the ice maker thermostat.** Open the freezer door to access the ice maker. Remove the screws that secure the ice maker to the wall of the freezer (see Figure 22-14). Next, unplug the ice maker from the receptacle, and remove the ice maker (see Figure 22-15a). Now press the retaining tab, and pull the wiring harness from the ice maker head (see Figure 22-15b). Insert a flat-blade screwdriver between the shutoff arm and the white elongated hole on the ice maker head (see Figure 22-16). Remove the shutoff arm by prying the arm out of the ice maker head. Pry off the ice maker cover (see Figure 22-17) with a coin or a screwdriver. This will expose the ice maker module. To remove the ice maker module, remove the three screws (see Figure 22-18), and pull the module out of the ice maker head. Inspect the linkage, the cam follower, and the switch contacts. Next, remove the two screws that secure the ice maker head to the mold (Figure 22-20). Separate the ice maker head from the mold assembly, and you will see the thermostat on the mold side of the ice maker head.

5. **Remove the thermostat.** To remove the thermostat, use needle nose pliers to pull out the retaining clips (Figure 22-21). Remove the thermostat.

FIGURE 22-20 Removing the two screws.

FIGURE 22-21 Removing the retaining clips with a pair of needle nose pliers.

6. **Install a new thermostat.** To install the new thermostat, just reverse the order of disassembly, and reassemble the ice maker. Be sure that you properly index the pins on the thermostat. Before you assemble the ice maker head to the mold assembly, you must apply new thermal bonding material to the thermostat. This will allow the thermostat to make better contact with the ice maker mold. Reinstall the ice maker in the freezer, and test.

Type-1 Ice Maker: Mold and Heater Assembly

The typical complaints associated with failure of the ice maker mold and heater assembly are:

- Ice maker is not making any ice.
- Ice will not come out of the mold.
- There are black pieces of mold coating material in the ice cubes.

To handle these problems, perform the following steps:

1. **Verify the complaint.** Verify the complaint by checking the ice maker and the temperature in the freezer.

2. **Check for external factors.** You must check for external factors not associated with the ice maker. Is the ice maker installed properly?

3. **Disconnect the electricity.** Before working on the ice maker, disconnect the electricity to the refrigerator/freezer. This can be done by pulling the plug from the receptacle. Or disconnect the electricity at the fuse panel or circuit breaker panel. Turn off the electricity.

4. **Gain access to the ice maker mold assembly.** Open the freezer door to access the ice maker. Remove the screws that secure the ice maker to the wall of the freezer (see Figure 22-14). Next, unplug the ice maker from the receptacle, and remove the ice maker (see Figure 22-15a). Now press the retaining tab, and pull the wiring harness from the ice maker head (see Figure 22-15b).

   Insert a flat-blade screwdriver between the shutoff arm and the white elongated hole on the ice maker head (see Figure 22-16). Remove the shutoff arm by prying it out of the ice maker head. Pry off the ice maker cover (see Figure 22-17) with a coin or screwdriver. Insert a Phillips screwdriver into the access ports in the module (Figure 22-22) to loosen the screws. Separate the ice maker head assembly from the mold assembly. Also remove the ejector, the fill cup, the stripper, and the shutoff arm (see Figure 22-7). Do not remove the heater; the heater and mold come as a complete assembly.

5. **Install a new ice maker mold assembly.** To install a new ice maker mold assembly, just reverse the order of disassembly, and reassemble. When assembling the ice maker, apply a thin film of silicone grease to the end of the ejector that goes into the fill cup. This will prevent the ejector from freezing to the fill cup. Also apply a thin film of silicone grease to the other end of the ejector. Before installing the stripper, apply a heavy film of silicone grease to the top surface of the mold that is covered

Figure 22-22
The entire ice maker head can be removed by loosening the two screws through the access ports.

Shutoff arm

Mold attachment screw access ports (2)

by the stripper. This will prevent the water from wicking over the ice maker mold every time it cycles. Before you assemble the ice maker head to the mold assembly, you must apply new thermal bonding material to the thermostat. This will allow the thermostat to make better contact with the ice maker mold. Reinstall the ice maker in the freezer, and test it.

Water Valve

The typical complaints associated with failure of the water valve are:

- Ice maker is not making any ice.
- Ice maker fill tube is frozen.
- Ice maker mold fills with very little water.

To handle these problems, perform the following steps:

1. **Verify the complaint.** Verify the complaint by checking the ice maker and the temperature in the freezer.

2. **Check for external factors.** You must check for external factors not associated with the ice maker. Is the ice maker installed properly? Check the water supply to the water valve.

3. **Disconnect the electricity.** Before working on the water valve, disconnect the electricity to the refrigerator/freezer. This can be done by pulling the plug from the receptacle. Or disconnect the electricity at the fuse panel or circuit breaker panel. Turn off the electricity.

4. **Gain access to the water valve.** To access the water valve, pull the refrigerator/ freezer away from the wall. Remove the back access panel. Shut off the water supply to the ice maker.

5. **Remove the water valve.** Remove the water inlet tube from the water valve. Then remove the screws from the water valve bracket that secure the valve to the cabinet (Figure 22-23). Next, remove the ice maker fill line. Finally, disconnect the wiring harness from the solenoid coil of the valve.

PART V

FIGURE 22-23 Disconnect the water inlet line and the outlet line. Disconnect the wiring harness from the solenoid coil.

FIGURE 22-24 Testing the solenoid coil for resistance.

6. **Test the water valve.** Using your ohmmeter, set the scale on R × 10, and place the probes on the solenoid coil terminals (Figure 22-24). The meter should show resistance between 200 and 500 ohms. If not, replace the water valve. If you determine that the water valve is good but there is little water flow through the valve, inspect the inlet screen. If this screen is filled with debris, it must be cleaned out. To accomplish this, use a small flat-blade screwdriver and pry out the screen (Figure 22-25). Then wash out the screen, making sure that all of the debris is removed. After cleaning out the debris, reinstall the screen, and test the water valve.

FIGURE 22-25 Removing the inlet screen. Clean it out with warm water and soap.

7. **Install a new water valve.** To install the new water valve, just reverse the disassembly procedure, and reassemble. Also reconnect the ground wire from the water valve bracket to the refrigerator/freezer cabinet. Secure the water supply line to the cabinet. This will prevent the supply line from leaking when the refrigerator/freezer is pushed back against the wall. Check for water leaks before you push the refrigerator/freezer back against the wall. Also inspect the fill tubing for any cracks, etc. (Figure 22-26).

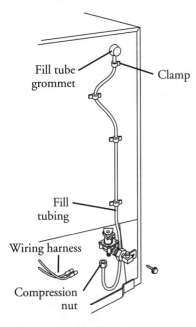

Fill tube grommet

Clamp

Fill tubing

Wiring harness

Compression nut

FIGURE 22-26   Checking for defective tubing. Check the clamp and compression nuts for leaks.

Type-2 Ice Maker: Head and Tray

The typical complaints associated with failure of the type-2 ice maker head are:

- Ice maker is not making any ice.
- Ice will not come out of the mold.

To handle these problems, perform the following steps:

1. **Verify the complaint.** Verify the complaint by checking the ice maker and the temperature in the freezer.

2. **Check for external factors.** You must check for external factors not associated with the ice maker. Is the ice maker installed properly?

3. **Disconnect the electricity.** Before working on the ice maker, disconnect the electricity to the refrigerator/freezer. This can be done by pulling the plug from the receptacle. Another way to disconnect the electricity is at the fuse panel or circuit breaker panel. Turn off the electricity.

4. **Gain access to the ice maker.** Open the freezer door to access the ice maker. Remove the screws that secure the ice maker to the wall of the freezer. Next, unplug the ice maker from the receptacle and remove the ice maker.

5. **Disassemble and remove the ice maker head assembly.** After testing the ice maker and determining that you must replace the ice maker head assembly, it's time to disassemble it. First, remove the retaining clip from the end of the tray, and then slide the tray off the ice maker drive shaft (Figure 22-27). Next, remove the fill cup

FIGURE 22-27
Always reinstall the ice cube tray in the same position as when it was removed. If the ice tray is installed upside down, water will flood the ice bucket every time the ice maker cycles.

Ice maker head

Fill cup

Tray drive shaft

Ice maker head

Shutoff arm

Retaining clip

PART V

FIGURE 22-28
Gently pry off the
shutoff arm.

and screw. On the back of the ice maker head is a clamp that holds the tray drive shaft in place; remove it. Pull the shaft out of the ice maker head. Then place a flat-blade screwdriver behind the shutoff arm, and pry it loose (Figure 22-28).

6. **Install a new ice maker head.** To install the new ice maker head, just reverse the order of disassembly, and reassemble. When replacing a defective ice maker head, only replace it with an exact replacement with the same amount of cycle operation time. Reinstall the ice maker in the freezer, and test.

Type-2 Ice Maker Tray

The typical complaints associated with failure of the type-2 ice maker tray are:

- Ice maker is not making any ice.
- Ice will not come out of the mold.
- Water in the tray freezes into a solid block of ice.

To handle these problems, perform the following steps:

1. **Verify the complaint.** Verify the complaint by checking the ice maker and the temperature in the freezer.

2. **Check for external factors.** You must check for external factors not associated with the ice maker. Is the ice maker installed properly?

3. **Remove the ice maker tray.** To remove the ice maker tray, remove the retaining clip from the end of the tray, and then slide the tray off the ice maker drive shaft (see Figure 22-27).

4. **Install a new ice tray.** To install the new ice tray, just reverse the disassembly procedure, and reassemble. Always reinstall the tray in the same position as that from which it was removed for proper operation. Test the ice maker. Let it run through a complete cycle and harvest. Then inspect the ice maker tray.

Self-Contained Ice Cube Maker

This type of ice cube maker is a freestanding, self-contained refrigeration appliance that produces a slab of ice, which is then cut into ice cubes (Figure 22-29). The production of ice cubes is all done automatically, and the entire mechanism is stored within the ice cube maker cabinet. The self-contained ice cube maker can also be installed under the counter. The thickness of the ice cubes can be adjusted by the thickness control, located on the control panel. This ice cube maker can produce up to 50 pounds of ice cubes in a 24-hour period. The amount of ice cubes will vary, depending on where it is installed, the room temperature, and the supply water temperature.

Principles of Operation

In the freeze cycle, water flows constantly, and it is recirculated over the evaporator freeze plate until a slab of ice is formed (Figure 22-30). When the ice slab reaches a predetermined thickness, the evaporator freeze plate temperature is sensed by the thermostat and the freeze cycle is terminated. At that point, the defrost cycle will begin to release the ice slab from the evaporator freeze plate. The ice slab will slide down onto the cutter grid, which cuts the ice slab into ice cubes. At the same time, the ice maker has automatically switched back into the freeze cycle.

FIGURE 22-29
The self-contained ice cube maker.

Controls

Ice cube storage bin

Compressor & condenser & fan motor

FIGURE 22-30   A pictorial view of the water system.

When the defrost cycle begins, the remaining water in the water tank reservoir is discarded through the overflow tube. At that point, fresh water will enter through the water valve and go into the water tank reservoir.

The refrigeration cycle (Figure 22-31) in this type of ice maker is similar to that in a conventional refrigerator. The compressor pumps the refrigerant into the condenser coil, which is cooled by a fan and motor. The refrigerant leaves the condenser coil as a high-pressure liquid, passes through the dryer, and enters the capillary tube. The refrigerant is next metered through the capillary tube and then enters the evaporator freeze plate. The refrigerant gas then leaves the evaporator freeze plate and returns to the compressor.

When the ice maker goes into the defrost cycle, it energizes the hot gas solenoid, which reverses the refrigeration cycle, during which the condenser fan motor and water pump will stop. The hot gas passes through the evaporator freeze plate, heating it up enough to release the ice slab. The thermostat senses the temperature of the evaporator freeze plate again and activates the freeze cycle. The hot gas solenoid valve will then close, the water valve will close, the condenser fan motor will start, the water pump will start, and the freeze cycle will begin to manufacture a new slab of ice.

Figure 22-31
A pictorial view of a refrigeration system.

High-pressure gas

High-pressure gas condensing to liquid

High-pressure liquid

Low-pressure liquid

Low-pressure gas

Step-by-Step Troubleshooting by Symptom Diagnosis

When servicing the self-contained ice cube maker, don't overlook the simple things that might be causing the problem. Step-by-step troubleshooting by symptom diagnosis is based on diagnosing malfunctions, with possible causes arranged into categories relating to the operation of the automatic ice maker. This section is intended only as a checklist to aid you in diagnosing a problem. Look at the symptom that best describes the problem you are experiencing with the automatic ice maker, and then correct the problem.

Compressor Will Not Run and There Is No Ice in the Storage Bin

- Is the ice cube maker located in an area where the temperature is below 55 degrees Fahrenheit?
- Test for proper voltage supply.
- Check for loose or broken wires.
- Test the compressor, relay, and overload protector.
- Check the controls for the proper setting.
- Test the bin thermostat for continuity. If contacts are open, replace the thermostat.
- Test the compressor.

Compressor Runs, but There Is No Ice in Storage Bin

- Check water supply.
- Check water valve.
- Check evaporator thermostat.
- Check the hot gas solenoid. It might be stuck "open."
- Check for sealed-system problems.
- Check for excessive use of ice cubes.
- Test cutter grid.
- Check wiring against wiring diagram.
- Is the water inlet tube from the water valve inserted in the return trough?
- Check condenser fan motor.

Ice Storage Bin Is Full of Ice and the Compressor Runs Continuously

- Check the calibration on the bin thermostat.
- Test the bin thermostat for continuity. Are the contacts stuck shut?
- Check wiring against the wiring diagram.

Low Ice Production

- Is the ice cube maker located in an area where the temperature is below 55 degrees Fahrenheit?
- Inspect the storage bin. Is water falling on the ice cubes?
- Check the calibration on the bin thermostat.
- Check the thickness control. Ice cubes produced should be between 1/2 and 5/8 inch thick.
- Check the hot gas solenoid. It might be stuck partially open.
- Check for sealed-system problems.
- Check the water supply in the reservoir. There might not be enough water circulating over the evaporator freeze plate.

Excessive Water Dripping on the Ice Cubes

- Is the water tank overflowing? Check for a blocked overflow tube.
- Is the water trough installed properly?
- Is the water inlet tube from the water valve inserted in the return trough?
- Check cutter grid for ice jam.
- Check the water deflector position.
- Check water valve for leaks.

Ice Cubes Are Too Thin

- Check thickness control setting.
- Check if there is enough water being circulated over the evaporator freeze plate.
- Check for restrictions in the water system.
- Check the water pump, motor, and the distributor tube.
- Check the thermostat calibration.

Ice Cubes Are Too Thick

- Check thickness control setting.
- Check the thermostat calibration.

The Condenser Fan Will Not Run During the Freeze Cycle

- Check the fan blades for binding on the shroud.
- Test the condenser motor for continuity.
- Check for open circuits against the wiring diagram.
- Check for a defective evaporator thermostat.

Water Pump Will Not Run

- Check the pump for binding in the housing.
- Check for open circuits against the wiring diagram.
- Test the pump motor for continuity.
- Check for a defective evaporator thermostat.

Water Tank Is Empty

- Water does not enter tank until the first defrost cycle is initiated.
- Check for an open circuit to the water valve solenoid.
- Check the water line for complete restriction.
- Check for a defective evaporator thermostat.
- Check the water valve; it might be stuck shut.
- Test the water valve solenoid for continuity.
- Check the water inlet tube from the water valve. It might not be directing the water into the tank.
- Check for a clogged water inlet screen in the water valve.

Treating the Water

In the freeze cycle, as the water passes over the evaporator freeze plate, the impurities in the water are rejected and only the pure water will stick to the plate. The more dissolved solids that are present in the water, the longer the freezing cycle. Bicarbonates, which are found in the water, are the most troublesome of all impurities. These impurities can cause:

- Scaling on the evaporator freeze plate
- Clogging of the water distributor head
- The water valve and many other parts in the water system may clog up

If the impurities become too concentrated in the water system, they can cause cloudy cubes and/or mushy ice.

All of the water system parts that come in direct contact with the water might become corroded if the water supply is high in acidity. The water might have to be treated in order to overcome problems with the mineral content. The most economical way to treat the water supply is with a polyphosphate feeder. This feeder is installed in the water inlet supply to prevent scale buildup. This will require less frequent cleaning of the ice maker. To install one of these feeders, follow the manufacturer's recommendations in order to treat the water satisfactorily.

Cleaning Instructions for the Ice Cube Maker

The manufacturer of this type of ice cube maker recommends that the ice maker be cleaned occasionally to help combat lime and mineral deposit buildup.

To clean the water system parts and the evaporator freeze plate, turn off the ice maker with the cycle switch. Open the bin door, and remove the cutter grid by removing the two thumb screws. Unplug the cutter grid, and remove it from the storage bin (Figure 22-32). A drain plug is located under the water tank. Remove it to drain the water out of the tank (Figure 22-33). After all of the water has been removed, reinstall the plug. Pour half a gallon of hot water into the tank, and set the switch to "clean." The hot water will circulate

FIGURE 22-32   Removing the thumbscrews, disconnecting the electrical harness, and pulling the cutter grid out of the bin.

FIGURE 22-33   The water tank is located in the storage bin.

FIGURE 22-34   Front grille held in place with two screws located on the toe panel section.

FIGURE 22-35   The condenser section of the ice maker.

through the water pump assembly and over the evaporator freeze plate, including all the water system components. Let the water circulate for five minutes, and then drain the water out of the tank. Replace the plug. Mix ice machine cleaner with half a gallon of hot water, and pour it into the water tank. If you use a recognized ice machine cleaner, follow the instructions on the label for best results. If you would rather prepare your own solution, add six ounces of citric acid and phosphoric acid to half a gallon of hot water, and pour into the water tank. Turn the switch to "clean," and circulate this solution for 20 minutes or longer; then drain the water. Follow with two clean water rinses that circulate for five minutes, and then drain the water again.

Remove the splash guard, the water dispenser tube, and the plastic water pump tank. Place them in a solution of mild laundry bleach for five minutes, and then rinse. Use one ounce of bleach to one gallon of hot water. Be sure the water temperature does not exceed 145 degrees Fahrenheit—it could damage the plastic parts. Finally, sanitize the ice bin, door, ice cube scoop, grid panel, and grid with a bleach solution.

Reinstall all parts in the reverse order of disassembly, and test the ice machine operation. After the cleaning treatment, apply a release agent to the evaporator plate. This will retard any future buildup of scale and mineral deposits, and it will make the plate more slippery, which will provide for better ice slab release.

To clean the condenser coil, remove the screws that secure the front grille, and then remove the grille (Figure 22-34). Vacuum all lint and dust from the coil and from the surrounding area (Figure 22-35). Reinstall the grille. The frequency of cleaning will be determined by the surrounding conditions.

Prevention of Water-Utilizing System Explosions

In certain water-utilizing refrigeration systems, water can leak into the refrigerant side of the system. This can lead to an explosion of system components, including but not limited to, the compressor. If such an explosion occurs, the resulting blast can kill or seriously injure anyone in the vicinity.

Systems at Risk of Explosion

Water-utilizing systems that have single-wall heat exchangers may present a risk of explosion. Such systems may include:

- Water-source heat pump/air conditioning systems
- Water cooling systems, such as ice makers, water coolers, and juice dispensers

Water-utilizing systems that have single-wall heat exchangers present a risk of explosion, unless they have one of the following:

- A high-pressure cutout that interrupts power to *all* leads to the compressor
- An external pressure-relief valve

How an Explosion Occurs

If the refrigerant tubing in the heat exchanger develops a leak, water can enter the refrigerant side of the system. This water can come in contact with live electrical connections in the compressor, causing a short circuit or a path to ground. When this occurs, extremely high temperatures can result. The heat buildup creates steam vapor that can cause excessive pressure throughout the entire system. This system pressure can lead to an explosion of the compressor or other system components.

Service Procedures

In light of the risk of explosion, be especially alert for signs of water leaking into the refrigerant side of the system. Whenever servicing or troubleshooting a water-utilizing system, always check to see if it has either a pressure-relief valve or a high-pressure cutout as previously described. If the system does not have at least one of these, *disconnect all electrical power*, and look for indications that water has leaked into the refrigerant side of the system. These indications may include:

- Observation of or a report of a blown fuse or tripped circuit breakers.
- Signs that water has leaked to the outside of the system.
- Reports that the system has made gurgling or percolating noises.
- A history of loss of refrigerant charge without a leak being found in the system. *Note:* Common leak-detection methods will not detect a water-to-refrigerant leak in the system's heat exchanger(s).
- Observation of or a report of the compressor giving off an unusual amount of heat.

If *any* of these indications are present, do the following checks to determine if water has leaked into the refrigerant side:

Step 1: Check for a Ground Fault (a Short to Ground)

- Check the compressor for a ground fault (also known as a short circuit to ground).
- If a ground fault does not exist, go to step 2.

- If a ground fault does exist, keep the power off. *Warning:* To avoid electric shock, electrocution, and terminal venting with ignition, do not energize a compressor that has a ground fault. Mark and red-tag the compressor to indicate that there is a ground fault. Do not reconnect the power leads. Tape and insulate each power lead separately. Proceed to step 2. Do not replace the compressor or energize the system before performing step 2.

Step 2: Check for Water in the System

Once the compressor is cool to the touch, open the system process valve slightly to see if any water comes out of the system. *Warning:* Opening the system process valve while the compressor is hot can cause severe burns from steam coming out of the valve. If any water comes out of the process valve, the entire system *must* be replaced. See the section "Replacing a Single-Wall Water-Utilizing System."

If water does not come out of the process valve, there is still a possibility that some water has leaked into the refrigerant side of the system. To address this possibility, determine if the system has a history of losing refrigerant charge without a leak being found or repaired.

If you find *any* indication of a history of losing refrigerant charge without detection of a leak, this is a sign that refrigerant has leaked into the water inside the heat exchanger. The entire system *must* be replaced. See the section "Replacing a Single-Wall Water-Utilizing System."

If you do not find any indication of a history of loss of charge without detection of a leak, you still need to install one of the following:

- A high-pressure cutout that interrupts power to *all* leads to the compressor
- An external pressure-relief valve

Also, if you found a ground fault in the compressor in step 1, replace the compressor before applying power to the system.

Replacing a Single-Wall Water-Utilizing System

When replacing a single-wall water-utilizing system, replace the system with one that has one of the following:

- A double-wall heat exchanger(s)
- A high-pressure cutout that interrupts power to *all* leads to the compressor
- An external pressure-relief valve

Repair Procedures

Each repair procedure is a complete inspection and repair process for a single ice cube maker component, containing the information you need to test and replace components. The actual repair or replacement of any sealed-system component is not included in this chapter. It is recommended that you acquire refrigerant certification (or call an authorized service company) to repair or replace any sealed-system component. The refrigerant in the sealed system must be recovered properly.

Compressor, Relay, and Overload Protector

The typical complaints associated with failure of the compressor, relay, and overload protector are:

- Ice maker does not run at all.
- No new ice production.
- Ice cubes in the storage bin are melting rapidly.
- Compressor won't run; it only hums.

To handle these problems, perform the following steps:

1. **Verify the complaint.** Verify the complaint by asking the customer to describe what the ice maker is doing.

2. **Check for external factors.** You must check for external factors not associated with the appliance. Is the appliance installed properly? Does the ice maker have the correct voltage?

3. **Disconnect the electricity.** Before working on the ice maker, disconnect the electricity. This can be done by pulling the plug from the receptacle. Or disconnect the electricity at the fuse panel or at the circuit breaker panel. Turn off the electricity.

4. **Gain access to the compressor.** Access the compressor. To access the compressor, remove the front grille. Remove the two screws in the condensing unit base, and pull the unit toward you. Be careful to not damage any refrigerant lines. Next, remove the compressor terminal cover (Figure 22-36) by removing the retaining clip that secures the cover. Remove the terminal cover.

5. **Test the compressor relay.** To test the compressor relay, remove the relay by pulling it from the compressor terminals without twisting it (Figure 22-37). On the relay body is stamped the word TOP. Hold the relay so that TOP is in the up position. Next, place the probes of the ohmmeter on the relay terminals marked S and M. Set the meter scale

FIGURE 22-36   A pictorial view of compressor, overload, and relay.

FIGURE 22-37   Pull the relay off the compressor. Be careful not to break the compressor pins.

on R × 1. The reading will show no continuity. Next, remove the probe from the terminal marked M, and place the probe on the side terminal marked L. The reading will show no continuity. Now, move the probe from terminal S, and place it on the terminal marked M. The reading will show continuity. With the probes still attached, turn the relay upside down (see Figure 21-34b), and perform the same tests. By turning the relay over, the switch contacts in the relay will close. When you retest the relay, you should get the opposite results. You should have continuity between terminals S and M and between S and L. The meter will not show continuity between M and L. If the relay fails this test, replace it.

6. **Test the overload protector.** To test the overload protector, remove the wires from the overload protector and compressor terminals. Then remove the overload protector (from the compressor) by removing the retaining clip that secures the overload protector (Figure 22-38). Next, place the probes of the ohmmeter on the overload protector terminals. Set the meter scale on R × 1. The reading will show continuity. If not, replace the overload protector.

7. **Test the compressor.** To test the compressor, remove the relay and the overload protector. This will expose the compressor terminals. The compressor terminals are marked C, S, and R: C indicates the common winding terminal, S indicates the start winding terminal, and R indicates the run winding terminal. Next, place the probes of the ohmmeter on the terminals marked S and R (Figure 22-39). Set the meter scale on R × 1, and adjust the needle setting to indicate a zero reading. The meter reading will show continuity. Now place the meter probes on the terminals marked C and S. The meter reading will show continuity. Finally, place the meter probes on the terminals marked C and R. The meter reading will show continuity. The total number of ohms measured between S and R is equal to the sum of C to S plus C to R.

To test the compressor for ground, place one probe on a compressor terminal and the other probe on the compressor housing or on any good ground (Figure 22-40). Set the meter

FIGURE 22-38 Disconnect the overload protector.

FIGURE 22-39 Testing the compressor windings.

FIGURE 22-40 Testing the compressor for ground.

scale to R × 1000. The meter reading will show no continuity. Repeat this for the remaining two terminals. The meter reading will show no continuity. If you get a continuity reading from any of these terminals to ground, the compressor is grounded and must be replaced.

Bin Thermostat

The typical complaints associated with bin thermostat failure are:

* The ice maker runs all the time, making too much ice.
* The ice maker doesn't run at all.

To handle these problems, perform the following steps:

1. **Verify the complaint.** Verify the complaint by checking the sensing tube and bulb and the control settings.

2. **Check for external factors.** You must check for external factors not associated with the appliance. Is the appliance installed properly? Explain to the user how to set the controls.

3. **Disconnect the electricity.** Before working on the ice maker, disconnect the electricity. This can be done by pulling the plug from the receptacle. Or disconnect the electricity at the fuse panel or at the circuit breaker panel. Turn off the electricity.

4. **Gain access to the bin thermostat.** To access the bin thermostat, remove the screws from the escutcheon (Figure 22-41), and remove the panel. Next, remove the screws from the control bracket (Figure 22-42). Pull back on the control bracket, exposing the controls.

5. **Test the bin thermostat.** To test the bin thermostat, remove the wires from the thermostat terminals and place the ohmmeter probes on those terminals (Figure 22-43). Set the range scale on R × 1, and test for continuity. The meter should read continuity between the contacts if the temperature of the capillary tube is above 42 degrees Fahrenheit. The meter should not read continuity between the contacts if the temperature of the capillary tube is below 36 degrees Fahrenheit.

FIGURE 22-41 Removing the control panel to expose the controls.

FIGURE 22-42   A pictorial view of the control panel.

FIGURE 22-43
The bin thermostat.

FIGURE 22-44
Peel back the liner to
gain access to the
capillary tube.

FIGURE 22-44
Peel back the liner to
gain access to the
capillary tube.

6. **Remove the bin thermostat.** Remove the bin thermostat from the control bracket by removing the two screws (see Figure 22-41). Remove the bin thermostat well from the left wall of the liner. Next, remove the five clips (under the gasket) from the left side of the liner (Figure 22-44). Now, bend the liner flange forward, and remove the capillary tube and thermostat control.

7. **Install a new bin thermostat.** To install the new bin thermostat, just reverse the order of disassembly, and reassemble. Then test the control. Remember to reinstall the capillary tube in the same location from which it was removed. If you do not, the ice maker will not cycle properly.

Evaporator Thermostat

The typical complaints associated with failure of the evaporator thermostat are:

- Unable to control ice cube thickness.
- Ice maker water not circulating.
- The ice maker runs, but there is no ice in the bin.

To handle these problems, perform the following steps:

1. **Verify the complaint.** Verify the complaint by checking the sensing tube and bulb and the control settings.

2. **Check for external factors.** You must check for external factors not associated with the appliance. Is the appliance installed properly? Explain to the user how to set the controls.

3. **Disconnect the electricity.** Before working on the ice maker, disconnect the electricity. This can be done by pulling the plug from the receptacle. Or disconnect the electricity at the fuse panel or at the circuit breaker panel. Turn off the electricity.

Figure 22-45
The evaporator
thermostat.

Multitester

4. **Gain access to the evaporator thermostat.** To access the evaporator thermostat, remove the screws from the escutcheon (see Figure 22-41), and remove the panel. Next, remove the screws from the control bracket (see Figure 22-42). Pull back on the control bracket, exposing the controls.

5. **Test the evaporator thermostat.** To test the evaporator thermostat, remove the wires from the thermostat terminals, and place the ohmmeter probes on terminals 1 and 2 (Figure 22-45). Set the range scale on R × 1, and test for continuity. The meter should show continuity between the contacts if the temperature of the evaporator freeze plate is 30 degrees Fahrenheit or warmer. The meter should not read continuity between the contacts if the temperature of the evaporator freeze plate is +10 to –3 degrees Fahrenheit. By disconnecting the water pump at the terminal board and operating the ice maker without the pump, the evaporator thermostat action can be easily observed. This will cause the thermostat to cycle in a matter of a few minutes.

6. **Remove the evaporator thermostat.** Remove the cutter grid by removing the thumbscrews (see Figure 22-32). Remove the evaporator thermostat from the control bracket by removing the two screws (see Figure 22-42). Remove the clamp from underneath the evaporator freeze plate, which secures the capillary tube to the evaporator. Next, remove the five clips, under the gasket, from the left side of the liner (see Figure 22-44). Now, bend the liner flange forward, and remove the capillary tube and thermostat control.

7. **Install a new evaporator thermostat.** To install the new evaporator thermostat, just reverse the order of disassembly, and reassemble. Then test the control. Remember to reinstall the capillary tube in the same location from which it was removed. Also, the capillary tube must be taped to the hot gas restrictor tube. If you do not do these things, the ice maker will not cycle properly.

PART V

FIGURE 22-46 The hot gas defrost valve. This valve will reverse the flow of refrigerant to the evaporator in the defrost cycle.

Solenoid coil

Hot gas valve

Hot Gas Solenoid Valve

The typical complaints associated with failure of the hot gas solenoid valve are:

- Ice maker runs, but there is no ice production.
- Evaporator freeze plate will not heat up to release ice slab.
- Ice maker runs continuously.

To handle these problems, perform the following steps:

1. **Verify the complaint.** Verify the complaint by checking the ice maker cycles.

2. **Check for external factors.** You must check for external factors not associated with the appliance. Is the appliance installed properly? Explain to the user how to set the controls.

3. **Disconnect the electricity.** Before working on the ice maker, disconnect the electricity. This can be done by pulling the plug from the receptacle. Or disconnect the electricity at the fuse panel or at the circuit breaker panel. Turn off the electricity.

4. **Gain access to the hot gas solenoid valve.** To access the hot gas solenoid valve (Figure 22-46), remove the grille (see Figure 22-34). Next, remove the deflector from the condenser (see Figure 22-35).

5. **Test the hot gas solenoid valve.** Test the hot gas solenoid valve for continuity. Remove the wires from the solenoid coil. Place the ohmmeter probes on the solenoid coil terminals (see Figure 22-46). Set the range scale on R × 1, and test for continuity. To test the hot gas valve itself, connect a 120-volt fused service cord (Figure 22-47) to the solenoid coil. Listen for a click sound as the plunger rises up. Now, disconnect the service cord, and you will hear the plunger drop back. If you cannot hear a distinct click sound from the hot gas valve, it will need to be replaced by an authorized service company (the sealed system might be under warranty from the manufacturer) or by a licensed refrigerant technician.

   The solenoid coil is a separate component that can be replaced without replacing the entire hot gas valve assembly. Another way to test the hot gas valve is to leave the wires off the solenoid coil and reconnect the service cord to the solenoid coil. This test requires the electricity to be turned on.

CAUTION *Tape the solenoid coil leads that were removed so that they will not touch the chassis when you plug in the ice maker for this test. Be cautious when working with live wires. Avoid getting shocked!*

With the ice maker plugged in and running, feel the hot gas defrost tube—it should feel warm or hot when the valve is energized.

Figure 22-47
A 120-volt fused
service cord test.

Three rubber insulated
NEC color-coded
alligator clips

Grounded to
electrical box

24-inch
test leads

On/Off
switch

Electrical box

120-volt fused
service cord

20-amp
resettable fuse

6-foot
service cord

Male plug
with ground

120-volt
receptacle

6. **Remove the hot gas solenoid coil.** To remove the hot gas solenoid coil, remove the spring clip from the top of the coil, and then remove the coil (be sure the electricity is off).

7. **Install a new hot gas solenoid coil.** To install the new solenoid coil, just reverse the order of disassembly, and reassemble. Then test the valve.

Water Valve

The typical complaints associated with failure of the water valve are:

- Ice maker runs, but there is no ice production.
- No water is circulating across the evaporator freeze plate.
- Water is flooding the storage bin, causing the ice to melt.

To handle these problems, perform the following steps:

1. **Verify the complaint.** Verify the complaint by checking the ice maker cycles.

2. **Check for external factors.** You must check for external factors not associated with the appliance. Is the appliance installed properly? Explain to the user how to set the controls. Is the water supply turned on?

3. **Disconnect the electricity.** Before working on the ice maker, disconnect the electricity. This can be done by pulling the plug from the receptacle. Or disconnect the electricity at the fuse panel or at the circuit breaker panel. Turn off the electricity.

Figure 22-48
The water valve.

Multitester

4. **Gain access to the water valve.** To access the water valve, remove the top insulated panel. The water valve is located in the upper-right front corner.

5. **Remove and test the water valve.** In order to test the water valve solenoid coil, the water valve must be removed from the storage bin. Shut off the water supply to the ice maker. Now disconnect the water line from the valve. Next, remove the screws from the water valve bracket. Pull on the valve to release it from the receptacle in the liner. Place the ohmmeter probes on the solenoid coil terminals (Figure 22-48). Set the range scale on R × 1, and test for continuity. If there is no continuity, replace the water valve.

6. **Install a new water valve.** To install the new water valve, just reverse the order of disassembly, and reassemble. Then test the valve. Don't forget to turn on the water supply.

Condenser Fan Motor

The typical complaints associated with failure of the condenser fan motor are:

- The ice maker has stopped producing ice.
- The condenser fan motor runs slower than normal.
- The condenser fan motor does not run at all.
- The compressor is sometimes noisier than normal.

To handle these problems, perform the following steps:

1. **Verify the complaint.** Verify the complaint by asking the customer to describe what the ice maker is doing. Is the condenser fan motor running during the freeze cycle?

2. **Check for external factors.** You must check for external factors not associated with the appliance. Is the appliance installed properly? Are there any foreign objects blocking the condenser fan blades?

3. **Disconnect the electricity.** Before working on the ice maker, disconnect the electricity to the ice maker. This can be done by pulling the plug from the receptacle. Or disconnect the electricity at the fuse panel or at the circuit breaker panel. Turn off the electricity.

4. **Gain access to the condenser fan motor.** To access the condenser fan motor, remove the front grille. Remove the two screws in the condensing unit base, and pull the unit toward you. Be careful not to damage any refrigerant lines.

5. **Test the condenser fan motor.** To test the condenser fan motor, remove the wires from the motor terminals. Next, place the probes of the ohmmeter on the motor terminals (Figure 22-49). Set the meter scale on R × 1. The meter should show some resistance. If no reading is indicated, replace the motor. If the fan blades do not spin freely, replace the motor. If the bearings are worn, replace the motor.

6. **Remove the condenser fan motor.** To remove the condenser fan motor, you must first remove the fan blades. Unscrew the nut that secures the blades to the motor. Remove the blades from the motor. Then remove the motor assembly by removing the mounting bracket screws (Figure 22-50).

7. **Install a new condenser fan motor.** To install the new condenser fan motor, just reverse the order of disassembly, and reassemble. Remember to reconnect the ground wire to the motor. Reconnect the wires to the motor terminals, and test.

Water Pump

The typical complaints associated with failure of the water pump are:

- Ice maker runs, but there is no ice production.
- No water circulating across the evaporator freeze plate.

FIGURE 22-49 The condenser fan motor.

FIGURE 22-50 A pictorial view of the condenser fan motor assembly.

To handle these problems, perform the following steps:

1. **Verify the complaint.** Verify the complaint by checking the ice maker cycles.

2. **Check for external factors.** You must check for external factors not associated with the appliance. Is the appliance installed properly? Explain to the user how to set the controls.

3. **Disconnect the electricity.** Before working on the ice maker, disconnect the electricity. This can be done by pulling the plug from the receptacle. Or disconnect the electricity at the fuse panel or at the circuit breaker panel. Turn off the electricity.

4. **Gain access to the water pump.** To access the water pump, remove the bin door, the front insulated panel, and the inner bin door. A drain plug is located under the water tank. Remove it to drain the water out of the tank (see Figure 22-33). Next, remove the thumbscrews that secure the water tank, and remove the tank.

5. **Test the water pump.** To test the water pump motor, isolate the motor, and place the probes of the ohmmeter on the motor terminals (Figure 22-51). Set the meter scale on R × 1. The meter should show some resistance. If no reading is indicated, replace the water pump. Next, check the motor with a 120-volt fused service cord (see Figure 22-47).

6. **Remove the water pump.** To remove the water pump (Figure 22-52), remove the screws from the water pump bracket that secure the pump to the liner. Disconnect the discharge hose from the pump. Remove the water pump.

7. **Install a new water pump.** To install a new water pump, just reverse the order of disassembly, and reassemble. Reconnect the wires to the motor terminals, and test.

Multitester

Figure 22-51   A bottom view of the water pump. When checking the pump, be sure the inlet is free of debris.

Cutter Grid

The typical complaints associated with failure of the cutter grid are:

- Ice slabs lie on top of the cutter grid.
- Cutter grid is not cutting ice slab into cubes evenly.

To handle these problems, perform the following steps:

1. **Verify the complaint.** Verify the complaint by checking the ice slab and cutter grid fuse.

2. **Check for external factors.** You must check for external factors not associated with the appliance. Is the appliance installed properly? Explain to the user how to set the controls.

3. **Disconnect the electricity.** Before working on the ice maker, disconnect the electricity. This can be done by pulling the plug from the receptacle. Or disconnect the electricity at the fuse panel or at the circuit breaker panel. Turn off the electricity.

Motor
fan

Motor

Sealing washer

Mounting
bracket

Grommets

Pump
impeller

Figure 22-52   A pictorial view of the water pump.

4. **Gain access to the cutter grid.** To access the cutter grid, open the bin door and remove the cutter grid by removing the two thumbscrews. Unplug the cutter grid, and remove it from the storage bin (see Figure 22-32).

5. **Test the cutter grid.** Examine the cutter grid for broken wires, and check the connecting pins for corrosion (Figure 22-53). As you inspect the cutter grid, look for cracked or broken insulators in the frame. Next, place the probes of the ohmmeter on the cutter grid plug terminals (Figure 22-54). Set the meter scale on R × 1. The meter should show continuity. If no reading is indicated, one or more grid wires or insulators are defective.

Figure 22-53
A cutter grid cuts an
ice slab into strips
and then into cubes.

Figure 22-55   Side view of the cutter grid, showing the clips, buss bars, and insulators.

Figure 22-54   A pictorial view of the cutter grid circuit.

6. **Repair the cutter grid.** If the cutter grid frame and insulators are broken, it would be advisable to replace the entire cutter grid. Using a C-clamp, compress the spring clip to relieve the tension (Figure 22-55). Next, use a pair of pliers to compress the adjacent spring clip, and remove the buss bar. Do the same procedure for the other side of the cutter grid. The insulators, clips, and grid wires can now be removed and replaced. If any of the grid wires break, it is time to replace all of the grid wires.

7. **Install a new cutter grid.** To install a new cutter grid, just reverse the order of disassembly, and reassemble. Reconnect the wires to the cutter grid, and test it.

Cutter Grid Transformer and Fuse

The typical complaints associated with failure of the grid transformer or fuse are:

- Ice slabs lie on top of the cutter grid.
- Cutter grid is not cutting ice slab into cubes evenly.

To handle these problems, perform the following steps:

1. **Verify the complaint.** Verify the complaint by checking the ice slab and cutter grid.
2. **Check for external factors.** You must check for external factors not associated with the appliance. Is the appliance installed properly?

FIGURE 22-56 The cutter fuse, located behind the control panel.

3. **Disconnect the electricity.** Before working on the ice maker, disconnect the electricity. This can be done by pulling the plug from the receptacle. Or disconnect the electricity at the fuse panel or at the circuit breaker panel. Turn off the electricity.

4. **Gain access to the cutter grid transformer and fuse.** To access the cutter grid transformer and fuse, remove the screws from the escutcheon (see Figure 22-41), and remove the panel. The fuse is located on the control bracket on the left side. To remove the fuse, push in and twist—it will pop out of the holder (Figure 22-56). Next, remove the screws from the control bracket (see Figure 22-42). Pull back on the control bracket, exposing the controls.

5. **Test the cutter grid transformer and fuse.** To test the transformer, disconnect the wires from the transformer to isolate it from the circuit. Use a 120-volt fused service cord, and connect it to the primary side of the transformer. *Note:* This test requires the electricity to be turned on. Be cautious when working with live wires. Avoid getting shocked!

You might have to look at the wiring diagram for assistance in identifying the primary side and for the proper color-coding of the wires. Using the volt meter, connect the probes to the secondary side of the transformer. Plug in the 120-volt fused service cord; the meter should read 8.5 volts. Unplug the service cord.

To test for resistance, disconnect the power cord, and set the ohmmeter scale on R × 1. Place the probes on the primary wires of the transformer. The meter should show resistance. Next, place the probes on the secondary wires of the transformer (Figure 22-57). The meter should show resistance. If the transformer fails either test, replace it.

FIGURE 22-57 The cutter grid transformer.

FIGURE 22-58
The cutter grid fuse.
If this fuse blows, the
cutter grid will not
function. Check for a
short.

Multitester

To test the fuse, place the probes on each end of the fuse (Figure 22-58). Set the ohmmeter scale on R × 1. The meter should show continuity. If not, replace the fuse.

6. **Remove the transformer.** To remove the transformer, remove the screws that secure it to the control bracket (see Figure 22-42).

7. **Install a new transformer or fuse.** To install a new transformer or fuse, just reverse the order of disassembly, and reassemble. Reconnect the wires to the components, and test it.

Cycle Switch

The typical complaints associated with failure of the cycle switch are:

- Unable to turn on the clean cycle.
- Unable to turn on the ice maker.

To handle these problems, perform the following steps:

1. **Verify the complaint.** Verify the complaint by checking the control settings.

2. **Check for external factors.** You must check for external factors not associated with the appliance. Is the appliance installed properly? Explain to the user how to set the controls.

3. **Disconnect the electricity.** Before working on the ice maker, disconnect the electricity. This can be done by pulling the plug from the receptacle. Another way to disconnect the electricity is at the fuse panel or at the circuit breaker panel. Turn off the electricity.

4. **Gain access to the cycle switch.** To access the cycle switch, remove the screws from the escutcheon (see Figure 22-41), and remove the panel. Next, remove the screws from the control bracket (see Figure 22-42). Pull back on the control bracket, exposing the controls.

5. **Test the cycle switch.** To test the cycle switch for continuity between the switch contacts, refer to the wiring diagram for the correct switch positions (Figure 22-59). Disconnect the wires from the switch, and turn the switch to the "on" position. Place the probes of the ohmmeter on the cycle switch terminals marked 2 and 3. Set the meter scale on R × 1. The meter should show continuity. Now place the probes of the ohmmeter on the cycle switch terminals marked 1 and 6. The meter should show continuity. Next, turn the cycle switch to the "clean" position. Place the probes of the ohmmeter on the cycle switch terminals marked 5 and 6. The meter should show continuity. If the switch fails these tests, replace it.

6. **Remove the cycle switch.** To remove the cycle switch, remove the two screws that secure it to the control bracket (see Figure 22-42).

7. **Install a new cycle switch.** To install a new cycle switch, just reverse the order of disassembly, and reassemble. Connect the wires to the switch terminals according to the wiring diagram, and test it.

Figure 22-59
A sample wiring diagram of a self-contained ice maker.

PART V

Self-Contained Ice Cube Maker Performance Data

There are a number of factors that can affect the overall performance of the ice cube maker.

Installation

- Installation in a dusty environment will cause the ice cube maker to operate continuously and have a low ice production rate.
- Installation in a greasy atmosphere will cause the ice cube maker to operate in high temperature conditions with low to no ice production.
- Installation next to a high temperature appliance will cause the ice cube maker to operate in high temperature conditions with low to no ice production.

Water Temperature

- Higher water temperatures will increase the operating time of the ice cube maker.
- Decreased ice production in a 24 hour period.

Ambient Temperature

- High temperatures will decrease ice production.
- Ice storage bin melting ice above 25 per cent.

Maintenance

- Restricted air movement through the condenser coil and front grille will cause low to no ice production.

Other

- If the ice cubes are more than ½ inch thick the ice production will decrease and the operating time will increase.

See Table 22-1, Table 22-2, and Table 22-3 for additional technical data.

Thermostat Position	Cut-in Temperature	Cut-Out Temperature
Evaporator thermostat - warm	38 degrees F ±2 degrees	11 degrees F ±4 degrees
Evaporator thermostat - coldest	38 degrees F ±2 degrees	−3 degrees F ±3 degrees
Bin thermostat	41 degrees F ±2 degrees	35 degrees F ±2 degrees

TABLE 22-1   Ice Thickness Controlled by Evaporator Thermostat.

Ambient Temperature °F	24-Hour Ice Production in Pounds			
100	36	36	35	34
90	41	40	39	37
80	46	44	43	41
70	51	48	46	44

To obtain the maximum ice production rate you will need to have the ideal conditions. The capacities that are shown above are the average production rates and they can vary depending on conditions.

TABLE 22-2 Water Temperature

Temperature	Cycle Time in Minutes
Ambient 70°F Water 60°F	25–29
Ambient 90°F Water 60°F	27–33
Ambient 110°F Water 60°F	38–46
Ambient 70°F Water 80°F	28–34
Ambient 90°F Water 80°F	32–40
Ambient 110°F Water 80°F	40–48

TABLE 22-3 Ice Cube Operating Cycle Times

Endnotes

1. The jumper wire should be made of 14-gauge wire and should be approximately 8 to 10 inches long, with 5/8 inch of the insulation removed from both ends. Do not short any contacts other than those specified in this test procedure. Why? Because you can damage the ice maker, injure yourself, or both.

2. You can purchase this ice maker test cord from your parts distributor.

Room Air Conditioners

Room air conditioners (RACs) are available in different sizes (rated in BTUs) and cabinet styles.[1] These units can be installed in a window or through the wall, or are available as a portable unit on wheels (Figure 1-16). Manufacturers have designed RACs with electromechanical controls or electronic controls. Consumers can purchase an RAC that is straight cool only, cool with heat (electric), heat pump (reverse cycle), or a combination of cool and heat (electric heat and reverse cycle).

The room air conditioner consists of the following:

- The base pan, fan housing, divider section (bulkhead), and outer cabinet
- The sealed system, which consists of the evaporator coil, the condenser coil, the compressor, and the connecting tubing
- The electrical circuitry, including the fan motor and other electrical components

Principles of Operation

The room air conditioner, when installed and running properly, will circulate the air in a room or area, removing the heat and humidity; some models will heat the air in the winter months. At the same time, the air filter, located behind the front grille, will filter out dust particles. Most models have a fresh air intake feature, which allows fresh outside air to enter the room when the unit is running. The thermostat will control the comfort level in the room or area, and cycle the air conditioner on and off according to the temperature setting.

Before continuing, take a moment to refresh your memory of Chapter 7. These two chapters combined will make servicing room air conditioners a breeze.

Safety First

Any person who cannot use basic tools or follow written instructions should *not* attempt to install, maintain, or repair any room air conditioners. Any improper installation, preventive maintenance, or repairs could create a risk of personal injury or property damage. If you do not fully understand the installation, preventive maintenance, or repair procedures in this chapter, or if you doubt your ability to complete the task on your room air conditioner, please call your service manager.

This chapter covers the electrical components and how to diagnose the sealed system. The actual repair or replacement of any sealed-system component is not included in this chapter. It is recommended that you acquire refrigerant certification (or call an authorized service company) to repair or replace any sealed-system component, as the refrigerant in the sealed system must be recovered properly.

Before continuing, take a moment to refresh your memory on the safety procedures in Chapter 2.

Room Air Conditioners in General

Much of the troubleshooting information in this chapter covers room air conditioners in general, rather than specific models, in order to present a broad overview of service techniques. The illustrations that are used in this chapter are for demonstration purposes only, to clarify the description of how to service these appliances. They in no way reflect on a particular brand's reliability.

Electrical Requirements

One of the most common problems that I have encountered over the years is when a consumer plugs the room air conditioner into an existing 115-volt receptacle only to find out that the circuit breaker keeps tripping. Most people do not take into account that this receptacle is connected to the other receptacles and/or lights in the same room and on the same branch circuit. This means that there is a limit to how many items can be connected to one branch circuit. Before a consumer purchases a room air conditioner, he or she needs to add up the total amperage or wattage of the items that are plugged into the branch circuit where the air conditioner is going to be plugged in. For example, on a 15-amp circuit, the total wattage on a 120-volt circuit should not exceed 1800 watts (Table 23-1). For safety reasons,

Circuit Breaker Size Rated in Amps	Volts	Watts	BTUs
15	120	1800	6138
20	120	2400	8184
25	120	3000	10,230
30	230	6900	23,529
40	230	9200	31,372
50	230	11,500	39,215
60	230	13,800	47,058
70	230	16,100	54,901
80	230	18,400	62,744
100	230	23,000	78,430

Conversion formulas: Watts = Volts × Amps; BTUs = Watts × 3.41; Amps = Watts/Volts

TABLE 23-1   Conversion Formulas for Amps, Volts, Watts, and BTUs.

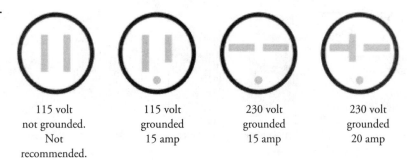

Figure 23-1
An electrical receptacle guide illustrating the type of receptacle needed for the amperage rating.

115 volt
not grounded.
Not
recommended.

115 volt
grounded
15 amp

230 volt
grounded
15 amp

230 volt
grounded
20 amp

all of the items, including the air conditioner, should stay below the amperage rating on the circuit breaker by 20 percent. The 115-volt, 15-amp circuit is reserved for lights, computers, televisions, stereos, etc. The 115-volt, 20-amp circuit is reserved for a refrigerator, dishwasher, automatic washer, or a garbage disposer only. Most 230-volt, 15- to 60-amp receptacles are on a single branch circuit. Figure 23-1 illustrates the type of plug connections needed for a room air conditioner. All room air conditioners and portable air conditioners must have a properly polarized and grounded receptacle, with the correct operating voltage for the unit purchased, preferably on its own branch circuit (Table 23-2). If the consumer only has multiple receptacles on one branch circuit, advise him or her to calculate the total amperage or wattage of each item (for example: television, computer, printer, computer monitor, lamps, microwave oven, etc.). Advise the consumer not to exceed the rated branch circuit, circuit breaker, or fuse. If the consumer does not correct the problem, this could lead to a possible fire hazard if the home wiring overheats.

Location and Installation of Room Air Conditioner

Room air conditioners and portable models can be installed in various types of windows (Figures 23-2 and 23-3) or through the wall (Figure 23-4) for permanent installation. It is the installer's responsibility to make sure that the air conditioner is installed properly according to the manufacturer's specifications and local building codes. If the air conditioner is installed in a window, the installer must make sure that the air conditioner is secured to the window so that it will not fall out, causing injury, death, or property damage. Through-the-wall models

Operating Voltages	Minimum Voltages	Maximum Voltages
115	104	127
120	108	132
208	188	229
220	198	242
230	207	253
240	216	264

Table 23-2 The Minimum/Maximum Operating Voltages for an Air Conditioner.

PART V

FIGURE 23-2
Room air conditioners can be installed in these types of windows. When selecting an air conditioner to purchase, measure the window opening.

Types of windows

Double-hung Casement Sliding Awning Transom-above a doorway

should be secured to the wall to prevent the unit from being pushed out of the wall when the unit is installed or serviced. Portable air conditioners are much easier to install. The air intake and discharge hoses are installed through a small opening in the window. This type of unit can be moved around from room to room easily.

The following are some helpful tips when considering a location to install a window air conditioner:

- Make sure you have the correct size (BTUs) for the area being cooled.[2]
- When selecting an air conditioner, make sure that the air vents will be able to point to the center of the room for better air circulation.

FIGURE 23-3
A typical portable air conditioner installation.

PART V

FIGURE 23-4
(a) A typical wall installation. The air conditioner must be installed with a slight pitch so that the condensate water can drain to the rear of the unit. (b) If the outside vents cannot clear the wall structure, purchase an air conditioner with a sleeve without side vents.

The cabinet vents must clear the exterior wall for maximum air circulation.

(a)

Some models have no side vents. The outside air is drawn into the rear and discharged through the rear.

Through-the-wall air conditioner

Wall sleeve

Rear grille

(b)

- When purchasing a window air conditioner, purchase one that is designed for the type of window in the home (see Figure 23-2). This will make for an easier installation.

- Make sure the unit will fit the window opening. Measure the opening of the window.

- Check the electrical outlet voltage. The voltage should match the operating voltage of the unit being installed.

- The electrical outlet should not be farther away than the length of the service cord.

- For larger air conditioners, install the brackets that come with the air conditioner. These brackets will reduce the stress on the window.

- Make sure there is adequate airflow with no obstructions on the outside of the building.

- Do not install a room air conditioner in an area where the temperature will exceed 120 degrees Fahrenheit.

The following are some helpful tips when considering a location to install a through-the-wall air conditioner:

- Make sure you have the correct size (BTUs) for the area being cooled.[2]

- When selecting an air conditioner, make sure that the air vents will be able to point to the center of the room for better air circulation.

- Measure the depth of the wall where the air conditioner will be installed. Make sure that the outside louvers will not be blocked (see Figure 23-4).

- Purchase an air conditioner with a slide-out chassis or with a wall sleeve for easier installation.

- Check the electrical outlet voltage. It should match the operating voltage of the unit being installed.

- The electrical outlet should not be farther away than the length of the service cord.

- Make sure there is adequate airflow with no obstructions on the outside of the building.

- Do not install a room air conditioner in an area where the temperature will exceed 120 degrees Fahrenheit.

Step-by-Step Troubleshooting by Symptom Diagnosis

When servicing an air conditioner, don't overlook the simple things that might be causing the problem. Step-by-step troubleshooting by symptom diagnosis is based on diagnosing malfunctions, with possible causes arranged into categories relating to the operation of the room air conditioner. This section is intended only to serve as a checklist to aid you in diagnosing a problem. Look at the symptom that best describes the problem you are experiencing with the air conditioner, and then correct the problem.

Air Conditioner Will Not Run

- Is the unit service cord plugged in?
- Check for voltage at the receptacle.
- Are the controls set properly?
- Check for a blown fuse or tripped circuit breaker. Check for total wattage on the branch circuit.
- Check the selector switch or electronic control board.
- Check all wiring connections and wiring.

Air Conditioner Will Not Cool or Cools Slightly

- Check for a dirty filter.
- Check for a dirty or restricted evaporator coil.
- Check for a dirty or restricted condenser coil.
- Check fan speed setting. Are the louvers adjusted correctly?
- Check the thermostat setting.
- Check the refrigeration cycle for leaks or an undercharge.
- Is the air conditioner the right size for the area to be cooled?[2]
- Is the fresh air intake or exhaust door open?

Compressor Will Not Run

- Check for the correct voltage at the receptacle.
- Check the control settings.
- Check the wiring connectors and wiring.
- Check the thermostat setting.
- Check the selector switch or electronic control board.
- Test the compressor for an open, short, or grounded winding.
- Test the overload.
- Test the compressor capacitor for an open, short, or ground.

Fan Motor Will Not Run

- Check the fan selector switch, main selector switch, or electronic control board.
- Check the wiring connectors and wiring.
- Test the fan capacitor for an open, short, or ground.
- Test the fan motor windings for an open, short, or ground.

Evaporator Coil Freezes Up

- Check for a dirty filter.
- Check for a dirty or restricted evaporator coil.

- Check the thermostat.
- Check fan motor.
- Check the refrigeration cycle for leaks, undercharge, or capillary tube restriction.

Condenser Coil Frozen (Heat Pump Models Only)

- Check defrost thermostat (outdoor thermostat) switch contacts and the capillary tube placement.
- Check the solenoid coil on the reversing valve.
- Check the reversing valve.
- Check the control settings.

Compressor Runs Continually and Will Not Cycle Off

- Check for excessive heat load. Is the air conditioner the right size for the area to be cooled?[2]
- Check for a partial refrigeration restriction in the line. Is the evaporator coil partially iced up?
- Check for a refrigerant leak.
- Check the running amperage against the model number identification plate on the air conditioner.
- Check the thermostat for proper operation.

Thermostat Will Not Cycle Off the Air Conditioner

- Check for stuck thermostat contacts.
- Check the thermostat setting.
- Check the wiring connections. Is the thermostat wired correctly?
- Is the air conditioner the right size for the area to be cooled?[2]

Thermostat Will Not Cycle the Air Conditioner On

- Check the thermostat bulb and capillary tube for loss of charge.
- Test the thermostat.
- Check the wiring connections and wiring.
- Check the control settings.

Thermostat Will Short-Cycle

- Is the air conditioner the right size for the area to be cooled?[2]
- Check for a dirty or restricted evaporator coil.
- Check for a dirty filter.
- Check the positioning of the thermostat bulb. Test the thermostat differential.
- Check and make sure that the plenum gasket is sealing properly.
- Check for outside air leakage into the air conditioner.

Compressor Runs and then Cycles on Overload

- Check for low voltage at the receptacle.
- Check the compressor overload.
- Did the compressor restart before the sealed system had a chance to equalize? Advise the customer to wait about three or four minutes after the air conditioner cycles off before restarting the compressor.
- Check the wiring connections and wiring.
- Check the compressor capacitor for an open, short, or ground.
- Check for a dirty or restricted condenser coil.
- Check the amperage of the air conditioner.
- Check for a kinked discharge line.

Air Conditioner Is Noisy

- Check the air conditioner installation.
- Make sure the fan blade and blower wheel are not striking the chassis.
- Check the compressor mounts and tubing.
- Inspect for loose cabinet parts.

Water Leaks Inside the Home

- Check the evaporator drain pan and drain for a blockage.
- Check weather sealing around the outer cabinet. Reseal if necessary.
- Check for water droplets on the outside of the base pan. If so, the evaporator drain pan might be cracked. Inspect the evaporator drain pan.
- Check the backside of the discharge grille. If it is wet, the evaporator coil might be dirty.
- Inspect all gaskets between the unit, outer cabinet, and window panes for air leakage.
- Check the angle of slope on the air conditioner installation.
- Check for a dirty evaporator coil.

Water Leaks on the Outside of the Home

- Check for water droplets on the outside of the base pan. If so, the evaporator drain pan might be cracked. Inspect the evaporator drain pan.
- Check for water between the condenser fan shroud and the compressor. If the water has collected around the compressor, inspect the fan shroud. Is it detached from the condenser coil?
- Check for a dirty condenser coil.
- Inspect the condenser fan blade—is the slinger ring in the correct position?

Air Conditioner Will Not Heat or There Is Not Enough Heat

- On electric heat models, check the electric heater and thermostats.
- On heat pump models, check the reversing valve, reversing valve solenoid coil, and thermostats.
- Check for a dirty filter.
- Check for an open fresh air intake or exhaust vent.
- Is the air conditioner the right size for the area?[2]
- Check for a dirty evaporator coil.
- Check the wiring connections and wiring.
- Check selector switch or electronic control board.

Heat Pump Will Not Go Into Defrost

- Check the defrost control and thermostats.
- Check the reversing valve and solenoid coil.
- Check the wiring connections and wiring.
- Check the electronic control board.

Air Conditioner Is Cooling when the Controls Are Set for Heat

- Check the wiring connections and wiring.
- Check the reversing valve and solenoid coil.
- Check the selector switch or electronic control board.
- Check the thermostats.

Room Air Conditioner Maintenance

Room air conditioners (including portable models) have air filters that need cleaning every 225 to 360 fan hours of operation. The discharge grille area also needs vacuuming to remove the dust buildup. Twice a year, the following areas need to be inspected and cleaned:

- The evaporator coil
- The condenser coil
- The evaporator pan and base pan
- The indoor blower housing and blower wheel
- All the wiring connections and wiring
- The electrical and mechanical controls
- The voltage at the receptacle
- The inside and outside of the air conditioner
- All gaskets

- The drain system (clean it, too)
- The cabinet seal (clean the outer cabinet)
- The copper tubing

When cleaning the air conditioner, use an approved cleaner to wash the unit. Remember to protect the electrical components and fan motor with plastic to prevent the water from damaging the components. Refer to the use and care manual that comes with every air conditioner for further maintenance instructions on the model you are servicing. Do not plug in or run the air conditioner after using water to clean the unit. Wait a few hours, allowing the air conditioner to completely dry out. To prevent electrical mishaps, the air conditioner must be totally dry before you can plug it in.

Performance Data

After you have completed the maintenance or repair on the air conditioner, perform an electrical test by checking the amperage or wattage on the unit, and compare the readings with the information on the model number data tag. At the same time, perform readings on the following:

- The room temperature and outside temperature.
- The temperature differential of the intake and discharge air through the evaporator coil. Take a reliable thermometer, place it in front of the air intake (where the air filter is located), and take a reading. Then place the thermometer in the discharge grille, and take a reading of the air blowing into the room. The difference between the two readings will be the temperature drop. This reading will vary among manufacturers and models. The temperature drop should be between 18 and 31 degrees Fahrenheit.
- The temperature differential of the intake and discharge air through the condenser coil. Use the same reliable thermometer to take the readings.
- Use a sling psychrometer to measure the indoor and the outdoor wet bulb temperatures. The sling psychrometer will measure the relative humidity in the room and outside (Table 23-3).
- Measure the operating voltage.
- Measure the startup and cycling amperage or wattage of the unit.

Take the readings and match them against the manufacturer's performance data. You can locate the air conditioner performance data on the manufacturer's Web site or in the manufacturer's service manual. The data that you accumulated should match—if it doesn't, adjustments will have to be made to bring the air conditioner up to manufacturer's standards. You might have to replace a component, clean the unit, or correct the installation.

Repair Procedures

Each repair procedure is a complete inspection and repair process for a single room air conditioner component. It contains the information you need to test and replace components.

PART V

Temperature of the Room or Outside (Dry Bulb) in Degrees Fahrenheit	The Temperature Difference Between the Dry Bulb and the Wet Bulb Temperatures in Degrees Fahrenheit								
	4	5	6	7	8	9	10	11	12
	Relative Humidity % at Pressure = 30.00 Inches								
40	68	60	52	45	37	29	22	15	7
50	74	67	61	55	49	43	38	32	27
52	75	69	63	57	51	46	40	35	29
54	76	70	64	59	53	48	42	37	32
56	76	71	65	60	55	50	44	39	34
58	77	72	66	61	56	51	46	41	37
60	78	73	68	63	58	53	48	43	39
62	79	74	69	64	59	54	50	45	41
64	79	74	70	65	60	56	51	47	43
66	80	75	71	66	61	57	53	48	44
68	80	76	71	67	62	58	54	50	46
70	81	77	72	68	64	59	55	51	48
72	82	77	73	69	65	61	57	53	49
74	82	78	74	69	65	61	58	54	50
76	82	78	74	70	66	62	59	55	51
78	83	79	75	71	67	63	60	56	53
80	83	79	75	72	68	64	61	57	54
82	84	80	76	72	69	65	61	58	55
84	84	80	76	73	69	66	62	59	56
86	84	81	77	73	70	66	63	60	57
88	85	81	77	74	70	67	64	61	57
90	85	81	78	74	71	68	65	61	58
100	86	83	80	77	73	70	68	65	62

TABLE 23-3   Psychrometric Table.

Electronic Components

The typical complaints associated with the electronic components are:

- Unable to program the touchpad panel functions.
- The display board will not display anything.
- Unusual display readouts.
- Unable to control the temperatures.

- The compressor will not run.
- The fan motor will not run.

To prevent electrostatic discharge (ESD) from damaging expensive electronic components, follow the steps in Chapter 9.

To handle these problems, perform the following steps:

1. **Verify the complaint.** Verify the complaint by operating the air conditioner controls. Turn off the electricity to the air conditioner, and wait for two minutes before turning it back on. If a fault code appears, look up the code. If the air conditioner will not power up, locate the technical data sheet behind the control panel for diagnostics information.

2. **Check for external factors.** You must check for external factors not associated with the air conditioner. Is there electricity to the air conditioner? Is the electrical receptacle polarized and properly grounded?

3. **Disconnect the electricity.** Before working on the air conditioner, disconnect the electricity. This can be done by pulling the plug out of the wall receptacle. Or disconnect the electricity at the fuse panel or circuit breaker panel. Turn off the electricity.

WARNING *Some diagnostic tests will require you to test the components with the power turned on. When you disassemble the control panel or remove the outer cabinet, you can position the panel in such a way that the wiring will not make contact with metal. This will allow you to test the components without electrical mishaps.*

---

4. **Gain access to the electronic components.** You can gain access to the electronic components by removing the front grille and the screws on the control panel. Some units have either a one-piece or a two-piece grille with locking tabs and/or screws (Figure 23-5). Be careful not to break the tabs on the grille. On window models, you might have to remove the air conditioner from the window to gain access to the controls. On other models, you should be able to gain access by removing the screws that secure the control panel to the air conditioner frame. Next, tilt the control panel away from the air conditioner, making sure not to pull any of the wires off the controls.

WARNING *Do not touch the wiring or capacitor until it's discharged.*

WARNING *A capacitor will hold a charge indefinitely, even when it is not currently in use. A charged capacitor is extremely dangerous. Discharge all capacitors immediately any time that work is being conducted in their vicinity. Redischarge after repowering the equipment if further work must be done.*

---

5. **Test the electronic components.** If you are able to run the air conditioner test mode, check the different functions of the air conditioner. Use the technical data sheet for the model you are servicing to locate the test points on the wiring schematic. Check all wiring connections and wiring. Using the technical data sheet, test the electronic control and display board, input voltages, and output voltages.

(a)   (b)

Figure 23-5   On some models, the air filter can be removed by sliding it out of the unit or by removing the front grille. Be careful not to damage the tabs on the grille.

6. **Remove the defective component.** To remove the defective component, remove the screws that secure the printed circuit board to the air conditioner frame. Disconnect the connectors from the electronic control board and display.

7. **Install the new component.** To install a new component, just reverse the disassembly procedure, and reassemble. Reinstall all panels and the front grille, and restore the electricity to the air conditioner. Test the room air conditioner operation.

Thermistor

The typical complaints associated with failure of the thermistor are:

- The air conditioner will not cool or heat.
- Erratic temperature control.
- The air conditioner doesn't run at all.

To handle these problems, perform the following steps:

1. **Verify the complaint.** Verify the complaint by checking the control setting. Turn the control to the lowest setting for cool or the highest setting for heat; then turn it back to a normal setting to see if the air conditioner starts cooling or heating.

2. **Check for external factors.** You must check for external factors not associated with the air conditioner. Is the air conditioner installed properly? Is the exhaust or fresh air intake vent open? Explain to the user how to set the controls.

3. **Disconnect the electricity.** Before working on the air conditioner, disconnect the electricity to the unit. This can be done by pulling the plug from the receptacle. Or disconnect the electricity at the fuse panel or at the circuit breaker panel. Turn off the electricity.

WARNING *Some diagnostic tests will require you to test the components with the power turned on. When you disassemble the control panel or remove the outer cabinet, you can position the panel in such a way that the wiring will not make contact with metal. This will allow you to test the components without electrical mishaps.*

4. **Gain access to the thermistor.** To access the thermistor, remove the front grille and filter. Some units have either a one-piece or a two-piece grille with locking tabs and/or screws (see Figure 23-5). Be careful not to break the tabs on the grill. On window models, you might have to remove the air conditioner from the window to gain access to the controls. On other models, you should be able to gain access by removing the screws that secure the control panel to the air conditioner frame. Next, tilt the control panel away from the air conditioner, making sure not to pull any of the wires off the controls.

WARNING *Do not touch the wiring or capacitor until it's discharged.*

WARNING *A capacitor will hold a charge indefinitely, even when it is not currently in use. A charged capacitor is extremely dangerous. Discharge all capacitors immediately any time that work is being conducted in their vicinity. Redischarge after repowering the equipment if further work must be done.*

5. **Test the thermistor.** The thermistor is attached to the evaporator coil; trace the thermistor wire back to the electronic control board. Disconnect the thermistor connector from the board. Set the ohmmeter on R × 10K, and place the probes on the connector pin terminals. Measure the resistance of the thermistor. Using the technical data sheet, look for the reading and see if the results match. The reading can vary ±10 percent on the chart. Remove the thermistor from the evaporator coil, and submerge the thermistor in ice water for a few minutes—the resistance will increase. As the thermistor warms up to ambient temperature, the resistance should return to the original reading. If you suspect an erratic thermistor, replace it with a duplicate of the original.

6. **Install a new thermistor.** To install the new thermistor, just reverse the order of disassembly, and reassemble. Then test the thermistor. Remember to reinstall the sensor in the same location from which it was removed. If you do not, the air conditioner will not cycle properly.

Thermostat (Operating)

The typical complaints associated with failure of the thermostat are:

- The air conditioner will not cool enough.
- The room or area is too cold.
- The air conditioner runs all the time.
- The air conditioner doesn't run at all.

To handle these problems, perform the following steps:

1. **Verify the complaint.** Verify the complaint by checking the control setting. Turn the control to the lowest setting for cool or the highest setting for heat; then turn it back to a normal setting to see if the air conditioner starts cooling or heating.

2. **Check for external factors.** You must check for external factors not associated with the air conditioner. Is the air conditioner installed properly? Is the exhaust or fresh air intake vent open? Explain to the user how to set the controls.

3. **Disconnect the electricity.** Before working on the air conditioner, disconnect the electricity to the unit. This can be done by pulling the plug from the receptacle. Or disconnect the electricity at the fuse panel or at the circuit breaker panel. Turn off the electricity.

WARNING *Some diagnostic tests will require you to test the components with the power turned on. When you disassemble the control panel or remove the outer cabinet, you can position the panel in such a way that the wiring will not make contact with metal. This will allow you to test the components without electrical mishaps.*

4. **Gain access to the thermostat.** To access the thermostat, remove the front grille and filter. Some units have either a one-piece or a two-piece grille with locking tabs and/or screws (see Figure 23-5). Be careful not to break the tabs on the grille. On window models, you might have to remove the air conditioner from the window to gain access to the controls. On other models, you should be able to gain access by removing the screws that secure the control panel to the air conditioner frame. Next, tilt the control panel away from the air conditioner, making sure not to pull any of the wires off the controls.

WARNING *Do not touch the wiring or capacitor until it's discharged.*

WARNING *A capacitor will hold a charge indefinitely, even when it is not currently in use. A charged capacitor is extremely dangerous. Discharge all capacitors immediately any time that work is being conducted in their vicinity. Redischarge after repowering the equipment if further work must be done.*

5. **Test the thermostat.** To test the thermostat, disconnect the wires from the thermostat terminals. On two-wire thermostats, remove the wires from the control, label them, and place the ohmmeter probes on the terminals (Figure 23-6). Set the range scale on $R \times 1$, and test for continuity. With the thermostat set in the warmest (off) position, you should not read continuity. When the thermostat is set to the coldest (highest) position, you should read continuity. Inspect the capillary tube for any damage. If the thermostat capillary tube has lost its charge, the air conditioner will not function properly. For thermostats with more than two terminals (heat models) on the control, you must refer to the wiring diagram on the technical data sheet for the correct terminals to test (Figures 23-7, 23-8, and 23-9). If the thermostat is good, the problem must be elsewhere.

(a)   (b)

FIGURE 23-6   (a) Testing the thermostat. (b) Bi-metal defrost thermostat.

6. **Remove the thermostat.** With the thermostat exposed and the wires already removed, the capillary tube must now be removed from the evaporator coil. The capillary tube is held in place on the coil with clips. Remove the capillary tube from the clips. If the clips come off the evaporator coil, remember where the clips go back. The placement of the clips is crucial for the air conditioner to function properly. Now remove the screws that secure the thermostat body to the control panel.

7. **Install a new thermostat.** To install the new thermostat, just reverse the order of disassembly, and reassemble. Then test the thermostat. Remember to reinstall the capillary tube in the same location from which it was removed. If you do not, the air conditioner will not cycle properly.

Thermostat (Defrost)

The typical complaints associated with failure of the defrost thermostat (heat pump models only) are:

- Ice buildup on the outdoor coil.
- The air conditioner will not heat.

FIGURE 23-7 A typical wiring schematic for a straight-cool-only air conditioner.

To handle these problems, perform the following steps:

1. **Verify the complaint.** Verify the complaint by checking the control setting. Turn the control to the highest setting for heat, and then turn it back to a normal setting to see if the air conditioner starts heating.

2. **Check for external factors.** You must check for external factors not associated with the air conditioner. Is the air conditioner installed properly? Is the exhaust or fresh air intake vent open? Explain to the user how to set the controls.

FIGURE 23-8 A typical wiring diagram for a straight-cool air conditioner with electric heat.

FIGURE 23-9 A typical wiring diagram for a straight-cool air conditioner with reverse cycle (heat pump).

3. **Disconnect the electricity.** Before working on the air conditioner, disconnect the electricity to the unit. This can be done by pulling the plug from the receptacle. Or disconnect the electricity at the fuse panel or at the circuit breaker panel. Turn off the electricity.

WARNING *Some diagnostic tests will require you to test the components with the power turned on. When you disassemble the control panel or remove the outer cabinet, you can position the panel in such a way that the wiring will not make contact with metal. This will allow you to test the components without electrical mishaps.*

4. **Gain access to the defrost thermostat.** To access the defrost thermostat, remove the front grille and filter. Some units have either a one-piece or a two-piece grille with locking tabs and/or screws (see Figure 23-5). Be careful not to break the tabs on the grille. On window models, you might have to remove the air conditioner from the window to gain access to the defrost thermostat. On other models, you should be able to gain access by removing the screws that secure the control panel to the air conditioner frame. Next, tilt the control panel away from the air conditioner, making sure not to pull any of the wires off the controls. Some manufacturers might place the defrost thermostat control behind the bulkhead of the unit. The easiest way to locate that type of control is to follow the capillary tube from the condenser coil to the defrost thermostat body. The defrost thermostat is a nonadjustable thermostat without a stem or knob attached to it. On other models, the defrost thermostat might be a bi-metal thermostat that is clamped to the end of the condenser coil with two wires coming out of it that lead back to the control panel.

WARNING *Do not touch the wiring or capacitor until it's discharged.*

WARNING *A capacitor will hold a charge indefinitely, even when it is not currently in use. A charged capacitor is extremely dangerous. Discharge all capacitors immediately any time that work is being conducted in their vicinity. Redischarge after repowering the equipment if further work must be done.*

5. **Test the defrost thermostat.** To test the defrost thermostat, disconnect the wires from the defrost thermostat terminals or from the selector control switch, label them, and place the ohmmeter probes on the terminals (see Figure 23-6a). Set the range scale on R × 1, and test the thermostat for continuity. Refer to the technical data sheet for the model you are servicing for the position of the switch contacts (open or closed position in the heat cycle) so that you can get the correct ohmmeter reading. Inspect the capillary tube for any damage. If the thermostat capillary tube has lost its charge, the air conditioner will not function properly. For defrost thermostats with more than two terminals on the control, you must refer to the wiring diagram on the technical data sheet for the correct terminals to test (see Figures 23-7, 23-8, and 23-9). If the defrost thermostat is good, the problem must be elsewhere.

6. **Remove the thermostat.** With the thermostat exposed and the wires already removed, the capillary tube must now be removed from the condenser coil. The capillary tube is held in place on the coil with clips. Remove the capillary tube from the clips. If the clips come off the condenser coil, remember where the clips go back. The placement of the clips is crucial for the air conditioner to function properly. Now remove the screws that secure the thermostat body to the control panel. If you are servicing a model that has a bi-metal clamp-on defrost thermostat (see Figure 23-6b), remove the clamp that secures the thermostat to the condenser coil. Remember where the thermostat was secured.

7. **Install a new thermostat.** To install the new defrost thermostat, just reverse the order of disassembly, and reassemble. Then test the defrost thermostat by running the heat cycle. Just remember to reinstall the capillary tube or bi-metal clamp in the same location from which it was removed. If you do not, the air conditioner will not cycle properly.

Capacitor

The typical complaints associated with failure of the capacitor are:

- Fuse is blown or the circuit breaker trips.
- Fan motor will not run.
- Fan motor has a burning smell.
- Motor tries to start and then shuts off on overload.

To handle these problems, perform the following steps:

1. **Verify the complaint.** Verify the complaint by operating the air conditioner. Listen carefully, and you will hear if there are any unusual noises or if the circuit breaker trips. If you smell something burning, immediately turn off the air conditioner, and pull out the plug.

2. **Check for external factors.** You must check for external factors not associated with the air conditioner. Is the air conditioner installed properly? Does it have the correct voltage?

3. **Disconnect the electricity.** Before working on the air conditioner, disconnect the electricity. This can be done by pulling the plug from the electrical outlet. Be sure that you only remove the air conditioner plug. Or disconnect the electricity at the fuse panel or at the circuit breaker panel. Turn off the electricity.

WARNING *Some diagnostic tests will require you to test the components with the power turned on. When you disassemble the control panel or remove the outer cabinet, you can position the panel in such a way that the wiring will not make contact with metal. This will allow you to test the components without electrical mishaps.*

4. **Gain access to the capacitor.** Some models have the capacitor mounted on the fan motor; on other models, it is mounted behind the electrical controls. Remove the front grille and filter. Some units have either a one-piece or a two-piece grille with locking tabs and/or screws (see Figure 23-5). Be careful not to break the tabs on the grille. On window models, you might have to remove the air conditioner from the

window to gain access to the controls. On other models, you should be able to gain access by removing the screws that secure the control panel to the air conditioner frame. Next, tilt the control panel away from the air conditioner, making sure not to pull any of the wires off the controls.

WARNING *Do not touch the wiring or capacitor until it's discharged.*

WARNING *A capacitor will hold a charge indefinitely, even when it is not currently in use. A charged capacitor is extremely dangerous. Discharge all capacitors immediately any time that work is being conducted in their vicinity. Redischarge after repowering the equipment if further work must be done.*

5. **Test the capacitor.** Before testing the capacitor, it must be discharged. Use a screwdriver with an insulated handle to discharge the capacitor by shorting it across both terminals. Remove the wire leads, one at a time, with needle nose pliers. Set the ohmmeter on the highest scale, and then place one probe on one terminal and the other probe on the other terminal (Figure 23-10). Observe the meter action. While the capacitor is charging, the ohmmeter will read nearly zero ohms for a short period of time. Then the ohmmeter reading will slowly return toward infinity. If the ohmmeter reading deflects to zero and does not return to infinity, the capacitor is shorted and should be replaced. If the ohmmeter reading remains at infinity and does not dip toward zero, the capacitor is open and should be replaced.

   To test for a grounded capacitor, leave one meter probe on the capacitor terminal, and attach the other probe to the outer casing of the capacitor or air conditioner chassis ground. There should be no reading if the capacitor is not grounded. Next, remove the probe from the capacitor terminal, and place the probe on the other terminal. Again, there should be no reading indicated if the capacitor is not grounded.[3]

6. **Remove the capacitor.** Remove the capacitor from its mounting bracket.

7. **Install a new capacitor.** To install the new capacitor, just reverse the disassembly procedure, and reassemble. *Note:* A capacitor is rated by its working voltage (WV or WVac) and by its storage capacity in microfarads (μF). Always replace a capacitor with one that has the same voltage rating and the same (or up to 10 percent greater) microfarad rating.

Fan Motor

The typical complaints associated with failure of the fan motor are:

- No air is blowing out of the discharge grille.

- There is no cooling or heating.

- When the motor runs, there are loud noises.

- The fuse or circuit breaker trips when the air conditioner is started.

Figure 23-10 Testing a capacitor for an open or short circuit.

To handle these problems, perform the following steps:

1. **Verify the complaint.** Verify the complaint by operating the air conditioner. Listen carefully, and you will hear if there are any unusual noises or if the circuit breaker trips.

2. **Check for external factors.** You must check for external factors not associated with the air conditioner. Is the air conditioner installed properly? Does it have the correct voltage?

3. **Disconnect the electricity.** Before working on the air conditioner, disconnect the electricity. This can be done by pulling the plug out of the wall receptacle. Be sure that you only remove the air conditioner plug. Or disconnect the electricity at the fuse panel or at the circuit breaker panel. Turn off the electricity.

WARNING *Some diagnostic tests will require you to test the components with the power turned on. When you disassemble the control panel or remove the outer cabinet, you can position the panel in such a way that the wiring will not make contact with metal. This will allow you to test the components without electrical mishaps.*

---

4. **Gain access to the fan motor.** Before you begin to remove the fan motor, test the motor windings. Remove the front grille and filter. Some units have either a one-piece or a two-piece grille with locking tabs and/or screws (see Figure 23-5). Be careful not to break the tabs on the grille. On window models, you might have to remove the air conditioner from the window to gain access to the controls. On other models, you should be able to gain access by removing the screws that secure the control panel to the air conditioner frame. Next, tilt the control panel away from the air conditioner, making sure not to pull any of the wires off the controls.

WARNING *Do not touch the wiring or capacitor until it is discharged.*

WARNING *A capacitor will hold a charge indefinitely, even when it is not currently in use. A charged capacitor is extremely dangerous. Discharge all capacitors immediately any time that work is being conducted in their vicinity. Redischarge after repowering the equipment if further work must be done.*

---

5. **Disconnect the motor wire leads.** Disconnect the motor wire leads from the selector switch, and label them. Check the motor windings for continuity (Figure 23-11). Check for resistance from the common wire lead (white) to the high-speed (black) wire lead (Figure 23-11a). Then check the resistance from the common wire lead (white) to the medium-speed (blue) and the common wire lead (white) to the low-speed (red) winding (Figures 23-11b and 23-11c). If the fan motor has a capacitor wire (brown), check for resistance from the brown wire lead to the common wire lead (white).[4] To check for a grounded winding in the motor, take the ohmmeter probes and check from each motor wire lead to the motor housing (Figure 23-12). The ohmmeter will indicate continuity if the windings are grounded.

Wire coding identification

Black - high speed
Blue - medium speed
Red - low speed
White - common
Brown - capacitor

(a)

(b)

(c)

FIGURE 22-11   (a) Check from the common winding to the high-speed winding. (b) Check from the common winding to the medium-speed winding. (c) Check from the common winding to the low-speed winding.

FIGURE 23-12
Checking for a grounded motor.

Ohmmeter

Wire coding identification

Black - high speed
Blue - medium speed
Red - low speed
White - common
Brown - capacitor

Finally, spin the motor shaft—it should turn freely. If the shaft is hard to turn, replace the fan motor. Now move the motor shaft in an up and down motion perpendicular to the motor body. The shaft should have virtually no movement. If there is a lot of play in the end bell bearings, replace the fan motor.[5]

6. **Remove the fan motor.** To remove the fan motor, the air conditioner will have to be uninstalled or the chassis slid out of the outer case and placed on a solid table or workbench.

WARNING *This procedure will require two people to uninstall and move the air conditioner.*

Depending on which model you are servicing, the removal of the fan motor will vary from manufacturer to manufacturer. However, the basic steps are the same for every air conditioner: The fan motor is secured to the bulkhead (Figure 23-13).

1. Condenser fan blade with clamp or set screw	7. Bulkhead housing
2. Condenser fan housing seal	8. Screws
3. Condenser coil and condenser fan blade housing	9. Evaporator blower wheel
4. Condenser fan housing seal	10. Blower wheel clamp
5. Fan motor	11. Evaporator blower housing
6. Fan motor mounting nuts	12. Evaporator blower housing

FIGURE 23-13   An exploded view of the fan motor, condenser fan blade, blower wheel, and the housings.

In order to remove it, you must first remove the condenser fan blade. To gain access to the fan blade, remove the screws and brackets[6] that secure the condenser coil to the fan shroud. On some models, the upper housing has a cover—remove it. The condenser coil will have to be lifted up out of the base pan and gently moved out of the way to gain access to the fan blade. You only have to move the condenser coil enough to gain access to the fan blade and to provide room to remove the fan blade without damaging the condenser coil.

WARNING *When moving the condenser coil away from the fan blade, do not kink or break the copper tubing. Remember that there is high-pressure refrigerant in the lines.*

Remove the setscrew or clamp from the condenser fan blade. When removing the fan blade, be careful not bend or damage it. To gain access to the evaporator blower wheel (see Figure 23-13), remove the screws from the evaporator blower housing, exposing the blower wheel. Depending on which model you are servicing, you may have to remove part of the housing to gain access. Next, remove the clamp from the blower wheel. You will not be able to remove the blower wheel at this time until the fan motor is removed from the bulkhead. With the condenser fan blade removed and the evaporator blower wheel loose on the motor shaft, you are now ready to remove the motor. Remove the fan motor wiring from the control area, making sure to free up the fan motor wiring harness that may be secured to the chassis. Now remove the fan motor nuts or bolts that secure the fan motor to the bulkhead. Grab hold of the fan motor, and pull it out of the air conditioner while removing the evaporator blower wheel from the motor shaft. The blower wheel will remain in the evaporator blower housing.

NOTE *Replace the fan motor with a duplicate of the original for easier installation.*

NOTE *With the fan motor removed from the unit, it is the perfect time to chemically clean the remainder of the air conditioner. Everything will be exposed, and it will be easier to clean the unit.*

---

7. **Reinstall the fan motor.** To reinstall the fan motor, just reverse the instructions in step 6, and reassemble. When you reinstall the condenser fan blade, make sure to place it back on the motor shaft in the same position. Most room air conditioner condenser fan blades have a slinger ring that is attached to the paddles. The slinger ring will sit in a sump area in the base pan, allowing the fan blade to pick up the condensate water, flinging it against the condenser coil. Before reinstalling the outer cabinet or installing the air conditioner in the wall, test the air conditioner first. Make sure the fan blade and the blower wheel are not hitting against anything.

Compressor and Overload Protector

The typical complaints associated with failure of the compressor are:

- The air conditioner will not cool.
- The room temperature is warm.

- The air conditioner does not run at all.
- The compressor makes a humming or buzzing noise and stops.
- The circuit breaker trips when the air conditioner starts up.

To handle these problems, perform the following steps:

1. **Verify the complaint.** Verify the complaint by asking the customer to describe what the air conditioner is doing.

2. **Check for external factors.** You must check for external factors not associated with the air conditioner. Is the air conditioner installed properly? Does it have the correct voltage? Check for a voltage drop during air conditioner startup.

3. **Disconnect the electricity.** Before working on the air conditioner, disconnect the electricity. This can be done by pulling the plug from the receptacle. Or disconnect the electricity at the fuse panel or at the circuit breaker panel. Turn off the electricity.

WARNING *Before you begin to service any high-voltage components within the air conditioner, you must discharge the high-voltage capacitors first.*

WARNING *Some diagnostic tests will require you to test the components with the power turned on. When you disassemble the control panel or remove the outer cabinet, you can position the panel in such a way that the wiring will not make contact with metal. This will allow you to test the components without electrical mishaps.*

4. **Gain access to the compressor.** To access the compressor, pull the air conditioner out and away from the window or wall. Remove the outer cabinet (on some models). This will expose the compressor, fan motor, and electrical controls (Figure 23-14). Next, remove the compressor terminal cover (Figure 23-15) by removing the retaining nut or clip that secures the cover. Remove the terminal cover and washer.

FIGURE 23-14   An illustration of the component location on a slide-out chassis wall air conditioner.

FIGURE 23-15
An exploded view of a
rotary compressor,
overload protector,
and terminal cover.

Terminal cover nut and washer

Compressor
terminal cover

Terminal
cover gasket

Overload protector

Compressor terminals

Compressor

5. **Test the overload protector.** To test the overload protector, remove the wires from the overload and compressor terminals. Next, place the probes of the ohmmeter on the overload terminals (see Figure 23-15). Set the meter scale on R × 1. The reading will show continuity. If not, replace the overload protector.

6. **Test the compressor.** The compressor terminals are marked C, S, and R: C indicates the common winding terminal, S indicates the start winding terminal, and R indicates the run winding terminal. Refer to the wiring diagram for the model that you are servicing. Set the meter scale on R × 1, touch the probes together, and adjust the needle setting to indicate a zero reading. Next, place the probes of the ohmmeter

on the terminals marked S and R (see Figure 21-36a). The meter reading will show continuity. Now place the meter probes on the terminals marked C and S. The meter reading will show continuity. Finally, place the meter probes on the terminals marked C and R. The meter reading will show continuity. The total number of ohms measured between S and R is equal to the sum of C to S plus C to R. The compressor should be tested for proper resistance, as indicated on the wiring diagram.

To test the compressor for ground, place one probe on a compressor terminal, and attach the other probe to the compressor housing or to any good ground (see Figure 21-36b). Set the meter scale to R × 1000. The meter reading will show no continuity. Repeat this for the remaining two terminals. The meter reading will show no continuity. If you get a continuity reading from any of these terminals to ground, the compressor is grounded. Replace it.

WARNING *This procedure may require two people to uninstall and move the air conditioner.*

Electric Heater Element

The typical complaint associated with failure of the electric heater element is that the air conditioner will not heat at all. To handle this problem, perform the following steps:

1. **Verify the complaint.** Verify the complaint by checking the control setting. Turn the control to the highest setting for heat; then turn it back to a normal setting to see if the air conditioner starts heating.

2. **Check for external factors.** You must check for external factors not associated with the air conditioner. Is the air conditioner installed properly? Is the exhaust or fresh air intake vent open? Check for the correct voltage to the air conditioner. Explain to the user how to set the controls.

3. **Disconnect the electricity.** Before working on the air conditioner, disconnect the electricity to the unit. This can be done by pulling the plug from the receptacle. Or disconnect the electricity at the fuse panel or at the circuit breaker panel. Turn off the electricity.

WARNING *Some diagnostic tests will require you to test the components with the power turned on. When you disassemble the control panel or remove the outer cabinet, you can position the panel in such a way that the wiring will not make contact with metal. This will allow you to test the components without electrical mishaps.*

4. **Gain access to the electric heater element circuitry.** To access the electric heater element circuitry, remove the front grille and filter. Some units have either a one-piece or a two-piece grill with locking tabs and/or screws (see Figure 23-5). Be careful not to break the tabs on the grill. On window models, you might have to remove the air conditioner from the window to gain access to the controls. On other models, you should be able to gain access by removing the screws that secure the control panel to the air conditioner frame. Next, tilt the control panel away from the air conditioner, making sure not to pull any of the wires off the controls.

WARNING *Do not touch the wiring or capacitor until it's discharged.*

WARNING *A capacitor will hold a charge indefinitely, even when it is not currently in use. A charged capacitor is extremely dangerous. Discharge all capacitors immediately any time that work is being conducted in their vicinity. Redischarge after repowering the equipment if further work must be done.*

5. **Test the electric heater element.** At this point, you do not have to disassemble the air conditioner to determine if the electric heater element is defective. Locate the wiring diagram (see Figure 23-8), and trace the electric heater element circuit. You will notice that the heater element, thermal fuse, and heater limit control (thermostat) are wired in series with the selector switch and operating thermostat. These five components make up the heating circuit. Locate the electric heater element wiring, and disconnect the wires from the terminals. For example, in the wiring diagram shown in Figure 23-8, the orange and red wires from the electric heater element go to the capacitor common terminal and to the number 4 terminal on the selector switch. If you remove those two wires from their terminals, you have just isolated the electric heater element, thermal fuse, and the heater limit control. Set your ohmmeter to the R × 1 scale, and check for continuity between those two wires.[7] If the ohmmeter reads resistance in the circuit, then the problem is elsewhere. If your ohmmeter reads an open circuit, you will have to remove the electric heater assembly for further testing.

6. **Remove the electric heater assembly.** The electric heater assembly is located in the evaporator blower housing on the discharge side. It may be behind a metal screen on older models. For newer models, the heater element is located behind the evaporator coil. To remove the heater assembly, the air conditioner will have to be uninstalled.

WARNING *This procedure may require two people to uninstall and move the air conditioner.*

Remove the screws that secure the top panel on the evaporator blower housing (see Figure 23-13). On some models, part of the evaporator housing can be removed for easy access to the heater assembly. Once you gain access to the heater assembly, remove the screws that secure the assembly to the air conditioner. Replace with a duplicate of the original.

NOTE *With the electric heater assembly removed from the unit, it is the perfect time to chemically clean the remainder of the air conditioner. Everything will be exposed, and it will be easier to clean the unit. Just remember to protect the electrical components and fan motor.*

7. **Install the heater assembly.** To install a new electric heater assembly, just reverse the disassembly procedure, and reassemble. Reinstall all panels and the front grille, and restore the electricity to the air conditioner. Test the room air conditioner heat operation.

Reversing Valve Solenoid Coil

The typical complaint associated with failure of the reversing valve solenoid coil is that the air conditioner will not heat. To handle this problem, perform the following steps:

1. **Verify the complaint.** Verify the complaint by checking the control setting. Turn the control to the lowest setting for cool or the highest setting for heat; then turn it back to a normal setting to see if the air conditioner starts cooling or heating.

2. **Check for external factors.** You must check for external factors not associated with the air conditioner. Is the air conditioner installed properly? Is the exhaust or fresh air intake vent open? Explain to the user how to set the controls.

3. **Disconnect the electricity.** Before working on the air conditioner, disconnect the electricity to the unit. This can be done by pulling the plug from the receptacle. Or disconnect the electricity at the fuse panel or at the circuit breaker panel. Turn off the electricity.

WARNING *Some diagnostic tests will require you to test the components with the power turned on. When you disassemble the control panel or remove the outer cabinet, you can position the panel in such a way that the wiring will not make contact with metal. This will allow you to test the components without electrical mishaps.*

---

4. **Gain access to the reversing valve solenoid coil.** To access the reversing valve solenoid coil wiring, remove the front grille and filter. Some units have either a one-piece or a two-piece grill with locking tabs and/or screws (see Figure 23-5). Be careful not to break the tabs on the grille. On window models, you might have to remove the air conditioner from the window to gain access to the controls. On other models, you should be able to gain access by removing the screws that secure the control panel to the air conditioner frame. Next, tilt the control panel away from the air conditioner, making sure not to pull any of the wires off the controls.

WARNING *Do not touch the wiring or capacitor until it's discharged.*

WARNING *A capacitor will hold a charge indefinitely, even when it is not currently in use. A charged capacitor is extremely dangerous. Discharge all capacitors immediately any time that work is being conducted in their vicinity. Redischarge after repowering the equipment if further work must be done.*

---

5. **Test the reversing valve solenoid coil.** To test the solenoid coil, disconnect the wires from the selector switch terminals. Look on the wiring diagram for the correct wires to remove. Set the ohmmeter on the R × 10K scale, and place the meter probes on the solenoid coil wire terminals. You should read resistance in the coil. If not, replace the solenoid coil with a duplicate of the original.

6. **Replace the reversing valve solenoid coil.** To access the solenoid coil, pull the air conditioner out and away from the window or wall. Remove the outer cabinet (on some models). This will expose the compressor, reversing valve, fan motor,

and electrical controls (see Figure 23-14). The reversing valve is located near the compressor. Remove the retaining nut and washer from the pilot valve stem on the reversing valve, and remove the solenoid coil.

7. **Install the reversing valve solenoid coil.** To install a new solenoid coil, just reverse the disassembly procedure, and reassemble. Reinstall all panels and the front grille, and restore the electricity to the air conditioner. Test the room air conditioner heat operation.

Selector Switch

The typical complaints associated with failure of the selector switch are:

- The air conditioner will not cool or heat.
- The air conditioner doesn't run at all.

To handle these problems, perform the following steps:

1. **Verify the complaint.** Verify the complaint by checking the control setting to see if the air conditioner starts cooling or heating.

2. **Check for external factors.** You must check for external factors not associated with the air conditioner. Is the air conditioner installed properly? Is the exhaust or fresh air intake vent open? Is the voltage correct to the air conditioner? Explain to the user how to set the controls.

3. **Disconnect the electricity.** Before working on the air conditioner, disconnect the electricity to the unit. This can be done by pulling the plug from the receptacle. Or disconnect the electricity at the fuse panel or at the circuit breaker panel. Turn off the electricity.

WARNING *Some diagnostic tests will require you to test the components with the power turned on. When you disassemble the control panel or remove the outer cabinet, you can position the panel in such a way that the wiring will not make contact with metal. This will allow you to test the components without electrical mishaps.*

4. **Gain access to the selector switch.** To access the selector switch, remove the front grille and filter. Some units have either a one-piece or a two-piece grill with locking tabs and/or screws (see Figure 23-5). Be careful not to break the tabs on the grille. On window models, you might have to remove the air conditioner from the window to gain access to the controls. On other models, you should be able to gain access by removing the screws that secure the control panel to the air conditioner frame. Next, tilt the control panel away from the air conditioner, making sure not to pull any of the wires off the controls.

WARNING *Do not touch the wiring or capacitor until it's discharged.*

PART V

WARNING *A capacitor will hold a charge indefinitely, even when it is not currently in use. A charged capacitor is extremely dangerous. Discharge all capacitors immediately any time that work is being conducted in their vicinity. Redischarge after repowering the equipment if further work must be done.*

5. **Test the selector switch.** To test the selector switch, locate the selector switch circuit on the wiring diagram (see Figures 23-7, 23-8, and 23-9). Identify the terminals that turn on the different components to be tested. Set the ohmmeter on the R × 1 scale. Next, place the ohmmeter probes on those terminals. Then select the setting by either rotating the dial or by pressing the proper button on the switch (Figure 23-16). If the switch contacts are good, your meter will show continuity. Test all of the remaining component settings on the selector switch. Remember to check the wiring diagram for the correct switch contact terminals (those that correspond to the setting that you are testing).

6. **Remove the selector switch.** To remove the selector switch, remove the screws that hold the switch to the control panel, and remove the switch.

7. **Install the new selector switch.** To install the new selector switch, just reverse the disassembly procedure, and reassemble. Then reattach the wires to the switch terminals according to the wiring diagram. Reassemble the control panel. When you are reassembling the control panel, make sure that the wires do not become pinched between the control panel and the air conditioner frame.

FIGURE 23-16
The selector switch contact test points.

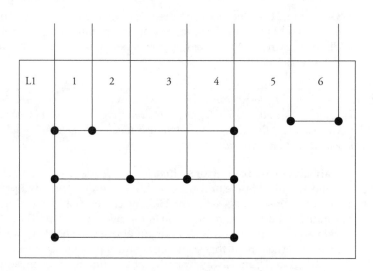

Selector switch	
Switch position	Contacts closed
Off	None
Cool	L1 to 1 & 4
Heat	L1 to 2, 3, 4, 5, & 6
Fan only	L1 to 4

Troubleshooting Sealed-System Problems

If you suspect a sealed-system malfunction, be sure to check out all external factors first. These include:

- Thermostats
- Compressor
- Overload on the compressor
- Fan motor
- Evaporator and condenser coils getting good air circulation
- Air conditioner installation
- Make sure the heater is not on at the same time as cooling
- Make sure that the reversing valve and solenoid coil are operating properly

After eliminating all of these external factors, you will then systematically check the sealed system. This is accomplished by comparing the conditions found in a normally operating air conditioner. These conditions are:

- Room temperature
- Wattage
- Condenser temperature
- Evaporator inlet sound (gurgle, hiss, etc.)
- Evaporator cooling pattern
- High-side pressure[8]
- Low-side pressure[8]
- Pressure equalization time

One thing to keep in mind: No single indicator is conclusive proof that a particular sealed-system problem exists. Rather, it is a combination of findings that must be used to definitively pinpoint the exact problem.

Low-Capacity Compressor

Symptoms of a low-capacity compressor in the sealed system are:

- Temperatures in the room or area will be above normal.
- The wattage and amperage will be below normal, as indicated on the model/ serial plate.
- The temperature of the condenser coil will be below normal.
- At the evaporator coil, you will hear a slightly reduced gurgling noise.
- The evaporator coil will show a normal cooling pattern.
- The high-side pressure will be below normal, and the low-side pressure will be above normal.[8]
- The pressure equalization time might be normal or shorter than normal.

Refrigerant Leak

Symptoms of a refrigerant leak in the sealed system are:

- Temperatures in the room or area will be below normal.
- The wattage and amperage will be below normal, as indicated on the model/serial plate.
- The condenser coil will be cool to the touch at the last pass, or even as far as midway through the coil.
- At the evaporator coil, you will hear a gurgling noise, a hissing noise, or possibly an intermittent hissing or gurgling noise.
- The evaporator coil will show a frost pattern in the lower rungs of the coil.
- The high- and low-side pressures will be below normal.[8]
- The pressure equalization time might be normal or shorter than normal.

Overcharged Air Conditioner

If the sealed system is overcharged, the symptoms are:

- The room temperature will be higher than normal.
- The wattage and amperage will be above normal, as indicated on the model/serial plate.
- The temperature of the condenser coil will be above normal.
- At the evaporator coil, you will hear a constant gurgling noise—generally, a higher sound level than normal.
- The evaporator coil will show a full frost pattern. If you remove the cover, you will possibly see the suction line frosted back to the compressor.
- The high- and low-side pressures will be above normal.[8]
- The pressure equalization time will be normal.

Slight Restriction

Symptoms of a slight restriction in the sealed system are:

- The room temperature will be below normal.
- The wattage and amperage will be below normal, as indicated on the model/serial plate.
- The temperature of the condenser coil will be slightly below normal.
- At the evaporator coil, you will hear a constant gurgling noise and a low sound level.
- The evaporator coil cooling pattern will be receded.
- The high- and low-side pressures will be below normal.[8]
- The pressure equalization time will be longer than normal.

Partial Restriction

Symptoms of a partial restriction in the sealed system are:

- The room temperature will be higher than normal.
- The wattage and amperage will be below normal, as indicated on the model/serial plate.
- The temperature of the condenser coil will be below normal more than halfway on the coil.
- At the evaporator coil, you will hear a constant gurgling noise and a considerably lower sound level.
- The evaporator coil cooling pattern will be considerably receded.
- The high- and low-side pressures will be below normal.[8]
- The pressure equalization time will be longer than normal.

Complete Restriction

Symptoms of a complete restriction in the sealed system are:

- The room temperature will be warm.
- The wattage and amperage will be considerably below normal, as indicated on the model/serial plate.
- The temperature of the condenser coil will be cool or at room temperature.
- At the evaporator coil, you will hear no sounds.
- The evaporator coil will not be cool.
- The high-side pressure will be equal to the pressure of refrigerant at room temperature.
- The low-side pressure will be in a deep vacuum.[8]
- There will be no pressure equalization time.

Moisture Restriction

Symptoms of a moisture restriction in the sealed system are:

- The room temperature will be above normal.
- The wattage and amperage will be considerably below normal, as indicated on the model/serial plate.
- The temperature of the condenser coil will be below normal.
- At the evaporator coil, you will hear a constant gurgle, low sound level, or no sound at all.
- The evaporator coil might have some frost on the evaporator inlet.
- The high-side pressure will be below normal.[8]
- The low-side pressure will be below normal or in a deep vacuum.[8]
- The pressure equalization time will be longer than normal or there will be no equalization at all.

Reversing valve–cooling

(a)

Reversing valve–heating

(b)

FIGURE 23-17   The refrigerant flow in a reverse-cycle valve. (a) The reversing valve solenoid coil is de-energized in the cooling mode. (b) The reversing valve solenoid coil is energized in the heating mode.

Reversing Valve (Heat Pump Models)

In a straight-cool air conditioner, the refrigerant flows from the compressor discharge through the condenser coil, capillary tube, evaporator coil, and back through the suction line to the compressor. The ability of an air conditioner to reverse the direction of the refrigerant flow is achieved with a reversing valve installed in the refrigerant circuit. The reversing valve is designed as a single-port, double-throw valve. It turns the function of the evaporator coil from a heat absorption coil into a heat dissipation coil (condenser coil) (Figure 23-17). When the solenoid coil is energized, the valve will reverse the refrigerant flow. For cooling, in a heat-pump air conditioner, the refrigerant flows from the compressor discharge through the reversing valve to the outdoor coil, through the capillary tube and indoor coil, and back through the reversing valve to the suction port on the compressor. For heating, the refrigerant will flow from the compressor discharge; through the reversing valve, indoor coil, capillary tube, and outdoor coil; and then back through the reversing valve to the suction port on the compressor (Figure 23-18).

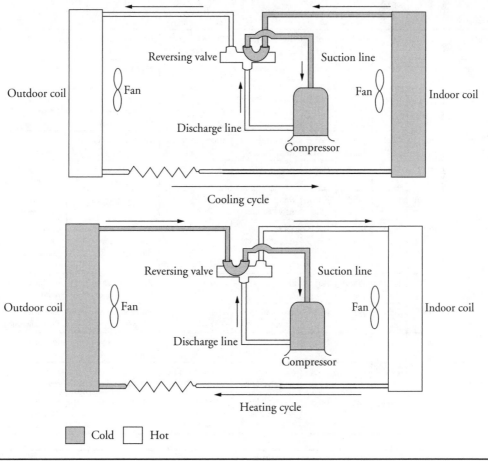

FIGURE 23-18   The refrigerant flow in a heat pump for cooling and heating.

The reversing valve itself is a non-serviceable component and must be replaced with a duplicate of the original if it fails. It is recommended that a qualified technician with refrigerant certification replace the reversing valve. The refrigerant in the sealed system must be recovered properly.

Diagnostic Charts

The following diagnostic flow charts, tables, and wiring diagrams will help you to pinpoint the likely causes of problems with room air conditioners (see Tables 7-4 and 23-4 and Figures 23-7, 23-8, 23-9, 23-19, and 23-20).

Reversing Valve Operating Condition	Discharge Line from the Compressor	Suction Line to the Compressor	Copper Line to the Inside Coil	Copper Line to the Outside Coil	Reversing Valve Left Pilot Capillary Tube	Reversing Valve Right Pilot Capillary Tube	Probable Cause	Solution
Normal cooling	Hot	Cool	Cool	Hot	Same temp as valve body	Same temp as valve body	Normal	None
Normal heating	Hot	Cool	Hot	Cool	Same temp as valve body	Same temp as valve body	Normal	None
Reversing valve will not shift from cooling to heating	Hot	Cool	Cool	Hot	Same temp as valve body	Hot	Bleeder hole blockage or leaking piston	Replace reversing valve
	Hot	Cool	Cool	Hot	Same temp as valve body	Same temp as valve body	Vent tube blockage	Replace reversing valve
	Hot	Cool	Cool	Hot	Hot	Hot	Both pilot valve ports are open	Replace reversing valve
	Warm	Cool	Cool	Hot	Same temp as valve body	Warm	Bad compressor	Replace compressor
Reversing valve will begin to shift and it will not complete the reversal	Hot	Warm	Warm	Hot	Same temp as valve body	Hot	Low on refrigerant or valve body damage	Check refrigerant charge or replace
	Hot	Warm	Warm	Hot	Hot	Hot	Both pilot valve ports are open	Replace reversing valve
	Hot	Hot	Hot	Hot	Same temp as valve body	Hot	Stuck valve or valve body damage	Replace reversing valve
	Hot	Hot	Hot	Hot	Hot	Hot	Both pilot valve ports are open	Replace reversing valve
Fluctuating heating	Hot	Cool	Hot	Cool	Same temp as valve body	Same temp as valve body	Internal valve leaking at piston	Replace reversing valve
	Hot	Cool	Hot	Cool	Valve body warmer than usual	Valve body warmer than usual	Internal valve leaking at piston	Replace reversing valve

TABLE 23-4 Reversing Valve Touch Test Chart. (*continued*)

Reversing Valve Operating Condition	Discharge Line from the Compressor	Suction Line to the Compressor	Copper Line to the Inside Coil	Copper Line to the Outside Coil	Reversing Valve Left Pilot Capillary Tube	Reversing Valve Right Pilot Capillary Tube	Probable Cause	Solution
Reversing valve will not shift from heating to cooling	Hot	Cool	Hot	Cool	Same temp as valve body	Same temp as valve body	High refrigerant pressure differential	Restart air conditioner and recheck
	Hot	Cool	Hot	Cool	Hot	Same temp as valve body	Bleeder hole blockage	Replace reversing valve
	Hot	Cool	Hot	Cool	Hot	Same temp as valve body	Internal valve leaking at piston	Replace reversing valve
	Hot	Cool	Hot	Cool	Hot	Hot	Defective pilot valve	Replace reversing valve
	Warm	Cool	Warm	Cool	Warm	Same temp as valve body	Bad compressor	Replace compressor

Note: Before replacing the reversing valve, turn the air conditioner off for a few minutes, and restart the cycle. On some occasions, the reversing valve may return to normal operation.

TABLE 23-4   Reversing Valve Touch Test Chart.

FIGURE 23-19
The diagnostic flow chart: The air conditioner is cooling when the heat function is selected.

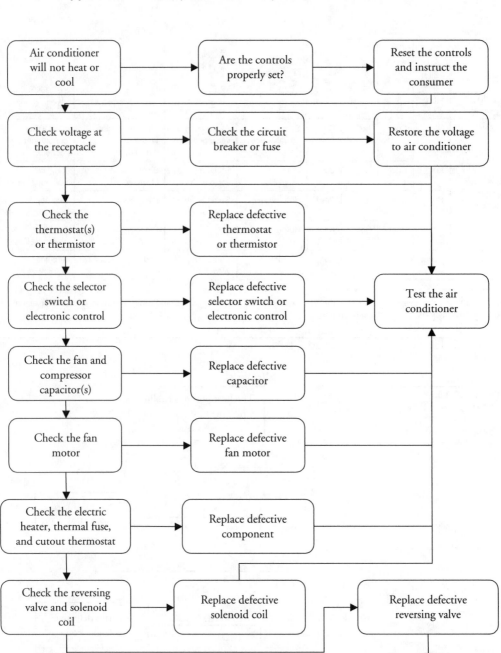

FIGURE 23-20   The diagnostic flow chart: Air conditioner will not heat or cool.

Endnotes

1. Room air conditioners (RACs) are available with top or side discharge air grilles.

2. Refer to Chapter 1 for correct sizing.

3. Capacitors with a metal outer case can only be checked for a ground. Start capacitors are encased in plastic and they cannot be checked for a ground.

4. Some fan motor manufacturers may use different wiring colors to indicate the various speeds. Check the technical data sheet for the model you are servicing for the correct fan motor winding wiring identification and color coding.

5. When the bearings wear out in the end bell of the motor, the motor will begin to make a vibrating sound or a growling noise.

6. On some models, there may be a brace that secures the condenser fan shroud to the bulkhead. This brace must be removed to allow access to the condenser fan blade.

7. If you did not read resistance, it does not mean that the electric heater element is defective. After removing the heater assembly, test the thermal fuse and the heater limit control (thermostat mounted to heater frame) for continuity separately. The thermal fuse is a non-resetable fuse—it will open the heating circuit if the heating temperature or amperage exceeds its rating, indicating that there may be other problems with the air conditioner (for example, a defective fan motor, dirty filter, or dirty evaporator coil). The thermal fuse acts as the last line of defense if the heater limit control fails. The heater limit control will also open the circuit, limiting the heating, if there is a reduction in airflow. On some models, this device will reset, automatically, allowing the electric heater to come back on again. Both of these devices are installed in the heating circuit to prevent the heating element from causing a fire in the air conditioner and home. These safety devices must remain in the circuitry. Do not bypass these safety devices. Replace the defective component with a duplicate of the original.

8. If you open up the sealed system, you will void your warranty. The sealed system must be repaired by an authorized service company.

Glossary

air shutter   An adjustable shutter attached to the primary air openings of a burner used to control the amount of primary air entering the burner.

allen wrench   An L-shaped tool that is used to remove hex screws.

aldehyde   A class of compounds that can be produced from incomplete combustion of gas fuels.

alternating current (AC)   Electrical current that flows in one direction and then reverses itself and flows in the opposite direction. In 60-cycle current, the direction of flow reverses every 120th second.

ambient temperature   Temperature of air that surrounds an object on all sides.

ammeter   A test instrument used to measure current.

ampere   The number of electrons passing a given point in one second.

armature   The section of a motor that turns.

automatic defrost timer   A device connected in an electrical circuit that shuts off the refrigeration cycle and turns on the electric heaters to melt the ice off the evaporator coil.

bake element   Lower heating element in an oven used for baking foods.

belt   A band of flexible material used to transfer mechanical power from one pulley to another.

bi-metal (strip or thermostat)   Two dissimilar metals joined together to form one unit with a differential expansion rating. It will bend if there is a temperature change.

bracket   A hardened structure used to support a component.

broil element   Upper heating element in an oven used for broiling foods.

BTU   (British Thermal Unit) The amount of heat required to raise the temperature of one pound of water one degree Fahrenheit.

burner   A device in which an air and gas mixture has entered into the combustion zone to be expelled and burned off.

burner head   The portion of the burner past the outlet mixing tube that contains the burner ports.

cabinet   The outer wrapper of an appliance.

cam A rotating surface with rises and falls that open and close the switches in a timer.

capacitor A device that stores electricity; used to start and/or run circuits on many large electric motors.

capillary tube A metering device used to control the flow of refrigerant in a sealed system. It usually consists of several feet of tubing with a small inside diameter.

Celsius The metric system for measuring temperature. The interval between the freezing point and the boiling point of water is divided into 100 degrees. This is also called the *centigrade scale.*

circuit A path for electrical current to flow from the power source to the point of use and back to the power source.

circuit breaker A safety device used to open a circuit if that circuit is overloaded.

closed (circuit) An electrical circuit in which electrons are flowing.

combustion The rapid chemical reaction of a gas with oxygen to produce heat and light.

component An individual mechanical or electrical part of an appliance.

compressor An apparatus similar to a pump that is used to increase the pressure of refrigerant that is to be circulated within a closed system.

condenser coil A component of the refrigeration system that transfers the heat that comes from within a closed compartment to the outside surrounding area.

condenser fan A fan motor and fan blades used to cool the condenser coil.

contact points Two movable objects, or contacts, that come together to complete a circuit or that separate and break a circuit. These contact points are usually made of silver.

continuity The ability of a completed circuit to conduct electricity.

control A device, either automatic or manual, that is used to start, stop, or regulate the flow of liquid, gas, or electricity.

current The flow of electrons from a negative to a positive potential.

cycle A series of events that repeat themselves in the same order.

defective Refers to a component that does not function properly.

defrost cycle That part of the refrigeration cycle in which the ice is melted off the evaporator coil.

defrost timer A device used in an electrical circuit to turn off the refrigeration long enough to permit the ice to be melted off the evaporator coil.

diagnosis The act of identifying a problem based on its signs and symptoms.

direct current (DC) Electric current that flows in one direction in a circuit.

end bell The end plates that hold a motor together.

energize To supply electrical current for operation of an electrical component.

evaporator coil A component of the refrigeration system that removes the heat from within a closed compartment.

evaporator fan A motor and fan blades used to circulate the cold air.

Fahrenheit The standard system for measuring temperature. The freezing point of water is 32 degrees Fahrenheit, and the boiling point of water is 212 degrees Fahrenheit.

flashback   Also known as extinction pop. Where the burner flames enter back into the burner head and continue to burn after the gas supply has been turned off.

fuse   A safety device used to open a circuit if that circuit is overloaded.

gasket   A flexible material (either airtight or watertight) used to seal components together.

ground   A connection to earth or to another conducting body that transmits current to earth.

ground wire   An electrical wire that will safely conduct electricity from a structure to earth.

Hertz (Hz)   A unit of measurement for frequency. One hertz equals one cycle per second.

hot gas defrost   A defrosting system in which hot refrigerant from the condenser coil is directed to the evaporator coil for a short period of time to melt the ice from the evaporator coil.

housing   A metal or plastic casing that covers a component.

idler pulley   A device that rests on or presses against a drive belt to maintain a specified tension on it.

insulation   Substance used to retard or slow down the flow of heat through a substance.

insulator   A material that does not conduct electricity. It is used to isolate current-carrying wires or components from other metal parts.

ladder diagram   A wiring schematic in which all of the components are stacked in the form of a ladder.

lead (wire)   A section of electrical wiring that is attached to a component.

lint   Fine pieces of cotton fiber that have broken away from a garment.

module   A self-contained device with a group of interconnecting parts designed to do a specific job.

nut driver   A tool used to remove or reinstall hexagonal-head screws or nuts.

ohm   A unit of measurement of electrical resistance.

ohmmeter   A test instrument used for measuring resistance.

open (circuit)   A break in an electrical circuit that stops the flow of current.

orifice   An opening in a hood cap (orifice spud) through which a gas is discharged and where the gas is metered.

overload protector   A device that is temperature-, pressure-, or current-operated that is used to open a circuit to stop the operation of that circuit should dangerous conditions arise.

parallel circuit   Components that are parallel-connected across one voltage source. All of the branches are supplied with the same amount of voltage.

port   An opening in the burner head where the air and gas mixture is discharged and lit by an ignition source.

pressure switch   A device that is operated by pressure and turns a component on or off.

pulley   A wheel that is turned by or driven by a belt.

primary air   The main air supply introduced into the burner which mixes with the gas before it reaches the burner head.

refrigerant   A chemical substance used in refrigeration and air conditioning that produces a cooling effect.

relay   A magnetic switch that uses a small amount of current in the control circuit to operate a component needing a larger amount of current in the operating circuit. A remote switch.

relief valve   A safety device that is designed to open before dangerous pressure is reached.

resistance   The opposition to current flow. The load in an electrical circuit.

run winding   Electrical winding of a motor, which has current flowing through it during normal operation of the motor.

safety thermostat   A thermostat that limits the temperature to a safe level.

schematic diagram   A line drawing that gives the electrical paths, layout of components, terminal identification, color codes of wiring, and sometimes the sequence of events of an appliance.

series circuit   A circuit in which all of the components are sequentially connected in the same line. If one component fails, all the components fail.

short circuit   An electrical condition in which part of a circuit errantly touches another part of a circuit and causes part or all of the circuit to fail (and trip the circuit breaker or blow a fuse).

solenoid   A cylindrical coil of insulated wire that establishes a magnetic field in the presence of current.

start winding   A winding in an electric motor used only during brief periods when the motor is starting.

surface element   The top cooking element on a range used for cooking foods.

switch   A device to turn current on or off in an electrical circuit.

temperature   A measure of heat energy or the relative lack thereof.

terminal   A connecting point in a circuit to which a wire would be attached to connect a component.

test light   A light provided with test leads that is used to test electrical circuits.

thermometer   A device used to measure temperature.

thermostat   A device that senses temperature changes and that usually operates a control relay.

thermostat (operating)   A thermostat that controls the operating temperature of a component.

transformer   A device that raises or lowers the main AC supply voltage.

venturi throat   A section of the mixing tube body that narrows down and then flares out again.

voltage   The difference in potential between two points; the difference in static charges between two points.

voltmeter   A test instrument used to measure voltage.

VOM (voltohm-milliammeter)   A test instrument used to measure voltage, resistance, and amperage.

water column (WC)   A unit used for pressure. A one-inch water column equals a pressure of 0.578 ounces per square inch.

watt   A unit to measure electrical power.

wattmeter   A test instrument used to measure electrical power.

Index